NEAR ABROAD

*PUTIN, THE WEST, AND
THE CONTEST OVER UKRAINE
AND THE CAUCASUS*

GERARD TOAL

Oxford University Press is a department of the University of Oxford. It furthers
the University's objective of excellence in research, scholarship, and education
by publishing worldwide. Oxford is a registered trade mark of Oxford University
Press in the UK and certain other countries.

Published in the United States of America by Oxford University Press
198 Madison Avenue, New York, NY 10016, United States of America.

© Oxford University Press 2017

First issued as an Oxford University Press paperback, 2019

All rights reserved. No part of this publication may be reproduced, stored in
a retrieval system, or transmitted, in any form or by any means, without the
prior permission in writing of Oxford University Press, or as expressly permitted
by law, by license, or under terms agreed with the appropriate reproduction
rights organization. Inquiries concerning reproduction outside the scope of the
above should be sent to the Rights Department, Oxford University Press, at the
address above.

You must not circulate this work in any other form
and you must impose this same condition on any acquirer.

CIP data is on file at the Library of Congress
ISBN 978–0–19–025330–1 (hardcover); 978–0–19–006951–3 (paperback)

*To my mother Bridie Toal,
and my daughters Sirin and Nives*

TABLE OF CONTENTS

Figures and Tables   ix
Acknowledgments   xi
Note on Place Names   xvii

Introduction: Near Abroads   1

1. Why Does Russia Invade Its Neighbors?   17

2. Geopolitical Catastrophe   55

3. A Cause in the Caucasus   93

4. Territorial Integrity   126

5. Rescue Missions   166

6. Places Close to Our Hearts   198

7. The Novorossiya Project   237

8. Geopolitics Thick and Thin   274

Notes   303
Index   371

FIGURES AND TABLES

Figures

1.1 Map of Russia and Its Neighborhood   37
1.2 Post-Soviet Space as a Contested Geopolitical Field   38
2.1 School Number 1, Beslan, North Ossetia. Putin described the September 2004 terrorist attack there as an international plot against the territorial integrity of the Russian Federation. Author photo.   91
4.1 Map of South Ossetia   136
4.2 Memorial to Ossetian refugees killed by Georgian irregulars on the Zar Road, May 29, 1992. Author photo.   137
4.3 Eduard Kokoity, Tskhinval(i), dedicating a statue to Pushkin, March 2010. Author photo.   141
5.1 Valery Gergiev conducts his Mariinsky Orchestra in front of the destroyed local assembly building in Tskhinval(i), August 21, 2008 [Maxim Shipenkov, EPA].   187
5.2 "Thank You Russia!" Billboard, Tskhinval(i), March 2010. Author photo.   189
5.3 Gutted houses, Kurta, former "capital" of Georgian Provisional Government in South Ossetia, March 2010. Author photo.   196
6.1 Map of Southeast Ukraine   200

6.2 Crimea's Choice as Nazism or Russia. Referendum Billboard, Sevastopol [Zurab Kurtsikisza, EPA]. 226
6.3 Russian president Vladimir Putin (2-R) with Crimean leaders (L-R), Prime Minister Sergei Aksyonov, parliamentary chair Vladimir Konstantinov, and Sevastopol mayor Aleksei Chaly sign the formal request for Crimea to join Russia in the Grand Kremlin Palace, March 18, 2014 [Alexey Druzhinyn, RIA Novosti, EPA]. 229
6.4 Putin addresses the Red Square music concert marking the one-year anniversary of Crimea's annexation, March 18, 2015 [Mikhail Klimevtyev, RIA Novosti]. 230
6.5 Monument to Annexation Troops as "Polite People" in Simferopol, Crimea. Photo courtesy of anonymous. 231
6.6 Self-Perceptions of Europeanness, SE6 and Crimea 233
6.7 Attitudes in SE6 and Crimea toward description of Euromaidan as a fascist coup 235
6.8 Attitudes in SE6 and Crimea toward Crimea Joining Russia 235
7.1 Pro-Russian protesters standing in front of an occupied police station, in Sloviansk, Ukraine, April 13, 2014 [Anastasia Vlasova, EPA]. 259
7.2 Support in SE8 for the Introduction of Russian Troops into Ukraine, Mirror/KIIS Survey, April 2014. 263
7.3 Support in SE8 for Secession from Ukraine. Mirror/KIIS Survey April 2014. 263
7.4 Attitudes toward Novorossiya: Historic Fact or Myth? 271
7.5 Attitudes toward Novorossiya: Basis for Separatism? 271
7.6 Attitudes toward Novorossiya: What is Novorossiya? 271
7.7 Attitudes toward Donbas Joining Russia Referendum, SE6 and Crimea 272

Tables

2.1 Competing Visions of Russia 72
2.2 Evolution of Russia's "Monroe Doctrine," March 1992–December 1993 83
8.1 Russia's Geopolitical Archipelago 284

ACKNOWLEDGMENTS

Geographies matter. I want to recognize how much this book owes to various research institutions, disciplinary networks, locations, and people. First, I would like to acknowledge sustained research support from the U.S. National Science Foundation (NSF). A small grant for exploratory research from the Geography and Regional Science program in late 2001 (award #0203087) allowed me and my colleagues to examine the 9/11 attacks and response in Russian geopolitical culture. A 2004 grant from the Human and Social Dynamics Initiative (award #0433927) supported research on the comparative dynamics of civil war outcomes in Bosnia-Herzegovina and the North Caucasus. A 2008 grant from the same initiative (award #0827016) supported a research project on Eurasian de facto states after the independence of Kosovo. Finally, a RAPID grant (award #14-1442646) from the Political Science Program supported research on attitudes and beliefs in Russian-supported de facto states and in southeast Ukraine in the wake of the annexation of Crimea. All of these grants supported social science surveys of public attitudes as well as travel to Russia, the Caucasus, Moldova, and Ukraine to conduct elite interviews and research ground-level realities. I am extremely grateful for this support, which enabled me and my colleagues to observe how unresolved territorial questions played into deteriorating U.S.-Russian relations since 9/11.

Second, I owe a great deal to my colleagues in these research endeavors. My largest debt is to John O'Loughlin (known as "JohnO"), my first graduate adviser more than three decades ago, and a research collaborator and friend since. Under JohnO's supervision I began to think like a political geographer and to place my upbringing in Ireland's borderlands into a larger context. I have always been grateful for his initial support, most especially as my theoretical interests at the time were different from his. I completed my graduate education in political geography at Syracuse University with John Agnew, another terrific mentor. At a political geography conference in Prague in August 1991, both JohnO and I met the distinguished Russian political geographer Vladimir Kolossov for the first time. The encounter was memorable because the coup against Gorbachev had just collapsed and political life was in great flux across the Soviet Union. JohnO and Vladimir subsequently collaborated on a series of research projects in Russia, Ukraine, and Moldova. It was only after a conference in September 2001 that the three of us started working together on the NSF-supported projects listed above. I am grateful to Professor Kolossov for his passion for field research, his spirit of collaboration, and his explanations and translations.

I also want to acknowledge his colleagues in the Institute of Geography at the Russian Academy of Sciences for providing an open welcome to Russian and Soviet-era networks that facilitated our travel to conflict regions. We visited North Ossetia and other ethnic republics in the North Caucasus in 2007, Transnistria in June 2009, Abkhazia in November 2009, South Ossetia in April 2010, Nagorny Karabakh in June–July 2011, and Ukraine in 2014. I would like to thank the following academics for discussions and occasional travel companionship during these visits: Andrei Gertsen, Mladen Klemencic, Nikolai Petrov, Jean Radvani, Vadim Saltikovsky, Olga Vendena, and Andrei Zubov. Thanks are due also to our research partners in the various sociological survey institutes we have used, most especially Alexei Grazdankin at the Levada Center in Moscow and Viktoriya Remmler in Krasnodar (who led our Abkhazia research), Volodymyr Paniatto and Natalia Kharchenko at the Kyiv International Institute of Sociology, Khasan Dzutsev in Vladikavkaz, Ion Jigau and CBS-AXA in Chișinău, Elena Bobkova in Tiraspol, and Gevork Poghosyan in Yerevan. Let me also acknowledge the many local officials and academics in these locations who met with us and provided their perspective and analysis. I am much wiser because of this. A special thanks to Vitaly Belezerov in Stavropol; Eldar Eldarov, Shakhmardan Muduyev, Sharafudin Aliyev, and Zagir Atayev in Makhachkala; Arthur Tsutsiev in Vladikavkaz; and Kosta Dzugaev and Alan Kharebov in Tskhinval(i).

As this book went to press, the Russian government decided to classify the Levada Center, Russia's leading independent polling agency, as a "foreign agent" because it worked with international academics, like JohnO and I, on social-scientific research. This move is part of a long-standing effort by the state to control independent knowledge about Russian attitudes and society. Unjust on its own terms, it is also counterproductive in that fictions and preconvictions now have an easier time not only within Russia but beyond it. At the very moment we need more social science facticity, we have post–factual politics.

Third, living and working in the Washington, DC, metro region shaped my decision to organize and write this book for a general audience rather than a strictly academic one. I am very grateful to my institutional home, the School of Public and International Affairs (SPIA) at Virginia Tech in the Washington metro region, for supporting my research down the years. In particular I would like to thank Timothy Luke for years of unstinting support; the current director of SPIA, Anne Khademian; and my colleagues in the Government and International Affairs program—Joel Peters, Giselle Datz, Patricia Nickel, and Ariel Ahram. I would also like to thank those students who provided research assistance—Megan Foran, Adis Maksic, Gela Merabishvili, Jeffrey Owen, and Emil Sanamyan—and those whose own research efforts helped inform my thinking: Julie Ademack, David Belt, Sonya Finley, Marc Jasper, Kevin Joyce, Walter Landgraf, Matthew Osterrieder, Mirian Popkhadze, Bryan Riddle, Christopher Lee Walker, and Heather King Westerman. Among research institutions, I found regular intellectual sustenance in the Kennan Institute; George Washington's Institute for European, Russian, and Eurasian Studies (IERES); and the Carnegie Endowment for International Peace. Thanks to them and others for hosting excellent events and speakers. I would like to thank the following individuals personally for helping me deepen my knowledge of the subject matter of this book: Kristen Bakke, Laurence Broers, Michael Cecire, Jon Chicky,

Colin Cleary, Alexander Cooley, Simon Dalby, Dan De Luce, Thomas de Waal, Valery Dzutsev, Geraldine Fagan, Eugene Fishel, Julie A. George, Dorota Gierycz, Giorgi Gogsadze, Thomas Graham, Magdalena Grono, Michael Haltzel, Fiona Hill, Edward Holland, Volodymyr Ishchenko, Stephen F. Jones, Kornely Kakachia, Richard Kauzlarich, Natalie Koch, Michael Kofman, Volodymyr Kulyk, Taras Kuzio, Marlene Laruelle, Philippe Leroux-Martin, Sergey Markedonov, Derek McCormack, Wayne Merry, Lincoln Mitchell, Steven Lee Myers, Niklas Nilsson, Donnacha Ó Beacháin, Olga Onuch, Caitriona Palmer, Paul Quinn-Judge, Matthew Rojansky, Angus Roxburgh, Eugene Rumer, Richard Sakwa, Dmitri Sevastopulo, Paul Stronski, Courtney Weaver, Andrew Weiss, and Elizabeth A. Wood. Academics working on contemporary conflict owe a great deal to the tremendous courage, dedication, and professional integrity of journalists, research analysts, and photographers covering these regions. Given the relentless information war enabled by 24/7 news channels and social media platforms that now envelop conflicts, high-quality professional journalism needs our financial support more than ever.

I want to acknowledge the help of the former diplomats, officials, and others from different countries who took the time to speak to me in interviews about their experiences and insights. There were some former officials I was unable to speak with for various reasons: I remain open to doing so in the interest of deepening the subject material of this book. One tragic absence is that of Ron Asmus. His untimely death prevented us from developing the conversation we started when he presented his book at Politics and Prose in my neighborhood in Washington, DC.

I want to recognize the openness shown by the Georgian Embassy in the United States. Ambassador Archil Gegeshidze and Deputy Chief of Mission George Khelashvili represent their country with dignity and refined skill. At Oxford University Press I would like to thank David McBride for feedback on an initial draft, and Emma Clements at Newgen. Thanks to Nancy Thorwardson at the University of Colorado for preparing the maps and Luis Liceaga for creating Figure 1.2. Finally, this book was greatly enhanced by the labor of friends and colleagues reading draft chapters. Dr. Ralph Clem read the first complete draft of the manuscript and provided valuable feedback. I thank Marlene

Laruelle for her comments on a draft of chapter 2; Laurence Broers for rich comments on chapters 3, 4, and 8; Arthur Tsutsiev for helpful observations on chapters 4 and 5; and Andrew Wilson for feedback on chapter 7. A forum on the finished manuscript draft, organized by Cory Welt at IERES in July 2016, was invaluable. I am deeply indebted to all those who participated—Susan Allen, Henry Hale, Edward Holland, Robert Orttung, Mykola Riabchuk, Sarah Wagner, Cory Welt, and Sufian Zhemukhov—and for their suggestions and criticisms not only then but over the years. I want to specially acknowledge Sarah, whom I first met in Srebrenica in 2004, as a great friend, insightful commentator, and accomplished soccer player. All these interlocutors helped sharpen the argument and push it toward greater clarity. None are responsible for any errors that may remain.

Finally, home is the most important place of all. This book would not have been possible without the love and support I receive from my family, my wife Sabine Durier and our two lively daughters Sirin and Nives. They provide joy on a daily basis and are power stations of happiness in times that are filled with dark events, distressing images, and disturbing trends. Our family life is greatly augmented by Mila Anos, whose household work helped make further time for writing possible. This book is dedicated to my mother, who made everything possible in the first place, and to Sirin and Nives, who are already writing their own stories.

NOTE ON PLACE NAMES

Geography literally means "earth-writing." Place-naming practices are part of how people write political geographies and live them. Names are often chosen to signify ownership of a territory or place, symbolically marking the political and cultural dominance of one group, and one geopolitical relationship, over others. In Ireland, naming a town Kingstown or Dún Laoghaire, Londonderry or Derry conveys power and identity. The lands of the former Soviet Union have similar postcolonial contentious dynamics over place names. Multiple claims and languages are at play, as are abundant historical legacies and memories. In most Union Republics, Russian was the dominant language, and the Russian administrative name for places tended to predominate locally and internationally. Since the Soviet collapse the newly independent states have made varying efforts to nationalize place names, renaming towns, streets, and other places to privilege new nation-state heroes and dates while removing Soviet symbols.

The place names in this book are, for the most part, simplified English language transliterations of official state language names. Thus, the names of oblasts and other locations in Ukraine are from Ukrainian not Russian. I use Kyiv instead of Kiev, Donbas instead of Donbass, and Kharkiv instead of Kharkov. Because language privileging, place naming, and territorial ownership are part of the contentious geopolitics examined in this book, I often provide competing names and forms of knowing places in the text. Thus, I use the Russian names Odessa, Kharkov, and Nikolaev when appropriate and when used by the speaker. Though Donetsk has a different spelling and pronunciation in Ukrainian and Russian, its common English transliteration misses this. Not so for Luhansk/Lugansk: I use transliteration from Ukrainian to name the oblast but from Russian to name the Lugansk People's Republic as this is its official name. Those with zero-sum mentalities will inevitably find problems with this but part of the pedagogic value of the strategy adopted in this book is to foreground place-contestation, and the life-world that is part of it, on the page.

The situation in the breakaway regions of Georgia is particularly complicated because one has trilingual geopolitical dynamics: titular nations (Abkhaz and Ossetians), a nationalizing state (Georgia), and a former imperial center that is now a privileged geopolitical player (Russia). This is to say nothing of nontitular minorities like the Armenians in Abkhazia. The politics of naming in South Ossetia begins with the existential question of whether that name is even recognized and acknowledged. The Georgian government abolished this region and name in December 1990. In Georgia, the north-central region of the country is named Shida Kartli (lit., "Inner Kartli"). The area around the capital of South Ossetia is called Samachablo (lit., "fief of the Machabeli clan"). The names are ownership claims that locate the area within Georgian nation-space imagining. Because South Ossetia nevertheless endured as a geopolitical fact on the ground the Georgian government began to refer to it euphemistically as the "Tskhinvali region." The name was a diminishing gesture in the face of the unilateral proclamation by those in power there that it was the Republic of South Ossetia. The practice I have followed in naming the two regional centers ("capitals") of the breakaway territories in Georgia is to use a form that signifies the contested name of these places, without privileging either. Thus, the Ossetian name Tskhinval and Georgian Tskhinvali is rendered Tskhinval(i). The same applies to Sukhum(i), the administrative center of Abkhazia.

As far as Georgians and most all the international community are concerned, South Ossetia and Abkhazia are Georgian territory. To the de facto regimes in these areas, however, Georgia begins at the de facto boundary lines Russia has helped them establish in the region. I follow the convention of using the terms "uncontested Georgia" and "Georgia proper" to refer to Georgian territory beyond the boundary lines of the de facto states. I recognize that these terms are objectionable to Georgia but they have the virtue of recognizing the material realities on the ground that persist in the face of imagined seamless maps of territorial integrity.

The Georgian government controlled enclave settlements within South Ossetia and Abkhazia until August 2008. The rule I adopt is to respect the place names that the majority of local residents use, with the alternative form in brackets. Thus, I use the name Tamarasheni (lit. "built by Tamar," a famous queen in the Georgian pantheon) for the first Georgian settlement immediately to the north of the town limits of Tskhinval(i) instead of the Ossetian name Tamares. In the last Soviet census of 1989, this settlement had both Ossetian and Georgian families, some intermarried. As a consequence of the violence of 1990–1992, Tskhinval(i) became predominantly Ossetian, and the settlements to its north overwhelmingly monoethnic Georgian communities. In saying this, however, we are summarizing a condition brought about by violence that forced people into either/or ethnic categories that may not reflect their kinship histories or actual beliefs and lived identities. I use the place name Akhalgori for the largely ethnic Georgian town to the southeast that was under Georgian government control until August 2008. In Soviet times it was known as Leningori. The de facto Republic of South Ossetia authorities privilege the Ossetian variant of this Soviet name: Leningor. Other Soviet names endure in South Ossetia. Tskhinval(i) today has both a Lenin Avenue and a Stalin Avenue, the latter figure an Ossetian folk hero (as he is also for some Georgians). Another Soviet name that endures is Roki tunnel, which is the Georgian form of the ethnic Ossetian village of Rouk on the southern slope of the Caucasus where it derives its name. I stick with the familiar Roki and not the Ossetian name. I use the Ossetian name Styr Gufta for the predominantly Ossetian settlement north of the Georgian enclave on the Transcaucasian Highway (TransKam). I also give the Georgian name, which is Didi Gupta. The strategic bridge outside the town is known as the Gufta bridge in Ossetian, and Gupta in Georgian. The central river in South Ossetia is Styr Liakhva in Ossetian, Didi Liakhvi in Georgian, and Bolshoi Liakhvi in Russian. I use the translation "Greater Liakhvi River" and reserve "Didi Liakhvi Valley" specifically for the Georgian enclave north of Tskhinval(i), as this helps signify its ethnic Georgian character. Its destruction was a concerted effort to erase that cultural identity. Campaigns of purification and erasure by the victorious, unfortunately, are all too common in the wake of episodes of ethnicized and geopoliticized violence.

In 2015 Ukraine's parliament launched a "decommunization" process that sought to erase Soviet names and symbols across Ukrainian territory. Thousands of Lenin statues have been toppled in Ukraine since its independence. A new wave started with the Euromaidan protests and continues, now legitimated by decommunization laws. Certain place names, like Dnipropetrovsk (now Dnipro), came too late to update maps in this work. Hundreds of place names were mandated for change in Crimea also, with a few replacement place names recognizing the heritage of the Crimean Tatar, heretofore ignored by Kyiv. A move by U.S. technology giant Google to implement these changes sparked outrage in Moscow and Crimea. One Russian lawmaker charged the U.S. company, cofounded by a Russian immigrant, with "topological cretinism." Place-naming controversies, and asterisks on maps indicating disputed territorial status, are expressions of a place remaking geopolitical contest that unfortunately looks likely to continue for some time. Geographies are inevitably political but they need not be about singular domination and control. They can be thought and lived differently.

Introduction
Near Abroads

ON MY THIRD EVENING in Russia, the world changed. I was in Stavropol, a city founded by Prince Gregory Potemkin at the time of the American Revolution as one of ten fortresses to defend the borders of the expanding Russian Empire. To the south were the Caucasus, formidable mountains and myriad peoples. Stavropol grew as an administrative center of tsarist and later Soviet power. It briefly fell to the Wehrmacht in 1942 as the invading army drove unsuccessfully toward the oilfields of Baku. Later, a popular young party secretary from the area got noticed in Moscow, joined the Politburo, and in 1985 became general secretary of the Communist Party of the Soviet Union. Mikhail Gorbachev's reforms would inadvertently lead to a geopolitical earthquake, the end of the Cold War in Europe, and the unthinkable—the collapse of the Communist empire built by Lenin and Stalin.

That evening the provost of Stavropol State University toasted the health of the international academics attending the conference starting the next morning. Many other benevolent toasts were exchanged, and a singularly somber one. A researcher with the Memorial Human Rights Center reminded us that a war raged nearby in Chechnya, an "inner abroad" of Russia. Here Russia's new president had approved the indiscriminate shelling of a Russian city and a dirty war against citizens redefined as "terrorists." Returning to our hotel that evening in a bus under armed guard, a Croatian friend and I were chatting when told to turn on the television. Russian television was broadcasting footage of airplanes crashing into skyscrapers in lower Manhattan on what seemed like a continuous loop. The full magnitude of what had happened was only apparent the next day. Like many, the Twin Towers were entwined with personal memories—first seeing them in rural Ireland on a pennant my uncle brought back from his vacation to New York, and later visiting the observation deck with my parents and friends. Furthermore, the attack on the Pentagon was only two miles from my home, a few more from where I worked, and all too close to some former students who worked in the building. The entangled experience of proximity and distance only deepened when a speaker began the conference by declaring that now the United States would understand what Russia had been fighting for years. Chechnya is everywhere, he warned. Russia and the United States, in his opinion, needed to form an alliance of civilized powers against international Islamic terrorism.

President Putin had said as much in his initial statement on the September 11 attacks. They went "beyond national borders." They were "a brazen challenge to the whole of humanity, at least to civilized humanity." The international community needed to pool its efforts in the struggle against terrorism, the "plague of the twenty first century."[1] In the weeks that followed, Russia offered the United States intelligence cooperation, use of Russian airspace, and no objections to it taking over former Soviet military bases in Central Asia to prosecute a war against the Taliban regime in Afghanistan. Such cooperation, controversial within Russia at the time, would not last.[2] Within a matter of months, disenchantment at the unilateralism of the Bush administration took hold. It only deepened as the United States legitimized preemptive war and then invaded Iraq on false pretenses and overthrew the country's longstanding regime. In August 2008, Russia went to war with Georgia, a country that had sent troops to support U.S. military efforts in Iraq and Afghanistan. In late February 2014 Russian troops seized Crimea, while pro-Russian activists destabilized Ukraine's east in response to what President Putin portrayed as a Western-sponsored "fascist coup" in Kyiv (Kiev in Russian). What began as a geopolitics of solidarity and unity in September 2001 devolved over the course of more than a decade into a geopolitics of mutual antagonism and paranoia.

This book examines the making of the geopolitical struggle between Russia and the United States over Georgia and Ukraine, two important Soviet successor states on the western and southern borders of the core successor state, the Russian Federation. The English phrase "near abroad" first emerged in 1992 as a consensus translation of the Russian phrase *blizhneye zarubezhye* (lit., "near beyond border"). That phrase was a popular geopolitical label Russian politicians, in the wake the Soviet Union's sudden dissolution, gave to the former Soviet Republics that were now independent sovereign countries in their own right.[3] The phrase acknowledged difference yet also enduring proximity. Russia's new borderland states were familiar parts of the same country for decades and, for most places, lands within the Russian Empire before then. It was difficult for many Russians to let go of memories of these places as parts of their country, just as it was difficult for some beyond Russia to reconcile themselves to living in a separate new country. Confusion and mixed emotions were common. Many in the West heard in the phrase "near abroad" reluctance on the part of Russia to acknowledge the full sovereignty of the new post-Soviet states.[4] "Near abroad," they argued, was a new term for an old desire on the part of Russia to have a sphere of influence in the lands next to its borders. But near abroad was not one essential thing. It simultaneously named a new arrangement of sovereignty and an old familiarity, a longstanding spatial entanglement and a range of geopolitical emotions.

The other successor states of the Soviet Union had their own near abroads, countries they looked to with aspirations and hopes. The Baltic states (Estonia, Latvia, and Lithuania) looked to Scandinavia. The new countries in Central Asia and the Caucasus looked to Turkey and later China, most with ambitions to reestablish historic linkages and acquire contemporary investment and support. Armenians looked to their successful diaspora in Europe and North America as did some Ukrainians. Those in Ukrainian Galicia looked toward neighboring Poland, with whom they had a contentious history. Georgia looked to Turkey but also to the United States where, after fragmentation and civil war, its well-known leader, Eduard Shevardnadze, was remarkably successful in attracting financial support for his beleaguered country. Between 1992 and 2001, the United States gave over $1 billion in financial assistance to Georgia, making it among the largest per-capita recipients of U.S. aid worldwide.[5] The United States also made it possible for some young Georgians to study in America. One of these students was Mikheil Saakashvili, who gained admirers in Washington and returned to Georgia with an ambition to transform his country into an America in the Caucasus. Though thousands of miles away, Georgia under Saakashvili adopted the United States as its "near abroad," and a relationship of fealty and moral debt, enabled by globalization and the global war on terror, blossomed. Married to a Ukrainian American who had worked in the U.S. State Department and Reagan White House, the leader of Ukraine's Orange Revolution, Viktor Yushchenko, also looked to Washington and Brussels for financial aid and geopolitical support. Both leaders cooperated to create a new interstate organization, the Community of Democratic Choice, as a vehicle for moving their states beyond Russia's orbit and into aspirational alignment with the European Union and the North American Treaty Organization (NATO).

Geographically concentrated minority groups within Ukraine and Georgia did not share this geopolitical ambition. Indeed, before it was even expressed, some started contesting the prospect of the then Soviet Republics acquiring greater power. In July 1990 a declaration of sovereignty by Soviet Ukraine's parliament heightened concerns among ethnic Russian Ukrainians in Crimea and elsewhere. In Odessa, a local professor formed a group called the Democratic Union of Novorossiya (New Russia).[6] As Soviet Georgia proclaimed a return to the constitution of the pre-Soviet Democratic Republic of Georgia, mobilized ethnopolitical factions in Abkhazia and South Ossetia sought to secede. In this they were following examples set by Armenians in the Nagorny Karabakh oblast within Azerbaijan and Soviet factory bosses and their Russophone workers in Transnistria, a region of Soviet Moldova. From regional rebellions in Azerbaijan, Georgia, and Moldova (those attempted in Ukraine were unsuccessful) came forth a new and surprisingly enduring geopolitical space: the post-Soviet de facto state. For these unrecognized polities, ethnic kin communities in nominally separate neighboring states and symbols of the Soviet Union were near and not abroad.

For most Americans, the geopolitical dramas of post-Soviet space unfolded in faraway places of which they knew little. But the United States had its own domestic entanglements with post–Cold War Europe and post-Soviet space that were to prove significant. Ethnic communities from these lands cared about anchoring them in the West, and they lobbied successfully for the extension of NATO membership to Poland, Hungary, and the Czech Republic. Both an imagined transnational community—a Judeo-Christian West—and personalized affective ties mattered in making this possible. Prominent European Americans like Zbigniew Brzezinski (born in Warsaw in 1928), National Security Advisor to President Jimmy Carter, and Madeleine Albright (born in Prague in 1937), the United States' first female secretary of state, were powerful advocates. George W. Bush's White House helped extend NATO farther east in March 2004, incorporating not only former Warsaw Pact territories (Bulgaria, Romania, and Slovakia) but also the Baltic states, remembered in Cold War America as "captive nations" forcefully incorporated by the Soviet Union in 1940 and again in 1945. Ukraine and Georgia, founding territories of the Soviet Union, did not have powerful ethnic lobbies in the United States but they did have friends in high places, and people interested in spreading freedom and democracy to the very borders of Russia. In Senators John McCain, Joe Lieberman, Lindsey Graham, Joe Biden, Carl Levin, and their like, men who came of age at the height of the Cold War, they had influential advocates.

When Mikhail Gorbachev first spoke of a "common European home" in 1987, the administration of George H. W. Bush—which took office two years later, in 1989—countered with the slogan, "a Europe whole and free."[7] Both notions were deliberately vague, but Bush's catchphrase expressed what the NATO alliance states wanted. The phrase rearticulated in positive terms a longstanding Cold War aspiration to roll back the Soviet Empire on the European continent. Europe was a continent of captive nations awaiting liberation from a repressive imperial machine headquartered in the Kremlin. The end of the Cold War in Europe and the Soviet Union's subsequent collapse seemed to fully affirm this vision. Furthermore, the violent dissolution of Yugoslavia provided NATO with the opportunity to present itself as an indispensible security organization for advancing a Europe whole and free.

There was considerable ambivalence about the place of Russia in this geopolitical vision. On the one hand, Russia was the historic enemy, the non-West, the main reason its former client states sought the protection of NATO membership. On the other hand, as Gorbachev's vision underscored, Russia was a European state, and furthermore it was culturally, geographically, and civilizationally part of the West. NATO's policy of containment was widely credited with "winning" the Cold War. Its architect, George Kennan, had no ambivalence about NATO expansionism, describing it as a "strategic blunder of potentially epic proportions."[8] "Enlargement," NATO's term of choice for expansion, went ahead nevertheless. Aspirant state leaders wanted it, while President Bill Clinton was keen to demonstrate the United States' indispensable leadership and to reap potential domestic political advantages.[9] Russia was presented with a fait accompli to which it would have to adjust. Robert Hunter, the Clinton administration's ambassador to NATO (1993–1998), oversaw the first round of NATO expansionism but subsequently became skeptical of further expansionism.[10]

In a new 1999 Strategic Concept document, NATO presented itself as "one of the indispensable foundations for a stable Euro-Atlantic security environment, based on the growth of democratic institutions and commitment to the peaceful resolution of disputes, in which no country would be able to intimidate or coerce any other through the threat or use of force."[11] Its open door toward new members said nothing about geostrategic location. The prevailing geopolitical descriptions were either simply "Europe" or "the Euro-Atlantic area," the latter specified as the territory of the participating states of the Organization for Security and Co-operation in Europe (OSCE), a group that now comprises over fifty states stretching across three continents.[12] In ignoring geography, foregrounding "democratic values," and expanding eastward toward Russia's borders, NATO looked like it was realizing the Cold War dream of rollback. Certainly many of its biggest supporters in the U.S. Congress saw it that way. Proponents spoke of a "blue blob" of democracy enlarging now that the "red blob" of totalitarianism had disappeared.[13] Enduring imaginary collectives like "the free world" and "the West" became commonplace phrases like "Europeanizing," "joining the West," "completing Europe," and "extending the borders of freedom"; all revealed the implicit geopolitical teleology at work. So also did a subtle divergence in the meaning of "Euro-Atlantic" from simply naming a three-continent space of plural political systems to naming also the aspirational space being created by NATO and European Union integration efforts. Russia and Central Asia were part of the first but not the second.

Many prominent officials from NATO's new member countries—famously dubbed "new Europe" by Donald Rumsfeld in 2003—amplified the idea of liberation, in the process touting their past victimhood and framing contemporary tensions with Russia as a civilizational struggle.[14] They made common cause with the Bush administration's democracy promotion efforts, which became known as the Freedom Agenda by 2004. In 2008 they strongly supported the efforts of the Bush administration to grant a NATO Membership Action Plan—ironically, for a policy ostensibly indifferent to location, abbreviated as MAP—to Ukraine and Georgia.

The Bucharest Declaration of April 2008 is central to this book. Though not approving a membership action plan to Georgia and Ukraine, the North Atlantic Council nevertheless declared: "we agreed today that these countries will become members of NATO."[15] It was an extraordinary pronouncement in many ways. First, it was a declaration of intent to expand NATO into territories that were intimately entwined with Russian and Soviet history, identity, and territory. This is not simply a matter of noting that Crimea, home of the extant Russian Black Sea Fleet, was formally part of Russia until 1954. No country that was part of the *original* Soviet Union, whose territorial borders were drawn by Bolsheviks from tsarist spatial legacies, had ever joined "the West," not to mention the U.S.-led Cold War transatlantic alliance arrayed against the Soviet Union for decades. Second, NATO sought to incorporate these states despite the Russian government communicating that its expansionism there crossed a "red line" as far is its national security was concerned. Third, it sought to incorporate these two states despite the fact that they both had longstanding internal divisions over NATO membership and significant ethnoterritorial polarization that was inevitably going to be exacerbated by the move.

The Bucharest Declaration was the moment the "near abroads" of NATO and Russia clashed head on. Indeed the year 2008 was a moment of significant rupture in the post–Cold War security order in Europe. Kosovo unilaterally declared itself independent from Serbia and was recognized as a new state by the United States and other European states (but not Russia). In August 2008, Georgia and Russia went to war over the breakaway territory of South Ossetia, recognized at the time as a region within Georgia even by Russia. Within weeks Russia had unilaterally recognized both secessionist regions of Georgia—South Ossetia and Abkhazia—as independent states. Tensions with Russia were high in Ukraine, but these eased somewhat when Russia's favored candidate, Viktor Yanukovych, won the presidential election there in 2010 (thanks, in part, to a political makeover supervised by the U.S. political consultant Paul Manafort). Six years earlier the Orange Revolution had deprived Yanukovych of this office. NATO membership aspirations were suspended. Instead, it was Russia that sought to lock Ukraine into a competing geopolitical project, the Eurasian Economic Union. Yanukovych's rejection of an alternative European Association Agreement in November 2013 brought people into the streets again to protest. Ukraine, as they saw it, faced a "civilizational choice" and should become European not Eurasian. As is well known, the protests spiraled out of control and became a challenge to Yanukovych's government. Fearing the worst, he fled, and a pro-Western government took power. This triggered Russia's invasion of Crimea and subversion of southeast Ukraine, a story that is still unfolding.

This book offers a critical geopolitical analysis of these events as moments in the making of contemporary European history and world order. *Critical geopolitics* is a form of scholarly criticism of the discourse and practice of geopolitics; it is to geopolitics what literary criticism is to literature, a structure of thought and knowledge existing in its own right, with some measure of independence from the speech acts, performances, and practices of the actors it examines.[16] Its institutional origins are within the field of Anglo-American political geography, and it draws inspiration from a broad range of Euro-American critical theory and social science. This book develops a critical geopolitical analysis by building out three conceptual foundations of this approach. The first is the notion of a *geopolitical field*. Classic geopolitics tends to understand the state as a naturalistic and contingent territorial achievement. It has long emphasized the importance of the geographical setting within which empires and states operate as they pursue security and prosperity. Critical geopolitics rejects the geo-determinism and naturalization of prevailing prejudice found in classic geopolitics. It operates with a more expansive and open conception of the geographical setting of statecraft, one concerned with how power structures (like states and markets) have produced spaces and places, territories and landscapes, environments and social agents. A geopolitical field is both the sociospatial context of statecraft and the social players, rules, and spatial dynamics constituting the arena.[17] The international state system rests on norms concerning the parcelization of space into agreed sovereign state territories. This delimiting and dividing is often deeply quarrelsome, for there are few natural borders—just socially agreed ones.[18] The borders of the successor states of the Soviet Union were particularly contentious, drawn and redrawn by a Communist ruling elite in response to considerations that only partially accommodated demographic realities and local sentiments. The nationalist wave that overwhelmed the Soviet Union had an early start in a particularly contentious border dispute in the southern Caucasus between Armenia and Azerbaijan over the predominantly Armenian enclave of Nagorny Karabakh, a district first established and located within Soviet Azerbaijan by Communist bureaucrats in 1923.[19] The Soviet Union contained many such contentious territories. Its collapse allowed the emergence of popular mobilization to challenge and change these boundaries.

This book argues that Russia's near abroad needs to be seen as a particular type of geopolitical field, one featuring a quintet of players and territories that were in discord with each other. The general field is postcolonial—metropolitan state and former colony—but regional concentrations of populations loyal to the former imperial metropole created conditions for state territorial fragmentation. A normative great power center ("the West") on the horizon and territorial separatists ("terrorists") within the metropole further complicate the nexus of interactions between the players and spaces. This book is informed by fieldwork in all four post-Soviet de facto states created by the initial dissolution of the Soviet Union (the process arguably is still unfolding). Visiting these places allows an understanding of how local elites view their condition, their aspirant state, and its future. Subsequent research on public attitudes in these contested regions provided further insights.[20] The U.S. National Science Foundation has generously supported my endeavors and those of my colleagues in this respect. In revealing the attitudes of ordinary residents of contested territories, we are not advocating on their behalf but contributing to a fuller understanding of the dynamic geopolitical field characterizing post-Soviet space.

A second conceptual foundation is thinking in terms of *geopolitical cultures*.[21] As chapter 1 explains, states are territorial power structures that come into being with distinct spatial identities and understandings of their position and mission in the world. This spatial identity and the ongoing debate about it define a state's geopolitical culture. In this book, geopolitics is not a perspective or an approach to international politics.[22] Instead, as a culture, it is how states see the world, how they spatialize it and strategize about the fundamental tasks of the state: security, modernization, the self-preservation of identity. The U.S. geopolitical culture is somewhat unusual in that geographic knowledge about the rest of the world is not necessarily central to how it works. There is, ironically, a geographical explanation for this absence of geography. Because it is a continent-sized country in its own hemisphere, the United States is its own world for most of its citizens. Knowledge of the wider world among the U.S. public is not particularly deep, nor is it needed in most instances. Consequently U.S. geopolitical culture has tended to be characterized by a moral rather than a geographic approach to the world beyond the shores of the United States. Ambrose Bierce's acidic remark a century ago—that "war is God's way of teaching Americans geography"—pointed to a persistent challenge for the U.S. foreign policy elite.[23] To explain international crises and, potentially, the need for U.S. intervention, leaders had to situate the places involved in larger frames, narratives, and emotions. Wars, in short, require active rhetorical location—the interpretative explanatory contextualization that gave war meaning and significance. For that they turned historically not to geography but to scripture and to messianic visions of an idealized American creed, the spreading of freedom, liberty, and democracy to all corners of the earth. Bierce did not quite have it right. War brought out a God-infused way of teaching geography; setbacks and defeats in wars taught Americans hard lessons about the actualities of world geography.

Critical geopolitics seeks to analyze geopolitical cultures in all their complexity, isolating organizing myths, favored narrative forms, prevalent conceits, and competing traditions within different cultures.

Here fieldwork also provides important insights. While we are awash in media-generated geopolitical culture all the time, I have had the good fortune of living and working in Washington, DC, for almost two decades. This has given me a unique vantage point into the making of U.S. geopolitical culture and how the geopolitical cultures of other states intersect with it. It has also provided me with access to some of the diplomats and leaders described in the following chapters. Informal conversations with a great variety of people—working and retired diplomats, think tank analysts, journalists, embassy officials, lobbyists, elected representatives and their staffs, quasi- and nongovernmental officials and activists—as well as attendance at public events over the years—has helped a great deal in the creation of this book. This inevitably brings some degree of bias, conscious and unconscious. It is worth noting that I am not a member of any Washington foreign policy think tank or association. I identify as both European and American, but my social distance from U.S. diplomats is a lot less than it is from Russian diplomats. I have strived in this work to preserve a scholarly distance from U.S. geopolitical culture, to describe the conditions of its production and operation and how visiting leaders seek to "play" it. While this study is deeply critical of U.S. geopolitical culture, this criticism nevertheless comes from a position within this culture.

I have sought to make an empathetic stretch toward Russian geopolitical culture throughout this work. It is vital to understand how that culture, and its leading articulator, Vladimir Putin, framed the question of Russia's relationship with post-Soviet space. Critical geopolitics takes discursive practices seriously and seeks to analyze the connections and interactions between different geopolitical cultures. Discursive practices are particularly important to study in conflict situations as they feature communication failures and breakdowns in the ability of parties to comprehend each other. Currently, any empathetic presentation of Russian geopolitical discourse in the West today faces social opprobrium. There is fear that using Putin's words or presenting his perspective will somehow legitimate his point of view—that it will infect thought and confuse or relativize what should be clear moral distinctions. Commentators of this persuasion tend to police articulations of Russian geopolitical discourse by framing it as "propaganda" or "information warfare." Some are quick to label those who present empathetic readings of Putin's discourse as "apologists" or, in the German context, as benighted "understanders" (*den Putin-Verstehern*). I believe this reflex serves U.S. and Western geopolitical culture poorly and inhibits our ability to understand the contemporary geopolitical crisis in relations between Russia and the West. It also displays a lack of confidence in the outcome of any presentation of divergent narratives. This book presents Putin's own words (as well as those of other actors) in the crises it examines as a means of deepening our understanding of how he framed these moments and set the agenda of Russian geopolitical culture. Readers should know that my own Russian-language skills are limited and that I have relied on official Kremlin translations in reproducing his words. Also, my sources are overwhelmingly English-language publications, so some will find my empathetic stretch inadequate; others may consider it too relativizing for their taste. Decentering our "natural" geopolitical narratives in search of deeper understanding is a contribution critical geopolitics can make, but it is inevitably limited by the positionality of the researcher.

Thinking critically about geopolitics as a culture requires that we break from prevailing practitioner understandings of the concept. An enduring one is the Kissingerian framing of geopolitics as great-power realpolitik and the conception of it as a cool and deliberative practice, a serious chess game in the pursuit of material interests. This notion is inadequate if not deeply misleading in that it can lead to a discounting of the actual terms used to describe and justify certain foreign policy decisions by leaders. Was there a nefarious, offensive realist plan by NATO to encroach upon Russia's sphere of influence in Bucharest? Or were certain leaders genuinely motivated by a desire to help the Georgian and Ukrainian people have a better future? Did Russia plan the August 2008 war, or were its leaders authentically outraged by the actions of the Saakashvili administration against innocent civilians in South Ossetia? This book builds upon a wealth of social scientific research that points to the necessary emotional foundations of rational thinking; the importance of embodied affect in human thinking; and the significance of the vast substratum of thinking that is automatic and unconscious, below or barely at the level of consciousness.[24] This literature, and a close reading of the empirical record of events, led me to appreciate the importance of civilizational and "rescue mission" storylines in the conceptualization of U.S. foreign policy toward Georgia and Russian foreign policy toward ethnic Russians and compatriots in Georgia and Ukraine. An affective geopolitics is at work in the foreign policy practices of both the United States and Russia, one that activates historic myths and heroic self-images while framing adversaries as equivalent to historic enemies (Hitler, Nazism, fascism). As this book will explain, it is a supreme irony of the current geopolitical crisis that both the United States and Russia draw upon structurally similar affective storylines in their geopolitical cultures to produce mutually incomprehensible interpretations of the same events. One reason for this is that geopolitics is as much, if not more, "hot" than it is "cold" in discourse and practice. It triggers powerful emotions—love, hope, and pride as well as outrage, contempt, and hate—and involves moments of ritualized enactment, such as pageant and parade, music and memory, celebration and dance, that are beyond words. That is why concerts organized by the Kremlin in wake of the 2008 war and the 2014 annexation of Crimea are discussed in this book. Affect and emotions, which find expression in talk of "values" and "ideals," are as central as strategic interests and material calculations.

The third conceptual building block is the *geopolitical condition*. This refers to an enduring concern in geopolitical writings with how emergent technological assemblages—military, transportation, and communication infrastructures—transform the way in which geopolitics is experienced, understood, and practiced. This question is complex and often separated into geostrategic questions about military technology and defense, and broader societal mobilization questions about news media, propaganda, and information control.[25] What is important is that these technological systems have a life of their own and create capacities and ecologies that transform how the game of geopolitics is played within geopolitical fields as well as how geopolitical cultures now operate. Distant conflicts can be rendered as proximate and immediate crises by global television networks. Local events can jump in scale to become global news. Agendas are set by visually arresting images or viral media memes. As a result, hierarchies of strategic interests, distinctions between vital and nonvital places, and understanding of location and distance can get thoroughly jumbled, confused, and lost.[26] In a hyperconnected postmodern geopolitical condition, actors can consciously create compelling visual spectacles for media consumption and global circulation to further their ends. State- or corporate-controlled media, especially television and tabloid newspapers, can shape how audiences see and experience geopolitical crises. Distant dramas can be rendered emotionally compelling and can inflame populations far from conflict zones. Other dramas can be ignored and forgotten. Certain lives can be privileged and mourned, others unseen or unheard.[27] These are aspects of the contemporary geopolitical condition, and they are part of the struggles described in this book.

While mindful of theory, this book is a series of narrative-driven essays that endeavor to weave together the different perspectives of the actors involved in the struggles described. The book begins with a question—why does Russia invade its neighbors?—that frontloads a widely accepted Western frame that I then seek to deepen. The first chapter reviews two predominant reactions in the United States to Russia's invasions of Georgia in 2008 and Ukraine in 2014 before discussing some of the concepts that are used in the rest of the work to provide thicker forms of geopolitical understanding than those that currently prevail. Chapter 2 considers the Soviet Union as a geopolitical complex and reviews how its collapse generated a Russian geopolitical culture with three distinct visions of the country. It presents Vladimir Putin's agenda as revanchist rather than necessarily revisionist. Chapter 3 examines the factors behind the development of a special relationship between the United States and post-Soviet Georgia. Chapter 4 then considers the government of Mikheil Saakashvili and how he sought to instrumentalize this special relationship to reclaim Georgia's two breakaway provinces. This attempt ended up precipitating the August 2008 war, for which Saakashvili bears considerable responsibility. Chapter 5 examines how this was represented and framed by different parties in the war—Ossetian, Georgian, Russian, and U.S. officials. Chapter 6 then considers the case of Ukraine and describes the circumstances that precipitated Russia's invasion of Crimea in late February 2014 and how its subsequent annexation was presented as geopolitical drama to the Russian and international publics. Chapter 7 turns to Russia's role in fomenting revolt and rebellion in eastern Ukraine at the same time. It documents how the Novorossiya project of Russian imperial nationalists in Ukraine and Russia failed to attract support beyond the Donbas and how the fate of that area diverged from the rest of Ukraine. Defeat and failure forced the Russian army to intervene militarily in late August 2014 to save the pro-Russian separatists they had fostered there. Finally, chapter 8 turns to U.S. geopolitical culture and examines whether three popular geopolitical frames used to understand Russian geopolitics in the near abroad are helpful or misleading. In questioning these frames, and a broader tendency toward "thin geopolitics" in U.S. geopolitical culture, the book offers an alternative analysis and set of conclusions.

Given the complexity of the subject matter of this book, and the vast literature already written about it, I have had to leave many aspects of the geopolitical contest described here unaddressed. I confine my account of Russia's near abroad overwhelmingly to Georgia and Ukraine, so there is insufficient discussion of Moldova, the Baltic countries, Armenia, and Azerbaijan—and no consideration given to Belarus and Central Asia. Also I have had to leave out discussion of the prologue and epilogue to the August War in Russia.[28] The European Union as an institution, and individual states like Germany, France, and Poland, were significant actors in the events discussed here. Other works discuss their role as well as the significance of the geopolitics of energy to Russian foreign policy and European interests.

As will be apparent, this book is not a work of advocacy, though it is not neutral about many of the events discussed here. Scholarship is inevitably enmeshed in the complexities of the world we live within, and cannot sit on the sidelines. I expect the narrative presented here will bring objections from most, if not all, sides in the struggles it describes because it does not affirm their privileged narrative. I understand the need of most Georgians and Ukrainians to begin with the injustice of the violence inflicted upon them by Russian military invasion. Ukrainian military forces on the frontlines with Russian-backed separatists live it daily, as do the community of Western journalists, policy analysts, and think tankers who follow the violence intently on social media. Violence polarizes and inflames. This book seeks to explain how this violence has a history and a geography, a context and an entanglement with other violence and experiences of victimhood. Some will fear this decenters their victimhood and induces moral equivocation. But the issues are too important, the current crisis too serious, to not strive for deeper intellectual and moral understanding. This is what this book endeavors to do.

I

Why Does Russia Invade Its Neighbors?

IT WAS SUPPOSED TO be China's coming-out party, a moment in the global spotlight affirming its arrival as an economic superpower. But hours before the opening ceremonies of the 2008 Beijing Olympics, news of a war in the Caucasus flashed across the world's TV screens. On the southern slopes of the Caucasus Mountains, the state of Georgia launched a military offensive against South Ossetia, a small breakaway territory beyond its control since the Soviet collapse. Georgia's offensive quickly brought Russia to the defense of its local Ossetian allies. As Soviet-era tanks rolled through the Roki tunnel, the only land connection between South Ossetia and Russia, Russian aircraft bombed Georgian targets in the region and beyond. For the first time since the Cold War ended, Russia was invading a neighboring state. Instead of glowing stories about China, speculation about a new Cold War filled the front pages of the Western press. Yet within a week the war was over and a ceasefire agreed. Thereafter a rapidly moving global financial crisis displaced what seemed a harbinger of geopolitical rupture to an afterthought. As quickly as it had flared, the Russo-Georgian war disappeared, and with it talk of a return to geopolitics past.

Six years later Russia was in the global spotlight as host of the XXII Olympic Winter Games in Sochi, located on the shores of the Black Sea at the western end of the Caucasus Mountains. Despite well-grounded fears of terrorism, the Olympics were a triumph for Russia and its leadership. Yet a few days later, the world recoiled in shock as Russia once again invaded a neighboring state. Responding to a perceived "fascist coup" in Kyiv, unmarked Russian military personnel seized control of the Ukrainian province of Crimea, once part of Soviet Russia and home to Russia's Black Sea Fleet. A hastily organized referendum followed, creating the appearance of legitimacy for Russia to formally annex the province, and the city of Sevastopol, in late March 2014. At the same time, Russia massed military forces on Ukraine's eastern border while covertly facilitating a rebellion across southeast Ukraine against Kyiv's interim government. Special forces previously active in Crimea and volunteer fighters flocked to the Donbas region of eastern Ukraine bordering Russia, the one region where street protests mutated into insurgency and developed as a more conventional war between Russian-backed separatist fighters and diverse Ukrainian military formations. On July 17, 2014, this localized conflict reached into the skies and engulfed a Malaysian Airlines Boeing 777 jet passing overhead on a flight path from Amsterdam to Kuala Lumpur. Rebel leaders, having downed Ukrainian military aircraft days earlier, shot the civilian plane from the skies in an apparent case of mistaken identity. The bulk of the plane crashed near the village of Grabovo in eastern Donetsk, killing all 298 people on board. Overwhelming circumstantial evidence suggests the plane was shot from the skies by a BUK missile vehicle that transited into rebel-controlled Ukraine from Russia.

The incident vaulted the conflict in eastern Ukraine beyond its local geographic confines; its victims now included innocents from across the globe. The Netherlands alone lost 193 of its citizens. Revulsion at the horror galvanized the leadership of the European Union and the United States into imposing further sanctions against individuals within Russia's governing elite and select sectors of the Russian economy. Russia's leadership nevertheless continued to see the conflict in the Donbas differently from North Atlantic Treaty Organization (NATO) countries. In late August, Russia intervened decisively with regular military formations to stop the advance of Ukrainian forces against pro-Russian separatists. Efforts to agree to a ceasefire made progress in September, but further discussions in February 2015 were required before fighting wound down. Progress on the negotiated agreement, however, has stalled, low-intensity fighting has resumed, and a return to full-scale war is possible.

Russia's multiple invasions of Ukraine in 2014 precipitated the greatest interstate security crisis on the European continent since the collapse of the Berlin Wall. Not since World War II had one sovereign state within Europe annexed the territory of another sovereign state. The invasion and forced incorporation of Austria into the German Third Reich in March 1938 was an ominous event. It was followed by further territorial aggrandizement and compromised sovereignty in Czechoslovakia. On August 23, 1939, Nazi Germany and the Soviet Union signed a pact delimiting respective spheres of influences in the event of a "territorial and political rearrangement in the areas belonging to the Polish state" and "to the Baltic States (Finland, Estonia, Latvia, Lithuania)."[1] Germany invaded Poland a week later, triggering near-total war across the European continent. The outcome of World War II—what became known as the Great Patriotic War in the Soviet Union—left the continent divided into opposing military blocs. It took decades before there was agreement on norms for states and territories. Among the principles approved by the thirty-five signatory states of the Helsinki Final Accords of 1975 were respect for the sovereign equality of states, the inviolability of frontiers, and the territorial integrity of states. Given that Soviet military conquest—occupation, repression, forced population movements, and redrawn borders—had profoundly reshaped the political geography of states in the eastern lands of the European continent, agreement on these principles was a victory for the Soviet Union. A declaration known as the "Charter of Paris for a New Europe" reaffirmed the principles in the wake of the fall of the Berlin Wall and Eastern-bloc dictatorships in November 1990. The Soviet Union and Yugoslavia joined other states to proclaim that the "era of confrontation and division of Europe has ended." "Europe whole and free," the document asserted (echoing President Bush's phrase the previous May), "is calling for a new beginning. We invite our peoples to join in this great endeavour."[2] Barely a year later, an agreement signed in a forest dacha by the presidents of Soviet Ukraine and Russia with the Belarusian parliamentary chair triggered the once-unthinkable: the dissolution of the Union of Soviet Socialist Republics. The legal principle applied to decolonization to ensure continuity of territorial units, *uti possidetis juris* ("as you possess"), was now applied to the Soviet Union and to Yugoslavia, also in dissolution. Intra–Soviet Union boundaries, drawn first by the Bolsheviks in 1922 and redrawn numerous times thereafter, were now international borders. Many groups were unhappy with this fiat of international law. Yugoslavia's dissolution spawned four wars and more than a decade of territorial crises. Gorbachev's reforms had reopened territorial disputes across the Soviet Union.[3] Dissolution only deepened the turmoil over borders and sovereignty. Local groups sought their own "whole and free" visions, and their endeavors left some new states wartorn and fractured. At the end of Europe's dark twentieth century, questions of sovereignty, borders, and freedom remained unresolved in certain locations. Comprehending these places is vital to understanding Russia's twenty-first-century invasions of Georgia and Ukraine.

Understanding is not justification. Given the high stakes involved—nothing less than the preservation of peace and avoidance of nuclear war in Europe—it is vital that we strive to deeply comprehend Russia's invasions of 2008 and 2014. Whether we realize it or not, we already inhabit and express theories of Russia's actions based on the very frames we use to describe events, actions, and processes. Russia, for example, does not hold that it "invaded" Georgia in 2008. Nor does it accept that the term "invasion" is an appropriate description for its actions in Crimea or eastern Ukraine. Its storylines are very different from those prevalent in Western capitals; yet, as we will see, they are constructed from familiar stories if not from the very same affective dispositions and narratives used there. The motivation for this book comes not from a desire to tell Russia's stories about these invasions, though it is important that we listen attentively to Russia's arguments so we can grasp how they made sense of events and justified their actions. Rather, this work emerges from deep dissatisfaction with the two storylines that are used to explain Russia's invasions within Western capitals, most especially within the United States. The first of these storylines is by far the most prevalent and popular among politicians and commentators: the argument that Russia's actions are the work of an imperialist power imposing a sphere of influence in its borderlands. Aspects of this storyline are indeed persuasive, but I will argue that it is wholly inadequate and even dangerous in some respects. The second storyline is less prevalent and public, but nevertheless it is influential behind closed doors in policy circles. It is a self-styled "realist" counterstory to what it frames as a dominant legalistic-moralistic approach to international problems in the United States and elsewhere.[4] Both storylines converge on the same conceptual explanation for why Russia invades its neighbors: geopolitics. How they understand this explanation, though, differs. The dominant liberal storyline views geopolitics as an anachronistic practice used only by revisionist great powers to challenge the universal liberal norms that are the necessary foundation of world order. Geopolitics is the practice of aggressive great powers. The political-realist counterstory holds that all great powers practice geopolitics and, furthermore, that all are sensitive to security challenges that are geographically proximate to their borders. Power not principle, and location not law, are what matters. Russia acted because there were threats in its backyard, just as the United States has done in the past.

The argument in this book is that both perspectives offer inadequate explanations as to why Russia invaded Georgia and Ukraine. Both rest on superficial conceptions of geopolitics—thin accounts that flatten the complex spatialities at work in international territorial conflicts and render them as forced dichotomies, moralized contrasts, and parsimonious schemas. What is developed here is a thicker account of geopolitics, one that examines multiple scales of action and clashing spatial imaginations. This richer understanding of geopolitics draws inspiration from the scholarly literature within critical geopolitics, a tradition that for more than two decades has sought to challenge geodeterministic, state- and ethnocentric, and, more broadly, overly dichotomized explanations of international conflicts.[5] Critical geopolitics is not only about recovering the hidden geographic complexities of the practice of geopolitics but also offering critical analysis of the prevailing forms of geopolitical discourse. In this sense, it examines the geopolitics of geopolitics.[6] To begin the argument, this chapter is divided into three sections. The first provides a summation of the dominant liberal understanding of Russia's invasions; the second discusses the realist counterstory; and the third introduces a series of critical geopolitical thinking tools that are used throughout the rest of the book.

The Return of Geopolitics Past

Russia had long functioned as a "dark double" to the United States in its self-definition as a world power.[7] The United States' Cold War culture rendered this as a mirror imaging of universalist creeds—a clash between freedom and totalitarianism, democracy and enslavement, Western civilization and the East. In his famous "Long Telegram" of February 1946, George F. Kennan presented a theory that enabled such dichotomizing. Soviet foreign policy, he argued, was "not based on any objective analysis of situation beyond Russia's borders . . . it arises mainly from basic inner-Russian necessities." The "Kremlin's neurotic view of world affairs" is at source a "traditional and instinctive Russian sense of insecurity." Originally the "insecurity of a peaceful agricultural people trying to live on vast exposed plain in neighborhood of fierce nomadic peoples," this became, after contact with Western powers, "fear of more competent, more powerful, more highly organized societies" on its borders. Because of this Russia's rulers have "learned to seek security only in patient but deadly struggle for total destruction of rival power, never in compacts and compromises with it."[8]

Kennan's theory on the sources of Soviet conduct was the intellectual basis for the West's subsequent policy of containment. Because the end of the Cold War in Europe and subsequent collapse of the Soviet Union were experienced as Western victories, there was no major rethink of the intellectual foundation of Cold War geopolitics. Rather, its primordialist theories, universalist creed, and dichotomizing habits endured. For example, U.S. columnist George Will claimed in 1996 that expansionism was "in the DNA of Russians."[9] This assertion was made to promote NATO expansionism and the Clinton administration's leadership in rebranding the Cold War alliance as a civilizing force, enlarging the realm of democracy in former Communist lands. An aging Kennan, and many other respected political figures, warned against the policy but were ignored. NATO's expansion to former Warsaw Pact members and then to the Baltic countries, forcefully incorporated into the Soviet Union in 1940, heightened the contradiction between holding, on the one hand, that Russia was an insecure expansionist empire and, on the other, that expanding a revamped Cold War military alliance to its borders was the best strategy for promoting peace and security in Europe.

While the modalities of the Russian invasions of 2008 and 2014 were different, the events provoked similar rhetorical responses from the Bush and Obama administrations. The first and most immediate reaction was condemnation of the Russian actions as violations of territorial integrity. The circumstantial events that triggered the Russian actions—Georgia's attack on South Ossetia and the ouster of president Victor Yanukovych—were rarely acknowledged in public statements lest this be seen as granting them legitimacy. The initial Bush administration statement from Beijing on the August War was relatively restrained, noting only that "Georgia is a sovereign nation and its territorial integrity must be respected."[10] But with Russian forces moving beyond South Ossetia, and Russian Foreign Minister Lavrov telling U.S. secretary of state Condoleezza Rice on the phone that Georgian president Saakashvili "must go," the White House soon feared that Russia was intent on overthrowing the Georgian government by force. A second, more expansive White House statement in Washington, DC, on August 11, 2008 declared: "Russia has invaded a sovereign neighboring state and threatens a democratic government elected by its people. Such an action is unacceptable in the 21st century."[11] A further statement declared that the "days of satellite states and spheres of influence are behind us" and that "[b]ullying and intimidation are not acceptable ways to conduct foreign policy in the twenty-first century."[12] Analogies to the Soviet invasions of Hungary in 1956 and Czechoslovakia in 1968 were common in the U.S. media. The influential U.S. neoconservative Robert Kagan saw an older analogy: "The details of who did what to precipitate Russia's war against Georgia are not very important. Do you recall the precise details of the Sudeten Crisis that led to Nazi Germany's invasion of Czechoslovakia? Of course not, because that morally ambiguous dispute is rightly remembered as a minor part of a much bigger drama."[13]

President Obama's first statement on the Crimea crisis acknowledged Russia's historic relationship with Ukraine and the military facility in Crimea, but warned that "any violation of Ukraine's sovereignty and territorial integrity would be deeply destabilizing."[14] When this became undeniable, his administration imposed sanctions on individuals and entities it held responsible. Russia's actions were described as violations of international law, including Russia's obligations under the United Nations (UN) Charter, the Helsinki Final Act, the 1994 Budapest Memorandum, its 1997 military basing agreement, and its 1997 Treaty of Friendship, Cooperation and Partnership with Ukraine. "In 2014," Obama remarked, "we are well beyond the days when borders can be redrawn over the heads of democratic leaders."[15] This same representation of Russian behavior as a throwback to a more primitive past was repeated by Obama's secretary of state John Kerry: "[Y]ou just do not, in the twenty-first century, behave in nineteenth-century fashion by invading another country on completely trumped up pretext" and "it's really nineteenth-century behavior in the twenty-first century."[16] In Poland on June 4, 2014, President Obama described Russia as resorting to the "dark tactics of the twentieth century."[17] Russia, in short, was represented as a state regressing to the primitive behavior of an earlier era.

The Bush and Obama administrations also represented the stakes involved in Russia's invasions in similar ways. For the Bush administration, Georgia was already a symbol of its Freedom Agenda, the mission outlined in Bush's second inaugural address for the United States to end tyranny across the world by spreading democratic institutions and liberty. President Bush cast Georgia as a loyal friend the United States was morally obliged not to abandon. "The people of Georgia have cast their lot with the free world, and we will not cast them aside."[18] Europe was moving to become "a continent that is whole, free, and at peace. And it is essential that America and other free nations ensure that an embattled democracy seeking to stand with us remains sovereign, secure, and undivided."[19] Yet despite the rhetoric, the Bush administration never seriously considered intervening militarily to defend Georgia, though some neoconservatives called for a more forceful response.[20]

President Obama's rhetorical response to Crimea, and to a subsequent covert war in eastern Ukraine, did not directly cite Cold War constructs like "the free world," but it universalized the stakes in Ukraine in similar terms. In an address in Brussels, Obama remarked that "once again, we are confronted with the belief among some that bigger nations can bully smaller ones to get their way—that recycled maxim that might somehow makes right." Russia's leadership, Obama declared, "is challenging truths that only a few weeks ago seemed self-evident—that in the 21st century, the borders of Europe cannot be redrawn with force, that international law matters, that people and nations can make their own decisions about their future." Obama noted that neither the United States nor Europe had any interest in controlling Ukraine: "What we want is for the Ukrainian people to make their own decisions, just like other free people around the world."[21]

The dominant liberal storyline on Russia's invasions in U.S. geopolitical culture has certain manifest weaknesses. Here we need only briefly note three. The first is a structural feature of all geopolitical cultures, namely the predisposition to regard one's own intentions and behavior as a positive force in the world. That this afflicts U.S. geopolitical discourse is hardly news. Charges of "double standards" against the United States are routine in international affairs. Most of the U.S. leaders (with the important exception of President Obama) characterizing Russia's invasions as outdated state aggression had just years earlier supported a U.S.-led invasion that violated the sovereignty and territorial integrity of Iraq. That action was deeply divisive and illegal under international law. These same leaders, including President Obama, are strong supporters of the state of Israel, even though in 1967 it illegally seized and occupied territory belonging to neighboring states. Since then it has returned some of this land but colonized other parts, effectively annexing this territory into the state of Israel. In the summer of 2014, as Russia waged covert war in eastern Ukraine, Israel launched a war against the Gaza Strip, an exceptional territorial space created and controlled by Israel. Though Israel's actions generated thousands of noncombatant deaths, its invasion received strong statements of support from the Obama administration. Later in the summer of 2014, the Obama administration determined that it was in its interest to ignore international borders and attack Islamic State militia forces in both Iraq and Syria. Ben Rhodes, the president's deputy national security advisor, told reporters at the time: "We're actively considering what's going to be necessary to deal with that threat and we're not going to be restricted by borders."[22] Liberal storylines that tout universal principles of international law and territorial integrity tend to ignore these inconsistencies. To critics of U.S. liberal hegemony, "territorial integrity" has an unacknowledged colonial geography: there is one imagined set of liberal norms for Europe and a different set for the Middle East.[23]

A second weakness is how the liberal storyline tends to explain Russian behavior. Because most analysts are foreign to the state and culture under consideration, and because they have limited or no direct access to the private deliberations of state leaders, there is a general tendency to attribute observable public behavior to innate dispositional features of that leader, regime, or country. What this essentializing tendency does is systematically downplay, marginalize, and ignore the situational, contextual, and spatial factors that may account for state behavior. While there are important exceptions, explanations of Russian behavior tend to fall back on Kennan's primordialist claims. Russia's behavior is a product of a base neurosis, an essential disposition toward territorial expansionism and control of its neighboring states. The synecdochic language that is common and unavoidable in geopolitical writing—where a capital city, "Moscow," or institutional building, "the Kremlin," stands in for an administration, regime, or essentialist disposition—enables this tendency and gives it a particular scalar form. Geopolitical explanatory stories, in other words, tend to be great power–centric stories that privilege its decision-making centers over all other sites. There is a tacit presumption of control, coherence, and command.

This general habit is important for what it neglects. First, it tends to discount the detailed, empirical, contextual circumstances of a crisis. Contra Kagan, the details of who did what to precipitate Russia's invasions are very significant. Situational and interactive explanations, however, tend to be marginalized by preconvictions about the sinister character of one's enemy. Second, it tends to discount the significance of actions and decision making at sites other than at the pinnacle of a perceived power-vertical. Thus, there tends to be little discussion of the localized geopolitical dynamics within South Ossetia, Crimea, and southeast Ukraine. Both spatial location and locality are marginalized by discourse that casts questions in terms of moralized abstractions. Third, the practice provides little to no room for situational interpretations of foreign policy actions or efforts to see events through an adversary's eyes. What leaders actually say about why they acted tends to be discounted or rewritten to conform to existing presumptions about their character, regime, and state. Foundational identity frames about whether a state is a friend (like Georgia) or a competitor great power (like Russia) are crucial determinants of storyline production. Analysts tend to overattribute hostile intentions to perceived competitors, emphasize negative dispositions, and discount ambivalent contextual factors. The reverse applies to states perceived as friends. Such cognitive predispositions create favorable conditions for the generation of convenient and self-fulfilling truths, a "truthy" (feels true) not truthful geopolitics.

A final difficulty concerns the tendency on the part of commentators and practitioners to presume that their motivations, intentions, and actions are transparent and clear to others. Social psychologists term this the "illusion of transparency."[24] This is particularly relevant for thinking critically about geopolitical discourses of legitimation, such as the rhetoric justifying NATO expansionism, European Union (EU) enlargement, and democracy promotion. Liberal analysts hold that NATO is manifestly not a sphere of influence because it is an alliance structure that member states must choose to join. Choice, not coercion, is at its heart. Furthermore, the alliance is held to have peaceful and stabilizing aims. The illusion of transparency leads to overestimation of the degree to which other states, most especially Russia, understand and accept NATO's benign intentions. The flip side of this is that not understanding is seen as evidence of paranoia, or as Kennan puts it in the "Long Telegram," as not based on "objective analysis." It is perfectly reasonable, however, for Russian leaders to be skeptical of NATO expansionism. Russian state leaders tend not to see NATO as a benign security force but as a longstanding existential threat to their state. Its expansion is seen as the creation of a hostile sphere of influence encroaching the very borders of Russia. Former Soviet president Gorbachev expressed a version of this attitude in justifying Russia's actions in August 2008: "By declaring the Caucasus, a region that is thousands of miles from the American continent, a sphere of its 'national interest,' the United States made a serious blunder. Of course, peace in the Caucasus is in everyone's interest. But it is simply common sense to recognize that Russia is rooted there by common geography and centuries of history. Russia is not seeking territorial expansion, but it has legitimate interests in this region."[25] The illusion of transparency, of course, also applies to Russian geopolitical discourse.

Geopolitics 101

The United States' geopolitical culture has always contained traditions of thought that have challenged liberal internationalist sentiments and storylines. The most significant of these in the professional foreign policy community is political realism. Less a coherent doctrine than an attitude, political realism is an engaged critique of the moral and legalistic approach to international problems that is characteristic of U.S. diplomacy, or at least the public rhetoric of U.S. diplomacy. George F. Kennan attributed this in part to the influence of lawyers and the legal profession on the practices of U.S. state institutions.[26] This produced an outlook on the world that was myopic and institutionally biased toward the existing order of states. An experienced U.S. diplomat himself, Kennan reflected in 1950: "We tend to underestimate the violence of national maladjustments and discontents elsewhere in the world if we think that they would always appear to other people as less important than the preservation of the juridical tidiness of international life."[27] Certain groups have positive aspirations that they regard as legitimate and more important than "the peacefulness and orderliness of international life."[28] Further, Kennan argued, the legal glorification of the principle of state sovereignty "envisages a world exclusively of sovereign national states with a full equality of status." In so doing, Kennan argued, this "ignores the tremendous variations in the firmness and soundness of national divisions: the fact that the origins of state borders and national personalities were in many instances fortuitous or at least poorly related to realities." By institutional setup and mentality, U.S. diplomacy "ignores the law of change. The national state pattern is not, should not be, and cannot be a fixed and static thing. By nature, it is an unstable phenomenon in a constant state of change and flux."[29]

Kennan's perspective grew out of his years of practical diplomatic experience in Eastern and Central Europe and the Soviet realm and represents a worldly critique of tendencies toward the legal fetishization of prevailing state borders. Less worldly and more academic has been the recodification of political realism in recent years by the Chicago political scientist John Mearsheimer. He identifies three core beliefs of political realism: (1) that states are the principal actors in world affairs; (2) that the behavior of great powers is influenced mainly by their external environment, not their internal regime type; and (3) that calculations about power dominate state thinking.[30] Mearsheimer's own version of political realism rests on five assumptions that recall the social Darwinian assumptions of classic geopolitics. These are that the international system is anarchic; that great powers inherently possess offensive military capability; that states can never fully trust each other; that survival is the primary goal of great powers; and, finally, that "great powers are rational actors. They are aware of their external environment and they think strategically about how to survive in it."[31] States are conceptualized as fearing, thinking, observing, and calculating entities in a struggle for survival with other states. From these foundational assumptions, Mearsheimer identifies three general patterns of behavior in international affairs: fear, self-help, and power maximization.

Few politicians articulate and robustly defend political realism in public in the United States, though many may hold such views in private. This is particularly the case after events perceived as egregiously violating prevailing norms of international law, like one state invading another. One source of political-realist commentary on the August 2008 war was the Nixon Center, a think tank founded in 1994 by former president Richard Nixon and associates (it was renamed in 2011 as the Center for the National Interest).[32] In an opinion editorial in the *Washington Post*, Paul J. Saunders, its executive director, placed blame for the August 2008 war not on Russia but on Georgia's leadership. He criticized Georgian president Saakashvili's recklessness in invading South Ossetia, and his mistaken calculation that Russia would not respond for fear of provoking the West.[33] Saunders saw a timeless strategic logic at work: "Throughout history, weak nations with powerful neighbors have energetically sought strong allies." Saakashvili, he argued, "has embraced this tried-and-true strategy with gusto, sending a substantial share of the country's small army to Iraq...and parroting Bush administration talking points on international issues—especially on promoting democracy—more than almost any other leader worldwide." Saakashvili learned well, with the help of lobbyists, how to "push America's buttons," but his record belied his interest in establishing democracy in Georgia. Georgia was attacked by Russia not for what it represented but for its escalatory violence in South Ossetia. By his actions, Saakashvili "has allowed Moscow to demonstrate quite clearly the limits of American interests in Russia's immediate neighborhood. The Kremlin has much more at stake there than Washington and is willing to act decisively and with overwhelming force." Geographical location, in short, matters greatly in explaining the August 2008 war, and the U.S. state does not need reckless friends in such a neighborhood. Some—and perhaps many—in the U.S. diplomatic community privately shared this position.

In the wake of the Crimea annexation and eastern Ukraine crisis, political-realist readings were more abundant if varied. An essay with the provocative title "Why the Ukraine Crisis Is the West's Fault" by John Mearsheimer, which appeared in the September/October 2014 issue of *Foreign Affairs*, was perhaps the most discussed of the political-realist interpretations of Russia's actions.[34] In the article, Mearsheimer argued that a triple package of Western policies—NATO enlargement, EU expansionism, and democracy promotion—represented a direct security threat to the national interests of Russia. When Ukrainian president Victor Yanukovych was driven from office in February 2014, the Russian leadership reacted in a manner to safeguard its vital national interests in a strategic state on its doorstep. Mearsheimer, like Saunders, sees a transparent strategic logic grounded in a relationship between geography and security in how the Putin administration responded to the governmental crisis in Kyiv:

> A huge expanse of flat land that Napoleonic France, imperial Germany, and Nazi Germany all crossed to strike at Russia itself, Ukraine serves as a buffer state of enormous strategic importance to Russia. No Russian leader would tolerate a military alliance that was Moscow's mortal enemy until recently moving into Ukraine. Nor would any Russian leader stand idly by while the West helped install a government there that was determined to integrate Ukraine into the West. Washington may not like Moscow's position, but it should understand the logic behind it. This is Geopolitics 101: great powers are always sensitive to potential threats near their home territory.[35]

Mearsheimer argued that the United States does not tolerate distant great powers deploying military forces anywhere in the Western Hemisphere, much less on its borders. Calling out the illusion of transparency in NATO expansionist discourse, he argues that it is the Russians, not the West, who get to decide what constitutes a threat to their state. Putin, he suggests, is not an irrational actor but "a first-class strategist who should be feared and respected by anyone challenging him on foreign policy."[36] His actions in Ukraine have been defensive responses to events, not offensive actions. Mearsheimer's policy prescriptions for the Ukraine crisis are for the West to abandon its efforts to westernize Ukraine and instead seek to convert it into a neutral buffer zone between the NATO states and Russia. NATO expansion should be explicitly ruled out for both Georgia and Ukraine. Western leaders and Western-oriented elites in both these countries should understand that "might often makes right when great-power politics are at play."[37] Geopolitical location is more significant than abstract universal principles of self-determination in international affairs.

Political-realist storylines on Russia's neighborhood invasions avoid much of the positive illusions characterizing liberal accounts. But, in Mearsheimer's account at least, these are replaced by other weaknesses. While location matters to realists, the dispositional essence of states, and the state system, still tend to be privileged over the contextual practice of world politics. Indeed, with academic political realism at least we have explicitly stated forms of cognitive parsimony. Recall Mearsheimer's three core beliefs about realism. These can be alternatively phrased as rigid state-centrism, the structural determinism of balance of power system relations, and the privileging of a narrow cognitive conception of how states perceive, calculate, and act. These biases account for the many explanatory weaknesses of political-realist stories about Russia and its relations with its neighboring states. Let us briefly examine just two in Mearsheimer's case.

First, Mearsheimer's great power–centrism tends to marginalize the geopolitical field and the various actors that define post-Soviet space. This field is, first and foremost, a contentious post-imperial space characterized by competing zones of power that are reacting to imperial legacies, and each other, in divergent ways. But the Ukraine crisis, in Mearsheimer's telling, is "in essence" a crisis of "two sides" operating with "different playbooks": "Putin and his compatriots have been thinking and acting according to realist dictates, whereas their Western counterparts have been adhering to liberal ideas about international politics."[38] This is a highly reductionist view of the Ukraine crisis that is manifestly at odds with the historical record. In Mearsheimer's world, superpowers are the only ones with real agency, smaller states are subordinate clients, and substate actors are proxies. An analysis of the structure of the conjunctures in August 2008 and February 2014, as we will see, reveals a much more complex picture concerning agency and "playbooks." An analysis of the broader context also reveals statements that contradict later polices adopted by Russia. Why Vladimir Putin stated on May 17, 2002, that Ukraine joining NATO was "a matter for those two partners," or on August 29, 2008, that "Crimea is not a disputed territory," needs to be addressed.[39]

Second, Mearsheimer argues that Russia acted as it did in February 2014 out of defensiveness and fear. Russia's national interest is described as overdetermined by primordial fear of land invasion. Recall that Mearsheimer believes that great powers are rational actors who are aware of their external environment and think strategically about it. Mearsheimer suggests Putin does this well in describing him as a "first class strategist." Yet this rationality is built upon a singular affect—fear—that has neither empirical checks nor a responsibility to context and circumstance. Fear acts as an explanatory deus ex machina. But any responsible account of the 2008 and 2014 invasions has to account for why fear should be seen as the fundamental motivation behind these invasions. Why fear of NATO encroachment in 2008 or EU encroachment in 2014 triggered wars, but did not result in wars in 2004—when NATO expansion incorporated the former Soviet Baltic Republics—needs explanation. The relevance of fear of a land invasion in August 2008 is, to say the least, unclear. And, as his critics have pointed out, the question of Ukraine's membership in NATO was not an active matter of debate in 2014 (though it is important background to the Ukraine crisis).[40] More significantly, the actual empirical circumstances of these invasions suggest the operation of a richer range of affective motivations than fear that is connected to the places themselves. The Caucasus and Crimea, after all, are richly storied locations in Russian geopolitical culture. Russia identified strongly with the co-ethnic and compatriot populations in South Ossetia and Crimea, populations it viewed as under direct attack. Acting less out of fear than from righteous indignation mixed with feelings of protection, pride, and glory, Russia rescued these populations. Mearsheimer's account marginalizes this. Primordial state-security motivations crowd out all other motivations, from affective in-the-moment explanations to regime-preservation calculations. In assuming that state actions are strategic calculations, Mearsheimer disregards the power of emotional ties to certain places in foreign policy practice. Further, it attributes competence to actions that were manifestly risky and latent with self-defeating potential. Finally, in its commitment to debunking normative discourse and translating events into strategic moves by state leaders in a game of power politics, political realism can create its own illusion of transparency. Not everything is about the game of state power politics, as realists understand it.

In summary, both the predominant liberal and alternative realist explanations of why Russia invaded its neighbors are inadequate. For a deeper explanation, we need a deeper understanding of geopolitics. Conventional understandings of it as geographic determinism (the early-twentieth-century understanding), a Nazi-like policy of territorial aggrandizement (the prevailing English-language understanding during and after World War II), or a realpolitik approach to international relations (the Kissingerian understanding that took hold in the 1970s), are narrow and misleading. Geopolitics requires us to address the ineluctable territoriality and spatial situatedness of state formations; the cultures state elites develop to acquire power, prosperity, and security in a dynamic world; and how world politics is conditioned by technological systems, capitalist dynamics, and time-space compression. To do so we need to introduce some new concepts and themes. The rest of this chapter introduces five conceptual themes that are thinking tools for the empirical chapters that follow.

Post-Soviet Space as a Contested Geopolitical Field

The term "post-Soviet space" is commonplace in discussion of the lands that made up the former Soviet Union. Some, like the Baltic countries, find the term objectionable because they consider their Soviet past an imperial occupation and wish to be rid of the association. Others see it as part of the lexicon of sphere-of-influence thinking.[41] The phrase, however, is an important reminder of the centrality of the Soviet Union as an imperial power complex in creating the borders of its successor states, as well as the regions where separatists have challenged these borders. Post-Soviet space is a space of decolonization, but not a straightforward one. Precisely what is imperial and what is not is fundamentally contested.[42] Decolonization in the name of recovery of an imagined core nation may actually be an unwelcome new form of imperialism to noncore nations caught up in the process. Diverse experiences with colonization and imperial rule have generated divergent aspirations for decolonization across post-Soviet space (to say nothing of the decolonization process within Russia itself). Focusing on interconnecting and interactive relations, Rogers Brubaker described the reconfiguration of political space by nationalism in Central and Eastern Europe after the collapse of Yugoslavia, Czechoslovakia, and the Soviet Union as a triadic nexus of competing nationalist projects and processes.[43] The first is the "nationalizing nationalism" of the newly independent states that emerged from the breakup of these multinational states. This is the project of the dominant nation in the successor state, the core group that views independence as legitimation for a historic (re)nationalization of state space. "The core nation is understood as the legitimate 'owner' of the state, which is conceived as the state *of* and *for* the core nation."[44] The operational conceit of the postcolonial nationalizing state is to reverse perceived legacies of imperial rule, to overcome its status as an "incomplete nation-state," and to thus realize the true destiny of the state as the homeland of the core nation. How state elites go about this can vary greatly, as Brubaker subsequently argued.[45] Not all nationalizing state elites pursue national building projects in the same way, because there are differences in how nationhood and core boundaries are defined and operationalized in policy processes.

A different form of nationalizing project is homeland nationalism. It is the nationalism of a successor state with a spatial identity that extends beyond the borders inherited at the time of the collapse of the multinational state. Homeland nationalism arises in direct opposition to and in dynamic interaction with nationalizing nationalisms. Perceived outrages and interference spur each other into potentially escalating cycles of competition and tension. A state becomes an external national "homeland," according to Brubaker, when its elites imagine residents of other states as co-nationals, part of "a single transborder nation, and when they assert that this shared nationhood makes the state responsible, in some sense, not only for its own citizens but also for ethnic co-nationals, who live in other states and possess other citizenships."[46]

National minorities, noncore groups within nationalizing successor states, make up the third part of Brubaker's triad. They have their own distinct regional as well as national identities, but they are in a subordinate position in the new political geography of successor states, former colonies, and imperial centers. How they act depends on a series of factors: how numerous and prosperous they are relative to the core group; whether they are geographically concentrated or dispersed; whether they enjoy a degree of local autonomy or have some prospect of it within nationalizing states; whether they have strong or weak local organizations with independent leadership; and whether there is an opportunity structure for mobilization or not. These factors also determine whether they are likely to align with certain homeland nationalist discourses championed within an external national homeland (if these exist) and are open to "capture" by this external state, or are sufficiently autonomous to chart an independent path. Formerly dominant national groups within multinational states—ethnic Russians in the Baltic countries, Ukraine, and Kazakhstan, for example—may have great difficulty conceiving of themselves as national minorities and adjusting to the status reversal accompanying this shift. Other national groups that may have secured special homelands and rights within a multinational empire—ethnic Abkhaz or Ossetians in their own autonomous areas within Soviet Georgia—may fear for their position within the successor states that follow the collapse of such empires, and may have longstanding ties to ethnic kin groups on the other side of new international borders, as is the case with both the Abkhaz and Ossetians.

Brubaker's nexus (a variant on Pierre Bourdieu's concept of a field) is an analytical tool for the description of competitive relational nationalisms in a space defined by the end of empire.[47] Some have taken issue with his characterization of nationalizing states, while others argue for adding a fourth dynamic to his nexus, the normative power center that ameliorates and modifies the nation-building policies of nationalizing states.[48] There is also a case for adding a fifth dynamic, which is that associated with potential secessionist areas within the metropolitan power itself, an "internal abroad" that creates anxieties about territorial integrity and control in the former imperial capital. Specified more explicitly as a geopolitical field, we can identify a quintet of place-based powers in post-imperial space, in this case post-Soviet space (Figure 1.1). These are:

(1) A metropolitan state that is striving to define for itself a stable post-imperial spatial identity.
(2) A state-challenging movement within the metropolitan power that seeks greater autonomy or independence for a geographic region where a noncore nation is in the majority. As a noncore nation space within, this area is an "inner abroad" to the core nation of the metropolitan power.
(3) Nationalizing states on the metropolitan state's borders seeking to break free from legacies of dependence and interdependence with the former imperial center, and to possibly join alternative security structures.
(4) National minority regional organizations within nationalizing states, places that may have held special territorial status in the former multinational empire. Depending on the strategies adopted by the nationalizing states, these are either latent or active secessionist regions. Many border the metropolitan power and have historic and contemporary ties with kindred regions within the metropolitan power.
(5) An external normative power complex seeking to expand its influence and reach into former closed imperial lands that are now open to new alignments.

Conceptualizing post-Soviet space as a contested geopolitical field (see Figure 1.2) with a quintet of players moves us toward a richer and thicker understanding of its geopolitics. We will have more to say about this, the questions it raises, and the competitive field of nationalisms it produced in the following chapters. Previous moments of imperial transition and decolonization reveal similar geopolitical fields at work (with variable place-based actors). The prospect of Home Rule for the British-dominated island of Ireland and, after World War I, full independence provoked organized resistance by the Protestant Unionist population concentrated in the north counties of the island.

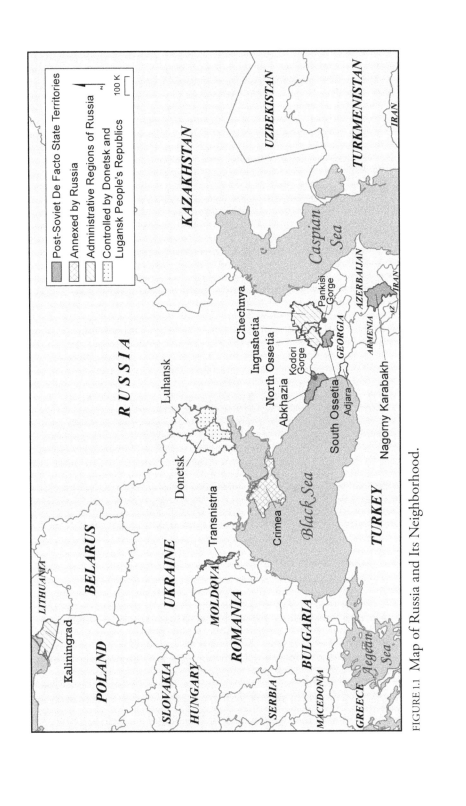

FIGURE 1.1 Map of Russia and Its Neighborhood.

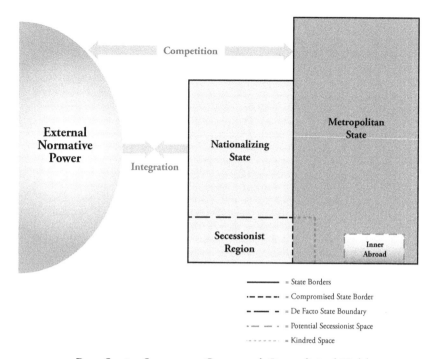

FIGURE 1.2 Post-Soviet Space as a Contested Geopolitical Field.

The resultant compromise was the establishment of a border between a Unionist-dominated polity in Northern Ireland and the Irish Free State on the rest of the island. Fear of Irish nationalist subversion (the origins of our contemporary association of "terrorism" with nonstate actors instead of states) within the United Kingdom shaped security policy on both the mainland and in Northern Ireland for decades.[49] Geopolitical engagement by external normative power centers (the United States and European Union) eventually facilitated opportunities for conflict management and resolution. Decolonization of the Indian subcontinent provoked group fears that were instrumentalized by local ethnopolitical entrepreneurs to create division and the formation of two separate polities—India and a two-part Pakistan—in previously shared space. Kashmir, as is well known, was torn apart by binary territorial logic, and its south remains a deeply troubled region within contemporary India. Turkey's relationship with its own "inner abroad"—the country's southeast, particularly the city of Diyarbakir—and the "near abroad" of its former Ottoman territories, and France's fraught relationship with Algeria, reveal similar geopolitical fields and territorial dilemmas.[50]

Geopolitical space needs to be thought of not only in terms of bordered territorial formations but also in terms of spatial flows. By this I mean the networks of movement, transit, and circulation that also characterize a geopolitical field as a space of flows.[51] Beyond its obvious character as an empire of nations, the Soviet Union was an empire of logistical networks, commodity chains, military-industrial factories, bases, command and control hierarchies, and pipelines delivering oil and gas across its vast space. This empire of connection is a second fundamental inheritance conditioning the geopolitical field of practice that was created by the Soviet collapse. Its power meant that nationalizing states seeking to break free from the metropolitan power were still bound to it by physical infrastructure and energy flows. These flow infrastructures would prove a particularly troublesome feature of post-Soviet space. This spatial field was very dynamic. Spaces were being made and remade but upon terrain made by the Soviet Union as an immense imperial complex of entwined territorial networks.

Geopolitical Cultures and Shared Discourses

Every state, and aspirant state, has a geopolitical culture. This can be defined as its prevailing sense of identity, place, and mission in the world.[52] Geopolitical cultures formulate answers to three fundamental questions facing all territorial states: who are we, how do we survive, and how do we prosper? A geopolitical culture is, first and foremost, about the identity of a territorial entity and the locational narrative it presents to itself and the world. Its specification involves boundaries of identity and difference, the broad civilizational realm within which it positions itself, the states it views as friends, and those it differentiates itself from and defines itself against. A geopolitical culture, in other words, is made up of a series of geographical imaginations about self and other in the world. A geopolitical culture is also about security and defense, about whom the territorial entity holds to be its enemies and the strategies it deems necessary to preserve its existence, identity, and capacity for maneuver. Certain entities may have been formed in opposition, reaction, and resistance to others and to the collapse of empires and world orders. Because they inherit a particular positionality in a geopolitical field, there may be fairly established and commonsense strategies they will need to pursue to achieve these ends. These may not necessarily be given though, and indeed there may be considerable debate within geopolitical cultures over what they should be. Similarly there may be considerable debate about what strategy of growth and modernization an entity should pursue to realize security and accumulate relative power within a broader geopolitical field. Certain states may be held as positive models of a desired future and others as negative exemplars to be avoided at all costs. This can and will vary depending upon the prevailing hegemonic order and its set of norms and understandings of "the modern."[53]

There is no objective relationship between the geographic location of a territorial entity and its geopolitical culture. Geography as earthly location and resource endowment is what a state makes of it. Instead, a state's geopolitical culture is delimited by its foundational myths and governing power structure, the latter operating as overlapping and intersecting networks of power.[54] Ideological power networks tend to generate cultural and civilizational discourses. Economic power networks tend to be the supports for modernization and accumulation-centric discourses. Security power structures organize around perceived threats to the state and can use insecurity and instability to entrench their power and position within the state. Together these networks generate civilizational (identity), modernization (accumulation), and state security (defense) forms of geopolitical thinking. Geopolitical cultures tend to feature competing visions that combine elements of each. Which one predominates and drives state foreign policy is the subject of struggle and entrepreneurship in the political arena. Here the political organization of structures of authority in a state is crucial. If decision-making authority is vested in a single central institution, such as a presidency, the power of that officeholder to determine the geopolitical orientation of a state is considerable. They have the power to creatively synthesize the different traditions in a state's geopolitical culture into specific geopolitical-policy storylines (see below). If power is distributed in a state across multiple sites of power, there is likely to be greater dissent and division within a geopolitical culture, and challenge to the power of an executive to have its storylines prevail over all others.

State elites debate geopolitical visions and orientations within an international arena characterized by competing as well as shared myths, norms, and discourse. A myth is a narrative form that creates coherence, structure, and identity for a community. Myths help establish boundaries within and between communities. Those who do not share the myth are held to not fully belong to the community.[55] Mythic archetypes refer to the recurrent structures, motifs, images, and symbols found in origin stories and historical-identity discourses. In Europe these often have strong Christian elements that reflect the historical power of Christianity as an interpretative system in that part of the world.[56] Normative discourses are also conditioned by Christian categories, but now mostly find expression in the realm of international law. In rhetoric an ideograph is a figure of thought, often a one-term summation of a putative national ethos or historical ideology. Words like "liberty" or "freedom" are examples, as are expressions like "rule of law" and "free world" as well as metaphors like "Mother Russia."[57] While these are commonplaces, they are also powerful transcendent signifiers that different political actors compete to appropriate and control for their own purposes. Not all uses will necessarily resonate. Some ideographs are metageographies, like "Europe" or "the West," and we will refer to them as "geo-graphs" here. They serve the useful purpose of spacing the world into distinct realms, usually representing different axiological systems. Others are visual, like a flag, statue, or map.[58]

Here we can do no more than simply note the discursive competition between the various actors within the post-Soviet geopolitical field over the reworking of longstanding Cold War categories and ideographs. Three of the most resonant are noted below.

Imperialism and Anti-Imperialism
Imperialism has always been a central discourse in U.S.-Russian relations and in the functioning of the Russian state. During the Cold War, each side charged the other with imperialist ambitions and policies.[59] In the moralized geopolitics of the first Reagan administration, the Soviet Union was "the evil empire." Russia's reputation as an empire preceded the Soviet Union and endured after it. Russia, to its critics, was never a "nation-state" and was too prone to imperialist behavior against its neighboring states. For the new nationalizing state elites, an imperial

condition shaped their past and menaced their future. Decolonization required forms of nation-building that undid colonial legacies. But the dichotomy of nation versus empire was self-serving and hid internal complexities. Just as Chechens charged the Russian state with internal imperialism, so also did Abkhaz and Ossetians charge Georgia with the same. The Soviet dissident Andrei Sakharov famously described Georgia as a "little empire" because of its initial ethno-exclusivist approach—"Georgia for the Georgians"—to nation-state building.[60] In 1991 an Ossetian leader explained to a visiting Western journalist: "We are much more worried by Georgian imperialism than Russian imperialism. It is closer to us, and we feel its pressure all the time."[61] Thus, not only for newly independent states but also for separatist forces within these states, anti-imperialism is a powerful normative discourse. "Freedom" in post-Soviet space meant different things to different peoples. Complicating matters further was that some within post-Soviet Russia championed the cause of what they saw as "stranded Russians" and Soviet-nation "compatriots" who had unjustly become "national minorities" beyond the borders of Russia. Russia had a responsibility, as a great power and former imperial center, to maintain order and protect the vulnerable beyond its new borders. Discourses on this varied. To some, Russia should follow the United States and become a "liberal empire" spreading capitalism.[62] To others appropriating Soviet rhetoric, Russia's mission was anti-imperialist. Russia's enemies beyond its borders were not only nascent "fascist" elites within the newly nationalizing states but supporters of such elites, led by the inveterate imperialist power, the United States of America.

Civilization, Europe, and the West

Evocation of civilizational distinctions and divides has always been central to the positioning of Russia in the world, both by itself and external powers. The Cold War was interpreted as an antagonistic "two ways of life" civilizational struggle, with deployment of "Western civilization" a crucial means by which the Federal Republic of Germany was recoded as part of the West and worthy of admission into NATO security structures.[63] This was the narrative adopted by the Baltic countries as the Soviet Union collapsed, and thereafter by other nationalizing successor states in post-Soviet space. They argued they were already part of "the

West" and of "Europe" (though they wanted to avoid the label "Eastern Europe") and were now "returning" after occupation and captivity by "the East" (or, in more pointed discourses, by "Asia"). Discourses of Europeanizing and Westernization worked as complementary articulations of civilizational geopolitical discourse.

Dichotomizing geo-graphs—"the West" versus "the East"—are not the only way in which civilizational discourses can be evoked and deployed. Civilization could also be conceptualized in singular terms, as a shared common identity and set of values in opposition to a barbarian outside world. The Gorbachev era saw the use of this construct in U.S.-Soviet relations, and it remained a flexible rhetorical resource used by Yeltsin and later Putin. This was Putin's usage on September 11, 2001, and he returned to it to justify Russia's interventionism in the Syrian civil war in 2015.

Genocide and Fascism

The genocide narrative is a product of the post–World War II era. The framing of Germany's exterminationist policies against European Jews as the crime of genocide was the result of pioneering legal entrepreneurship by Raphael Lemkin. Given the moral and normative power of this designation, it is hardly surprising that national groups with historical traumas involving mass killing and persecution would seek to reframe these traumas as genocide. The Armenians did so in the Soviet Union, with the construction of a memorial to mark the fiftieth anniversary of the persecutions of 1915 in Yerevan.[64] Other Soviet nations with histories replete with violent repression, such as the Abkhaz and Ossetians, reinterpreted their history within the terms of this discourse. In naming historic massacres as genocides, it powerfully constituted victims and persecutors in a highly moralized narrative. Genocide was the ultimate crime, the name for the unspeakable. No other word would do.

Genocide discourse is important not only to small national minorities, such as the Ossetians, but also to great powers. A mythic narrative of the Great Patriotic War was constructed in the Soviet Union to remember and memorialize World War II as a heroic struggle against fascist powers pursuing genocidal policies. This narrative is about much more than discourse. Songs, music, images, flags, and period-piece commonplace objects engender a surge of emotion and feeling. For example, in the Soviet Union, Shostakovich's Seventh Symphony, subtitled "Leningrad," automatically recalled the circumstances of its composition and the suffering it commemorated. The Great Patriotic War narrative obscured the Nazi-Soviet pact and crimes later committed by some Soviet troops as well as the establishment of Soviet client states in what became "Eastern Europe" or, more pointedly, the "Eastern bloc." "Fascism" was always a flexible rhetorical label for the Soviet Union and worldwide Communist movement. During the Cold War, it was repurposed to constitute the Western powers, and NATO, as successor fascist enemies to the Axis Powers.[65] As the Soviet Union came apart, this rhetoric was utilized to characterize the renewal nationalist movements that came to power in Georgia and Moldova.

It was complimented by a new post–Cold War discourse sponsored by Western powers that also had genocide at its center, namely a "responsibility to protect" creed that developed within the United Nations in the wake of the mass killing episodes in Bosnia, Rwanda, and Kosovo in the 1990s. This had its origins in the emergent Western consciousness of the systematic killing of Jews and other groups during World War II, what became known simply as "the Holocaust." A museum to this horrific chapter of European history dedicated in Washington, DC, in 1993 extended recognition and awareness of genocide as an affective storyline.[66]

One irony of the popular appeal of the genocide storyline is that both American neoconservatives and Putin administration officials demonstrate a preoccupation with Nazism and a rising fascist threat, but each view the other as embodying this viscerally experienced threat. Triggered by the deployment of the signifiers "fascism" and "genocide" are imperatives of rescue and protection. For many the matter is beyond debate, beyond words. Rescue missions, as we will see, are particularly significant forms of geopolitical imagining in post-Soviet space.

Affective Geopolitics

Geopolitics has historically been associated with deliberative, calculative, and instrumental modes of thinking. Poker and chess metaphors are common. High-profile geopolitical decisions over the last decade, such as the U.S. invasion of Iraq, however, have once again spotlighted factors other than deliberation at the center of decision making. An array of contemporary social sciences—feminist scholarship, cognitive neuroscience, and decision-making psychology—has also upended commonsense notions of human cognition as deliberative and rational. Antonio Damasio's work underscores how normal human reasoning is dependent upon emotional structures.[67] Indeed, ordinary reasoning is not possible without these capacities. The psychologist Daniel Kahneman's work emphasizes the importance of the operation and interaction of two "systems" in the human brain, a largely automated rapid reaction System 1 that he describes as "fast" thinking and a System 2 that he associates with more deliberative "slow" thinking.[68]

And then there is affect, which in itself has generated an enormous literature. Considerable debate surrounds its specification, and related concepts like feelings and emotions, in contemporary social science. For our purposes, *affect* is understood here as a largely biological phenomenon that concerns how human bodies register sensations in pre-cognitive and semiconscious ways. It is about the vast substratum of thinking that is automatic and unconscious, below or barely at the level of consciousness. When this gets organized and mobilized, it can be a particularly potent force. Affect is significant because it "retains an element of autonomy that pulls thinking beyond the steady control of intellectual governance."[69] High-profile events, like terrorist attacks or invasions, can generate affective waves that inflame populations into support for radical policies and actions.

Affective experience, however, is not all force and energy that enrolls and mobilizes. It can also be a slow phenomenon—for example, nostalgia for a lost order such as the Soviet Union. It can involve embodied experiences of vulnerability, passivity, suffering, fatigue, indolence, apathy, and hunger. These too have political implications. Feeling refers to how affect is consciously registered and personalized in human bodies; emotion refers to how such feelings are recognized and organized culturally. Both are subsets within the problematic of affect. If there is any theme that unites the different literatures that engage affect, it is taking biology seriously as a force in human culture and cognition. "A multilayered conception of culture and thinking is needed today," William Connolly writes, "one that comes to terms with how biology is mixed differentially into every layer of human culture."[70]

This has significant implications for the study of geopolitics. Feminist scholars have led the way in focusing attention on the significance of embodied practices and affect in the everyday experience of geopolitical events.[71] A few international relations scholars have highlighted intersections of geopolitics and emotion.[72] Attention is now being paid to how societies socialize bodies into automated ways of (re)acting, feeling, and thinking. It is in early parenting and schooling practices that biological and cultural inputs grow into and around each other. All this schooling, training, running, singing, playing, and working accumulate as an everyday life grounded in affective commitments to, or disaffections with, particular places, communities, countries, authorities, and ways of living. Regimes of spatial acculturation and socialization are the raw material of geopolitical cultures.[73] To speak of affective geopolitics, then, is to broaden and deepen our understanding of geopolitics as a form of thinking and acting that rests on bio-cultural foundations.

Two aspects of affective geopolitics are particularly relevant to understanding post-Soviet space. The first is how individuals and groups tacitly understand and acquire deep-seated dispositions about the territory of the state they live in and those that surround it. At a primordial level, this has to do with size and the experience of security, or insecurity, that it induces. As the largest country in the world, Russia's size would appear to offer security and reassurance in contrast to, for example, a country like Israel whose size, and dominant perception of being surrounded by enemies, perpetuates an already visceral sense of existential threat. However, Russia's great size has historically induced a sense of vulnerability and has been accompanied by discourses about plots and encirclement schemes by historic enemies, portraying Russia as a besieged fortress. Many schoolchildren learn about their country through a literature and poetry that frame it as an organic entity, a geo-body that is part of them. Its success is their success. Some learn to see the territory as a spiritual and cultural patrimony, as sacred space, a civilizational achievement. After state schooling, territorial images of the state circulate daily in newspapers and on television; they are part of the collective unconscious of communities. Color schemes and music are also part of this, such as the colors of flags and political movements ("color revolutions") and the music of celebration, remembrance, and defiance. The Russian national anthem, for example, is an adaptation of the anthem of the Soviet Union, with lyrics—"From the southern seas to the polar lands/Spread are our forests and fields/You are unique in the world, one of a kind/This native land protected by God!"—rewritten at the outset of Vladimir Putin's presidency in 2000. The repetitive and habitual performance associated with state holidays, religious services, and remembrance rituals create the nation-state as an embodied condition. Non-core nations reproduce themselves similarly but in contradistinction to the dominant nation-state.

We know from past historical cases that the breakdown of a particular state territorial order, especially if it is sudden and unexpected, can be experienced as fearful and traumatic by significant segments of the population. This is obviously relevant for understanding the collapse of the Soviet Union.[74] The sudden disappearance of a seemingly permanent "big country" was disorientating, especially for families with relatives scattered throughout the territory. Former popular holiday destinations and summer camps, affective spaces of leisure and pleasure like Black Sea destinations in Abkhazia and Crimea, were now in different countries. For millions of Soviet citizens, made differently but wired to the same central cultural referents and sites of power, this was a profound shock, a source of regret and later humiliation. The prominent Russian economist Yegor Gaidar described Russia as experiencing phantom limb pain.[75] Even if Soviet-era citizens mentally adjusted to the changes, and socialized new generations into different forms of belonging and knowing, their bodies were nevertheless made from Soviet raw material, from family memories and connections that were laid down in formative years. The depth and intensity of territorial affect is an empirical question. Many groups experienced the collapse of the Soviet Union as liberation, but for others it was a disaster of violence, displacement, and economic ruin. Many had conflicted feelings and mixed emotions. And within newly independent states, like Georgia, Moldova, and Azerbaijan, the Soviet collapse brought their own local experiences of territorial loss and dismemberment.

The second aspect that is relevant is how affective geopolitics conditions and shapes the leadership styles and foreign policy decisions of key players in post-Soviet space. Here the emphasis complements but is nevertheless different from that of feminist scholars who study leadership in terms of gendered performances of masculinity and strength.[76]

Four of the key figures in this book—Vladimir Putin, George W. Bush, Mikheil Saakashvili, and John McCain—exhibit distinctive forms of hegemonic masculinities. We can, however, nuance this analysis by examining how these figures were "cooked" by certain affective geopolitical contexts and went on to embody, consciously and unconsciously, gendered ideals and subject-positions. This approach allows us to have a degree of empathy for these politicians as figures sincerely fired up by certain beliefs and fighting for certain causes. This is both affective and gendered in that it is about politicians working within repertoires of heroic masculinity. This is best explained by an example concerning Putin. As is well known, Putin sought to join the KGB as a teenager, spontaneously volunteering in 1968 at its Leningrad headquarters. Biographers have noted the importance of the espionage drama (book, and then four-part TV series) *The Shield and the Sword* in motivating the impressionable teenager to volunteer in this fashion, and in subsequently making it his career choice.[77] Four decades later he not only remembered the film's sentimental theme song "Whence Does the Motherland Begin," and sang it with unmasked Russian spies kicked out of the United States in 2010, but also went to the trouble to learn how to play it on the piano.[78] Bush, Saakashvili, and McCain have equally sentimental formative influences and sincere commitments to causes much larger than themselves. Heroic scenarios and storylines—quests for liberation, responsibilities to protect and rescue—manifestly appeal to all these men. Both Bush and Putin viewed themselves as tough minded leaders of the civilized world in a war against the barbarianism of "international terrorism." Saakashvili saw himself as a new Georgian state-building hero (he named Atatürk, Ben-Gurion, and General de Gaulle as his role models) liberating his country from its Soviet past and restoring its territorial integrity.[79] His relationship with Putin quickly became toxic and competitive. This found expression in explicitly gendered ways, with Saakashvili symbolically diminishing Putin by rhetorically mocking him as "Lilli-Putin," while Putin famously threatened to hang Saakashvili "by the balls" in front of President Sarkozy of France in August 2008.[80]

Anger and resentment are undoubtedly aspects of affective geopolitics that need further research. Western sponsorship of the secession of Kosovo from Serbia, for example, is a clear anger point for the Russian leadership. Western interventionism in Libya and Egypt became anger points later. But affective geopolitics is also about positive desires to advance noble causes and save lives. Western desire to liberate the "captive nations" of Eastern Europe from Soviet occupation is a longstanding one. Representation of Russian behavior against its neighbors as "bullying" cues affective not strategic response strategies. The most consequential examples, of course, are Russia's invasions of 2008 and 2014, which were represented as rescues of co-ethnic and compatriot communities from "fascist" nationalism.

Geopolitics as Tabloid Storylines and State-Produced Drama

This brings us to the question of geopolitics and the media, an aspect of what I earlier described as the geopolitical condition. Governments employ communication strategies to circulate their storylines about events and actions to the public at large. We need to consider how this succeeds or fails. In an important work, François Debrix argues that the dominant style of geopolitics in the twenty-first century is based on tabloid culture.[81] The claim is surprising, for tabloid culture emerged more than a century earlier in the late-nineteenth-century newspaper and publishing industries in the United States and Great Britain. Indeed, the words "tabloid" and "geopolitics" were coined at roughly the same time and have important connections. William Randolph Hearst built his publishing empire by pioneering what became known as "yellow journalism," sensational stories of dubious veracity, which famously propelled the United States into war with Spain in 1898 over "atrocities" in Cuba. Alfred Harmsworth made his company Amalgamated Press into a publishing colossus in the United Kingdom by creating a popular working-man's newspaper, the *Daily Mail*, and pioneering what he termed the "tabloid newspaper." The *Daily Mail*'s chauvinistic representation of Anglo-German rivalry did much to shape popular hostility toward Germany and Kaiser Wilheim II, propelling the British people into the cataclysm of World War I. Tabloid newspapers built upon the nineteenth-century tradition of representing geopolitics through stock nation-state stereotypes (e.g., "John Bull," the "Russian Bear") and cartoon depictions of great international crises of the day. The Crimean War of 1853–1856 is viewed by some historians as the first time newspapers were a powerful force shaping public sentiment and the actions of state leaders and diplomats.[82]

What Hearst and Harmsworth helped create over a century ago were a set of principles and methods that remain significant today, though we live in a different media ecosystem. Cultural archetypes, mythic stories, and classic tropes are mobilized to render international affairs recognizable as particular forms of drama and spectacle. Truth is a function of the devices and procedures of presentation. Tabloid media are anti-elitist and anti-intellectual in stance, celebrating the "ordinary man"—a position that has generated notorious sexism over the years—while selectively exposing the "scandals" of the rich and powerful. Techniques of simplification and sensationalism drive presentation. Everyday life and political contestation are rendered through a dramatic narrative of good and evil. News is presented as a series of dramatic stories featuring easily grasped villains and heroes. (The tabloid form is partly built upon the nineteenth-century genre of melodrama.) Pulp novels and tabloid newspapers present the world as a struggle of opposites, of emotional drama and awe-inspiring heroism. As they developed, tabloid newspapers made visual imagery central to their appeal, with large headlines and dramatic photographs framing events. The importance of visuality only deepened with the advent of newsreels, television broadcasting, cable television networks, and today's interactive social media platforms.

Tabloid forms were initially driven by commercial considerations in capitalist states: the goal was to capture as large a mass audience as possible to facilitate profit-making. This dictated a populist strategy of appeal and presentation. State-controlled media in authoritarian states soon learned to use tabloid techniques for ideological control and propaganda. Totalitarian states developed elaborate infrastructures for domestic population control and indoctrination. As the Cold War took hold after World War II, it came to be defined as a "psychological war." Using experts from marketing and public relations industries, the Truman and Eisenhower administrations institutionalized a communications and media apparatus dedicated to "liberation" in Europe that endures in a revised form to this day.[83]

Tabloid techniques have endured even as the newspaper industry and government regulation have changed. Media corporations holding tabloid newspapers, such as News Corporation International, created tabloid-style television channels, the most celebrated of which is

Fox News in the United States. On television, tabloid techniques have generated what some media theorists term "hyperreality," a condition where persons and events are lifted into a self-referential world created by visual, cinematic, and digital modes of representation.[84] Through creative blending of tabloid storylines and visual production techniques, persons and events become hyperreal, given heightened dramatic form and an adrenalized presentation. As an apparatus-inducing, organizing, and directing political affect, Fox News has inspired admiration and imitation across the world, including within Russia. Under Putin, domestic television was brought under direct and indirect state control. In 2005, Russia launched Russia Today as part of an effort to improve Russia's image abroad. It actively presented a Russian and South Ossetian perspective on the August 2008 war against what it saw as "anti-Russian" bias in Western media organizations.[85] Russia's media operations were reorganized in 2013 under the leadership of Dmitry Kiselev, a prominent tabloid-television presenter on Russian television. Russia's state-controlled media complex has actively disseminated the Kremlin's point of view on international affairs.[86] It played a crucial role in 2014 in taking a governance crisis in Kyiv and turning it into a hyperreal story of a resurgent fascist threat to Russians and Russian-speakers in southeast Ukraine.

Two tabloid narrative forms that are now mainstream in all geopolitical cultures are conspiracy theories and nationalist heroics. The first is a product of the presentation of international politics as a competition for power and resources between two opposing world powers. Conspiracy theories provide various accounts of motives and ends in this struggle. The distinguishing feature of this form is the attribution of nefarious motives to a competing actor (leader, state, or great power) who represents a sinister other. Conspiracy theories are reductive and simplifying narratives, turning international affairs into a zero-sum drama over concealed ends and goals.[87] The narrative form features the "hidden hand" of the menacing other and accounts of injustices suffered by unwitting or duped victims. Stories concerning natural resources, military bases, secret agents, "behind-the-scenes" deals, and covert plots are stock tropes of conspiracy thinking. Indeed, a whole genre of popular geopolitical publishing, a type of pulp geopolitics, has developed within Russia and elsewhere dedicated to the subject.[88]

A second favored storytelling form is nationalist heroics. This features a series of classic tropes and flourishes—courage, defiance, and bravery—by ordinary or relatively powerless protagonists (individuals and small states), who confront evil and triumph. But it can also feature heroic behavior by powerful actors who accept a duty to protect the vulnerable and powerless. One particular heroic storyline of this ilk is the rescue fantasy, a scenario where a vulnerable and helpless small state (often feminized) is threatened by an evil empire with invasion (implicitly violation and rape). This form is old and in no way exclusive to contemporary post-Soviet space. It is, for example, the storyline of the U.S. intervention to "save" Cuba in 1898. This scenario is implicit in U.S. "captive nation" discourse accompanying the Cold War and post–Cold War aspirations to "rescue" states like Georgia and Ukraine from a Russian sphere of influence. As we will see, the Russian state version of this storyline explains interventionism as the protection and rescue of compatriots left stranded in threatening circumstances in the near abroad. The dramaturgy involved in the production of invasions as rescues requires our attention.

Escaping Geography through Globalization

The final theme that we need to introduce is the dynamic nature of proximity and distance in international affairs, a classic concern of those theorizing geopolitical conditions. Some twentieth-century geopoliticians presented geography as permanent and unchanging, but even early geopoliticians like Halford Mackinder were aware of how new technological systems of communication and transportation transformed spatial relations and connectivities. Mackinder warned the British ruling class in 1904 that railways were "transmuting" the conditions of power in Eurasia.[89] In the new post-Columbian epoch of closed political space and worldwide connectivities, land power would be ascendant, and sea-power states like the British Empire needed to modernize.

The end of the Cold War in Europe provided an impetus for dramatic changes in movement, connectivity, and accessibility across the continent. "Globalization" was the favorite term applied to this new round of time-space compression, of accelerating speeds of connectivities and communications combined with disappearing frictions of distance and borders. Greater airline connections and new communication systems like the Internet and mobile phones made places seem closer and more proximate than ever before. Unfortunately, these changes are often sloganized in glib ways—such as the "end of geography," the "death of distance," a "borderless" or "flat" world—that obscure the disparate consequences of technological change on our geopolitical condition.[90] One important geopolitical implication of greater ease of international travel is that it enabled elites within newly independent states to free themselves from the legacy of the Soviet Union, where formerly all communication lines tended to be routed through Moscow, and to build new communities of social proximity despite the physical distance of diasporas and far-off powers. In time, some of these elites built dense ties with Western states and tended to look toward the West for support in their struggles with Russia.

Mikheil Saakashvili is a case in point. In a memoir, his wife notes how, in 2000, "he became more and more focused on the media, on communications, and that he couldn't live without his mobile phone anymore—a luxury item."[91] Saakashvili already had a network of contacts in Washington and elsewhere. The U.S. ambassador to Georgia recalls that when Saakashvili assumed power after the Rose Revolution in December 2003, he was always on the phone, relentlessly cold-calling politicians, government officials, and lobbyists in Washington. Indeed, Saakashvili's late-night working habits were partially dictated by his proclivity to live by the time schedule of the United States rather than Georgia. From the outset it was apparent that Saakashvili's agenda was to propel Georgia out of its existing geographic and geopolitical context. His government redesigned the Georgian flag, placed the European flag next to the Georgian flag in all state buildings, and made NATO membership an explicit priority of the Georgian government. Slogans on streets and in the media broadcast the aspiration. Saakashvili spoke of Georgia as an "ancient European" country (even though that tradition of imagining had traveled to Georgia through Russia). Ultimately Saakashvili was ambivalent about Europe and looked to the United States as Georgia's future.[92] Georgia was to be

America in the Caucasus.[93] All of this willful geopolitical imagining (and imagineering) was enabled by information technology and associated dreams of transcendence and flight from geography. But it left Georgian actualities behind, most especially the country's grinding poverty; its Orthodox Christian culture; and its long history of ties to the immediate Caucasian region, to Russia, and to the Soviet Union, ruled for decades by a Georgian. It was a "luxury item" that, in the end, saw Saakashvili flee Georgia to the United States before he found a new cause in Ukraine.

The five thinking tools introduced here are not themes organized around the othering of Russia, the United States, or any other geopolitical power. Rather, these are a conceptual toolbox that builds out three core critical geopolitical concepts—field, culture, and condition—that are intermeshed and co-constituting in practice. The following chapters are informed by these conceptual thinking tools, but they are not case studies of the concepts. Rather, they are narrative arguments that draw upon these concepts lightly as they seek insight into the question of why Russia came to invade Georgia and Ukraine.

2

Geopolitical Catastrophe

ON APRIL 25, 2005, President Vladimir Putin addressed the Federal Assembly of the Russian Federation. In his lengthy speech Putin laid out a series of priorities for the Russian state in the coming decade. These priorities were not new—he had spoken about them in a similar address the year before—and their central aim was well known, "to build," as he put it in his address, "an effective state system within the current national borders." However, it was not Putin's discussion of democracy and corruption in state institutions that generated headlines in the Western media. Instead it was Putin's prologue for his reform agenda:

> Above all, we should acknowledge that the collapse of the Soviet Union was a major geopolitical disaster of the century. As for the Russian nation, it became a genuine drama. Tens of millions of our co-citizens and compatriots found themselves outside Russian territory. Moreover, the epidemic of disintegration infected Russia itself.
>
> Individual savings were depreciated, and old ideals destroyed. Many institutions were disbanded or reformed carelessly. Terrorist intervention and the Khasavyurt capitulation [the agreement that ended the first Chechen war] that followed damaged the country's integrity. Oligarchic groups—possessing absolute control over information channels—served exclusively their own corporate interests. Mass poverty began to be seen as the norm. And all this was happening against the backdrop of a dramatic economic downturn, unstable

finances, and the paralysis of the social sphere. Many thought or seemed to think at the time that our young democracy was not a continuation of Russian statehood, but its ultimate collapse, the prolonged agony of the Soviet system. But they were mistaken.[1]

Putin's rhetorical device was a conventional decline-and-renewal trope, describing the era of national decline and humiliation that set the stage for his heroic mission of restoring Russia's strength and capacity. However, Associated Press and BBC news service reports on the speech focused only on one phrase, to which they gave a different translation from that released by the Kremlin (cited above). Putin, they reported, had declared that the collapse of the Soviet Union was "the greatest geopolitical catastrophe" of the twentieth century.[2] From these summary dispatches, a damning Putin quote was born, one that would circulate for a decade in commentary on Russian foreign policy. In March 2014, for example, John Bolton, a former U.S. ambassador to the United Nations, cited it as explanation for Putin's response to the Ukraine crisis. Putin "wants to re-establish Russian hegemony within the space of the former Soviet Union. Ukraine is the biggest prize, that's what he's after. The occupation of the Crimea is a step in that direction."[3]

The enduring appeal of "the greatest geopolitical catastrophe" catchphrase is that it feels right. A form of truthy geopolitics, it reveals more about its users than about Putin's attitude toward the Soviet Union. Debate about translation of Putin's words is contentious.[4] The Kremlin's translation—"a major geopolitical disaster"—is disputed for it downplays what Putin critics see as his outrageous privileging of the collapse of the Soviet Union in a century full of dark geopolitical catastrophes. But it is fairly evident that his was a relativized view, most especially given the centrality of the Great Patriotic War to his geopolitical lexicon.[5] Putin's later clarification of the remark, in which he said "we do not regret" the breakup—he also cited the Russian epigram that "those who do not regret the collapse of the Soviet Union have no heart, and those that do regret it have no brain"[6]—is rarely, if ever, cited by Western political figures. Instead, the "greatest catastrophe" quote endures because critics view it as akin to an incriminating Freudian slip, one that reveals Putin as a crypto-commie imperialist whose deep desire is to re-create the Soviet Union.

This negative ascription displaced a more (in)famous embodied testimony attributing a straightforward and trustworthy character to Putin, by President George W. Bush after their first summit in June 2001. Bush declared he "looked the man in the eye" and "was able to get a sense of his soul."[7] Bush's then secretary of state Colin Powell let it be known that his response to Bush's remarks was: "Mr. President, I looked into President Putin's eyes and I saw the KGB." Bush's Republican rival, Senator John McCain, later appropriated that line. Running for president in 2007–2008, he implicitly critiqued Bush by telling audiences of how he "looked into Putin's eyes" and "I saw three things—a K and a G and a B."[8] Bush's second secretary of defense Robert Gates offered another variant when he stated he looked into Putin's eyes and saw "a stone-cold killer."[9]

The popularity of these embodied testimonies, sourced in cowboy movie masculinity, underscore the enduring framing power of Cold War referents—the Soviet Union, the KGB, cynical enemies—over perceptions of Russia in U.S. geopolitical culture. But the difficulty with such catchphrases is that they are too often substitutes for actual empirical analysis of Russian foreign policy (as is the case with Bolton above). Nostalgia for the Soviet Union and deep state-security structures *are* important background factors shaping Russian foreign policy. But equally vital is an understanding of post-Soviet space as a geopolitical field, an appreciation of the divisions within Russian geopolitical culture, Putin's actual affective dispositions, and the conditioning influence of the contemporary geopolitical condition (especially acts of terrorism as spectacles of disorder). The men known in Russia as the *siloviki*—present and former security and military officials who run Russia's power ministries—are central to the power structure of contemporary Russia, but analysis needs to dig deeper to grasp the broader structural context and culture within which they operated in the past and continue to operate today.

In this chapter I begin with a discussion of the Soviet Union as a distinctive imperial structure that collapsed rather suddenly. That collapse was head-spinning for millions of ordinary Soviet citizens, not only officials like Putin within the Soviet security apparatus at the time. It provoked an urgent identity crisis in Russia, and the emergence of a Russian geopolitical culture characterized by three competing traditions. While only one was explicitly revisionist, advocating different borders from those left by the Soviet collapse, there was widespread consensus in Russian geopolitical culture that it was victimized territorially by the Soviet collapse. Putin's relationship to these three traditions requires careful analysis, for it has evolved historically as state power, geopolitical circumstances, and his domestic political interests have changed. It is certainly misleading to claim that Putin wants to restore the Soviet Union, but it is not misleading to claim that Putin's project was revanchist from the outset. It is the nature of that revanchism and its unfolding entanglement with preexisting territorial disputes in the Caucasus and Ukraine that we need to understand.

Section one of this chapter provides an analysis of the Soviet Union as a particular type of empire whose collapse laid bare a series of fragile, unsettled territorial disputes between competing power centers across Soviet space. The second section provides a description of the new geopolitical environment that was post-Soviet space, a relational field characterized by the quintet of actors described in chapter 1. Section three provides a summary portrait of Russian geopolitical culture in the 1990s as it came to terms with the Soviet collapse and as different power networks offered competing visions of the country and its interests. Finally, section four outlines some enduring features of the geopolitical entrepreneurship of Vladimir Putin since his ascent to power in late 1999. It was always Putin's goal to restore Russia to the status of a great power in northern Eurasia. Pursuit of this end saw the adaptation and use of practices forged during the Soviet experience. But the end goal was not to re-create the Soviet Union but to make Russia great again.

Collapse of an Empire

While there was never much controversy about whether the Soviet Union was an empire within U.S. geopolitical culture, the question is much more complicated than has been generally acknowledged. Both the United States and Russia grew through a process of territorial expansion and centralization that was imperialist in character, with local peoples rendered subordinate to distant capitals and alien administrators. State- and empire-building were indistinguishable, and both were seen as progressive, modernizing endeavors until the twentieth century. Mobilizing against the imperialism of the great powers and their

"Great War," the Bolshevik vanguard seized power in Russia. Vladimir Lenin "sought neither to restore an empire nor to build a state, but rather to carry out an internationalist revolution that denied both the legitimacy of empires and the utility of states."[10] Championing the cause of national self-determination helped the Bolsheviks accumulate power across much of the collapsed Russian Empire. Through war, conquest, and repression they formed the Union of Soviet Socialist Republics (USSR) as "the first multiethnic state in world history to define itself as an anti-imperial state."[11] And, despite manifest contradictions, the Soviet Union always officially claimed to be the world's leading anti-imperialist power, championing this position against the United States during the Cold War. It was only at the end, crippled by governance and legitimation problems, that this tired ideological conceit fell away.

While the Soviet Union may not have begun as an empire, it did become one, perhaps the most uniquely complex of the twentieth century.[12] What made it so? For a start it inherited the largest land empire in the world, an imperial superstructure that knitted together a staggering heterogeneity of lands and peoples using a diverse set of strategies and techniques. At its demise the Soviet Union comprised fifty-three different ethnoterritorial units, a veritable matryoshka doll of nested governance: at the top were fifteen Soviet Socialist Republics, and below and within there were twenty different Autonomous Soviet Socialist Republics, eight autonomous oblasts, and, with the least status, ten Autonomous Okrugs.[13] David Laitin has identified three distinctive modes of Russian state control that spanned the tsarist and Soviet periods.[14] These modes reflected not only different geographic contexts and forms of incorporation but also divergent cultural positioning in a civilizational hierarchy that endured as a deep structure while Russia transitioned between political regimes. These ideal-type modes should be considered alongside a separate ideal-type spectrum of language repertoires. On one end is parochialism: people are monolingual in their own local language only. In the middle is unassimilated bilingualism (use of the local language in most social domains but Russian in limited official domains) and assimilated bilingualism (use of Russian in high-prestige domains and the local language in delimited social settings). The other end of the spectrum is full assimilation, where people are monolingual Russian speakers and do not know or use "their" local language (not addressed here).[15]

The first mode of incorporation is the classical colonial model in which the center subordinates a geographic region and recruits a local elite to broker their rule with the regional population. This local elite can enrich itself through its relationship with the center, but it can never acquire the same social status as those at the center. The language repertoire of this local elite is initially unassimilated bilingualism, but it can develop over time to assimilated bilingualism and even beyond, with local elites rarely speaking their local language. Laitin's example for this type is the Russian state's historic relationship toward Kazakhstan. Kazakh elites were able to build up considerable power within their own dominion but remained outside the circles of decision making and power at the Soviet center. Kazakhstan itself was long treated as a subaltern territory not equivalent to other regions like Ukraine or the Baltic states. Initially within the Soviet structure, Kazakhstan was a mere autonomous oblast, and while later acquiring nominal equivalence to the other Soviet Republics, its subordinate positionality endured. It and eleven other republics were not part of the founding meeting of the Commonwealth of Independent States (CIS).

The second mode of incorporation is what Laitin terms "integralist incorporation." Here a region is acquired that already has well-developed levels of modernization, literacy, and national identity. The relative cultural and economic strength of these regions gives them an identity that renders the idea of assimilation according to the mores of the center less appealing. The language repertoire of the local elite that takes on the role of brokering with the center remains largely stuck at unassimilated bilingualism. Local elites have limited opportunities to move into ruling circles at the center, and they also have sufficient local cultural confidence and demographic coherence to forgo or block russification. Laitin's examples here are the Baltic states where, he suggests, Russian rulers had "to adapt to the peripheral culture rather than the other way around."[16] He does not discuss the case of Georgia, but one can argue that it represents another example, albeit with a longer history of incorporation and distinctive features.[17] While many Georgian nobles gained entry into the Russian ruling circles, and a Georgian Bolshevik who adopted the moniker "Stalin" rose to become the most infamous ruler of the Soviet Union, Georgia maintained a separate cultural and linguistic world that saw limited Russian-language penetration. In aggregate statistical terms, relatively few Georgians made the leap from their country to the inner circles of the Politburo. A complication in the Georgian case, and indeed in all cases, is that certain territories within the Georgian domain had different modes of incorporation from that of the Soviet-created republic as a whole.

The third mode of incorporation is the "most favored lord" model. In this case, the elites of the incorporated territory are able to join the ruling class of the imperial state as near equivalents to its core identity. While some elites do not take advantage of this opportunity, many do so and assimilate into the hegemonic identity complex of the imperial state. Laitin's exemplar of this relationship is Ukraine, where members of the nobility were exempted from all government and military service by the tsar's "Charter to the Nobility" in 1785. Ukrainian Cossacks were able to ennoble themselves and acquire important positions across the breadth of the imperial administration. In the Soviet period, the close relationship between Russian and Ukrainian elites persisted, with some empirical studies revealing that Ukrainians received more top positions at the Soviet center than any other non-Russian nationality.[18] The process of co-optation had relatively low costs and potential high rewards for Ukrainian-speakers, many of whom embraced russification out of pragmatic necessity. From the Russo-Ukrainian borderlands came not one but two general secretaries of the Communist Party of the Soviet Union—Nikita Khrushchev and Leonid Brezhnev—both of whom identified as Soviet Russians. It was Khrushchev, once a metal worker in Donetsk, who presided over Crimea's transfer from Russia to Ukraine in February 1954. The occasion was the 300-year anniversary of the Treaty of Pereyaslav, wherein a Cossack parliament declared unity with Muscovy, with the transfer presented as a symbolic gift by an elder to a junior brother-nation. While there were supplemental economic and geographic reasons for the move, the surface symbolism about "friendship" between the two "brother" Soviet republics underscored the prevailing conceit of the Soviet Union as a postimperial state formation.

The ideology of the Soviet Union as an anti-imperial structure was strongest at the time of its formation, and it left an enduring legacy. Declared in 1922, the Soviet Union represented itself as a radical break with the past—as an empire of nations against the empire of great powers and Great Russian chauvinism. Faced with managing its internal diversities, the Bolsheviks fell back on established tsarist hierarchies of classification and delimitation before launching an ambitious ethnographic effort to organize and arrange all the distinct communities of the empire into national peoples, nationalities, and hierarchies based on size and perceived level of civilization.[19] Many have noted the irony of "the world's first state of workers and peasants" creating "the world's first state to institutionalize ethnoterritorial federalism, classify all citizens according to their biological nationalities and formally prescribe preferential treatment of certain ethnically defined populations."[20] But there were sound Marxist, and realpolitik, rationales for the innovative, elaborate, pyramidal ethnoterritorial governance structures created by the early Soviet Union. Nationalism was a powerful mobilizing force that had to be co-opted and disarmed by giving the diverse masses their "national form" but with the same "socialist content." Conversion to class consciousness had to take place through the local languages and cultural practices of the masses. A policy of *korenizatsiia* (indigenization) was launched in 1923 to train and promote local ethnic cadres as the representatives of their group within the Communist Party and the local governance apparatus. Modernization demanded the formation of nations as an evolutionary stage toward socialism and communism. *Korenizatsiia* gave this modernization process a mighty shove forward. Finally, given the history of colonial brutality and exploitation in the Russian Empire, smaller nations, or oppressed nations, generally deserved greater recognition and support, while the great power chauvinism of oppressor nations was a perpetual danger. Interestingly, Lenin and Stalin had initial disagreements about this latter principle, with Lenin restricting great power chauvinism to Russian nationalism in the USSR context, whereas Stalin viewed Georgian national chauvinism toward Ossetians and Abkhaz as a regional case of a broader great power chauvinism.[21]

Bolshevik theories of nations and their power relations profoundly shaped the Soviet Union and what came after. The Bolsheviks' conception of nations as resting on objective criteria of homeland, language, culture, and biology legitimated an exhaustive attempt to demarcate and delineate distinct ethnoterritories for those recognized as nations.

Certain groups were designated as the "titular nation" in particular territories and the forms of their culture were institutionalized there: a named polity with national symbols, national language schools, a cultural theater, an ethnographic museum, and so on. This was a federally supported and cultivated local nationalism built on affirmative action, exhibition, and display. It was a remarkable exercise in social and spatial engineering, one that collided with the multiethnic complexity of places and communities in many locations across the vast country. Most consequentially, over time these sponsored national forms became the material content of nationalist visions. Ethnoterritories became battlegrounds between the claims of titular peoples and the rights of other residents.

Layered upon the inherited imperial modes of incorporation and the empire of ethnoterritories created by the Bolsheviks was an empire of conquest that was an outcome of the traumatic geopolitics of World War II. Facing a rising Nazi threat in Central Europe, Stalin made common cause with the Nazis and expanded the Soviet Union through coercion and state terror from September 1939 to 1941.[22] In this way, Stalin temporarily reincorporated former tsarist territories in the Baltics and Bessarabia. The Nazi-Soviet pact, of course, proved temporary, but the military conquest of these territories was made permanent as the Red Army pushed back the Wehrmacht in the summer of 1944. The Soviet Union expanded through force of arms and built for itself in the following year an empire that extended all the way to Central Europe and Germany. While its efforts at co-optation and consensus were considerable, this was an empire underpinned by the reality of totalitarian state machinery and military interventionism should it come to that, as it did in Poland, Hungary, and Czechoslovakia. Finally, as noted in the previous chapter, the Soviet Union was a unique empire of infrastructural connections and logistical flows. Its top-down developmentalism created elaborate networks of connections and flows that depended on centrally coordinated economic governance and relatively friction-free movement corridors.

The Soviet Union was undone, in the end, by its own structural contradictions amid a rising tide of nationalism in most union republics. The very ideologies of national identity the USSR institutionalized and legitimated became convenient alternative sources of legitimacy for challenger elites amid cultural openness, economic restructuring, and ambitious constitutional change undertaken by Gorbachev in the late 1980s. What followed was a "parade of sovereignties" within and across the union republics, with Soviet concepts of "ethnogenesis" (primordial origin myths) and "objective" attributes of nationhood justifying the seizure of power by nationalist challengers or, in acts of deft maneuvering, the retention of power by Communist figures and Kremlin-anointed titular leaders.[23] Claims about original ownership and historic golden ages overrode the rights and requirements of those living in the present. This was the victory of national form over socialist content, with nationalism filling the legitimacy void with content that was, in the end, itself a theater of form, display, and myth.[24] Sealing the fate of the Soviet Union was the defection of its two pivotal republics, Russia and Ukraine, from its defense. The de-Sovietization of both, however, was to prove an extremely complicated matter.[25]

The emergent nationalizing projects in the post-Soviet successor states condemned the Soviet Union as an empire and sought varying degrees of distance from it. De-Sovietization was most manifest in the removal of certain symbols, markers, and political institutions of Soviet power, and their replacement by new national symbols and institutions.[26] But such a process was pursued using Soviet-made notions of nationality upon Soviet-made territories by Soviet-made politicians. Obfuscated by the rhetoric of decolonization were the territorial structures, balances, and compromises that made the Soviet Union function, most especially the founding formula that kept Russian power in check and an all-powerful party apparatus at the core of the state.[27] Nowhere was the process of decolonization more complicated than within the former center itself, within a Russia now nominally released from the bonds and burdens of Sovietness.

Post-Soviet Space

The collapse of the Soviet Union was not an outcome many expected or ever desired. The crucial period after the collapse of the August 1991 coup in Moscow saw officials around Russian president Boris Yeltsin launch a takeover of the Soviet "power" ministries. As Yeltsin consolidated power, leaders in most republics moved to do the same. While

Yeltsin's rivalry with Gorbachev rendered Russia's defection from the Soviet Union thinkable, it was Ukraine's leadership that forced the issue by scheduling a December 1 referendum that then decisively backed independence. Confronted with the prospect of a Soviet Union minus Ukraine, Yeltsin's team abandoned their vision of a Russian-dominated confederation. In its stead Russia, Ukraine, and Belarus declared on December 8, 1991, a Commonwealth of Independent States (CIS) as the successor to the Soviet Union. Soviet republics that never envisioned independence were suddenly faced with the prospect of it. Further, under the legal doctrine of *uti possidetis* ("as you possess"), the borders dividing republics were now unexpectedly borders between sovereign international states. Born of crisis and opportunistic moves by self-dealing elites, what were once faux Soviet sovereignty forms now had de jure content. This status, however, risked becoming a new order of legal fiction: nominally independent states with borders established and altered by administrative fiats, atop a complex spatial matrix shaped by seventy-plus years of central planning, economic integration, and nested matryoshka doll–like structures of ethnoterritorial governance. On December 21, 1991, eight other Soviet republics joined the original three in the CIS. Five days later the dissolution of the Soviet Union was formalized by a declaration of its Supreme Soviet.

Post-Soviet space had three salient features at the outset. First, the sovereignty and independence of the successor republics was extremely variable. The three Baltic states refused to join the CIS and argued theirs was a restored sovereignty, stolen by illegal occupation since 1939 and now reconstituted. Georgia also refused to join the CIS and restored the constitution and flag of the Democratic Republic of Georgia overthrown by a Red Army invasion in February–March 1921. The gesture, mirroring Baltic nationalist practice, signaled that Georgia was "occupied territory" during the whole Soviet period (in 2006 a Soviet Occupation museum modeled after ones in Tallinn and Riga opened in Tbilisi). These "restorationist" gestures compounded the border legitimacy problems, for earlier pre-Soviet states had borders different from their Soviet Republic borders. New state elites were, at the same time, claiming *uti possidetis* and rejecting it. Meanwhile, Belarus and the Central Asian states maneuvered tentatively so as to not imperil their longstanding dependence upon Russia. The post-Soviet states were "hardly states in any meaningful sense, merely shells of potential states."²⁸ Whether these shells could be filled with substantive identities and sovereignty was an open question.

The second salient feature of post-Soviet space was the emergence of "the Russian question." In purely statistical terms, using numbers from the last Soviet population census of 1989, the collapse of the Soviet Union meant that there were over 25 million people beyond the borders of Russia who had identified themselves as ethnic Russians in that census. Three-quarters of them were in just two successor republics: Belarus and Ukraine. The numbers were larger if the statistical category referred to those who declared their first language as Russian. Here the numbers rose to 36 million. Add those who were "Russian-language-leaning," namely those who named Russian their native or second tongue (after their titular language), and the figure rose further to 42 million. What these numbers meant was, at once, self-evident and uncertain. What was self-evident was the considerable disjuncture between the borders of the newly independent Russia and a putative Russian ethnospace and, beyond that, an adjoining Russophone sphere and neighboring Russian cultural world. Post-Soviet Russia occupied territory that was considerably smaller than the Russian Empire at the beginning of World War I. Indeed, Russia was smaller than it had been since the seventeenth century. It included largely nonethnic Russian areas in the Volga-Urals and in the North Caucasus but excluded traditional Russian ethnoterritories in northern Kazakhstan and Crimea. Alexander Solzhenitsyn was one of many public intellectuals prominently raising questions about what this means: "What exactly is Russia? Today, now? And—more important—tomorrow? Who, today, considers himself part of the future Russia? And where do Russians themselves see the boundaries of their land?"²⁹

Third, the immanent arrival of republican-level sovereignty triggered the unraveling of some of the many compromise structures of Soviet ethnoterritorialism. This began, as noted already, first in Nagorno-Karabakh, a predominantly ethnic Armenian autonomous oblast established within Soviet Azerbaijan in 1923. In February 1988, local activists agitated for the oblast's unification with neighboring Soviet Armenia. The crisis quickly spiraled outward, polarizing elites on either side and drawing the Kremlin in as crisis manager. The Soviet power vertical, however, was in crisis. On September 2, 1991, in the wake of the collapsed coup in Moscow and Azerbaijan's subsequent declaration of independence, the oblast's local parliament announced its secession from Azerbaijan and an independent Nagorny Karabakh Republic (NKR). A bloody war followed that left an estimated thirty thousand dead and over a million people displaced, as violent ethnic homogenization remade Karabakh, its neighboring provinces, and communities across Armenia and Azerbaijan.[30]

The coming to power of nationalist counterelites in Soviet Georgia and Moldova triggered the unraveling of their Soviet-era territorial structures. The murderous suppression of a Georgian nationalist demonstration by Soviet troops on April 9, 1989, had radicalized Georgian political life, creating permissive conditions for the election of the former dissident Zviad Gamsakhurdia as president of Soviet Georgia. Championing a sacralizing conception of Georgian territory as the homeland of a timeless Georgian ethnos, Gamsakhurdia's leadership of the Georgian national movement proved disastrous as threatened elites in the South Ossetian Autonomous Oblast, the Abkhazian Autonomous Soviet Socialist Republic, and the Adjara Autonomous Soviet Socialist Republic mobilized against him and the prospect of an independent nationalist Georgia. As we will examine later in greater detail, the upshot was a war that began over South Ossetia in December 1990, continued intermittently in 1991, and only ended, months after Gamsakhurdia fled office, with the signing of the Dagomys Agreement (also called the Sochi Agreement) between Russian president Yeltsin and the new Georgian president Eduard Shevardnadze in July 1992. A longer and bloodier fight followed thereafter over Abkhazia, which ended with the Georgian army's defeat, Shevardnadze's last-minute escape from the debacle, and the forced displacement of over two hundred thousand ethnic Georgians and others from that region by September 1993.

The situation was somewhat different in Soviet Moldova. Here what facilitated territorial fragmentation was not existing ethnoterritorial structures but historical fault lines. Formed from existing territories on the left bank of the Dniester River (the Moldovan Autonomous Soviet Socialist Republic established inside Soviet Ukraine in 1924) and on the right bank, Bessarabia (settled by the Russian Empire but lost to Romania in 1920, only to be recaptured in 1940 and again in 1944),

Soviet Moldova was destabilized by Gorbachev's reformist policies and the reemergence of sensitive historical topics like the Nazi-Soviet Pact and the circumstances of Soviet Moldova's creation. A convergence of interests between young Moldovan-language activists, aspirant Moldovan party elites, and urbanized Moldovan-language workers created the basis for mobilization around language reforms in the republic.[31] In August 1989, Moldova's Supreme Soviet declared Moldovan the state language, mandated a transition to the Latin alphabet, and implicitly recognized the unity of the Moldovan and Romanian languages. The reforms provoked countermobilization by Moldova's Russophone population that was centered mostly in industrial cities on the left bank (with the important exception of the city of Bendery). Moldova's Gagauz population (a Turkic-speaking recognized ethnic group) also mobilized and declared a Republic of Gagauzia the next year. The August 1991 coup radicalized cleavages that had ethnopolitical, linguistic, regional, and class dimensions. The Moldovan parliament's declaration of independence on August 27, 1991, provoked pro-Soviet elites on the left bank to proclaim a Pridnestrovian Moldavian Soviet Socialist Republic in response. Fighting in the spring of 1992 left the successor state divided into an internationally recognized Republic of Moldova and an unrecognized Pridnestrovian Moldavian Republic (PMR), known more commonly as "Transnistria" in English.

In all these cases, the territorial order established under Soviet rule within these republics failed to transition smoothly to a symmetrical post-Soviet territorial order. Indeed, separatism and territorial fragmentation began as soon as the political environment became more open, which was well before the Soviet collapse. The reason was simple. Soviet polities and borders had tenuous local legitimacy. The *uti possidetis* doctrine was mostly what successor state elites wanted to apply (though many nursed broader claims based on pre-Soviet polities), but in Georgia, Moldova, and Azerbaijan they never achieved internal territorial sovereignty: "as you possess" was only in legal theory and never on the ground. The dispute over Nagorny Karabakh launched Armenia and Azerbaijan into a war with each other that neither state has transcended. Karabakh remains an unresolved and dangerous conflict, one that saw its worst violence since the 1994 ceasefire in April 2016. Georgia ended up with two provinces it termed "breakaway territories" and had the added complication of a third, the former Adjara Autonomous Soviet Socialist Republic, run by a local strongman as a personal fiefdom. Moldova abjured unification with Romania and avoided widespread forced displacement in its brief war with the PMR. But, to this day, it only exercises sovereignty over part of its internationally recognized territory. The collapse of the Soviet Union gave birth to four de facto states (Abkhazia, South Ossetia, Nagorno-Karabakh and Transnistria), wannabe states that have managed to achieve practical internal sovereignty over a claimed territory but lack widespread international recognition. The term "frozen conflicts" is often applied, but it is only appropriate to the extent that these conflicts established unresolved territorial divisions that have endured. The dynamics of these conflicts are far from frozen.

Other Soviet Republics had federally established ethnopolities and contested territories nested within them also. While some like Chechnya sought to secede, none were able to successfully do so. Other territories were sites of countermobilizations against local independence, but these movements failed. These might be described as the territories that barked but did not take flight from the successor state. The most significant of these is obviously Crimea, which was officially nothing more than an oblast (province) within Soviet Ukraine since 1954. It was already a subject of considerable mobilization there and within Russia well before Ukraine voted for independence in December 1991. Other places within Ukraine saw pro-Russian mobilizations too. Why these mobilizations did not succeed at that time but were renewed over two decades later is an important question, one I will examine in chapters 5 and 6.

Russian Geopolitical Culture

The end of the USSR was made possible by the defection of Russia and Ukraine. But Yeltsin and his team were completely unprepared for the collapse and its ramifications. A personalized power struggle brought down a once-mighty superpower and left a naïve and deeply flawed populist at the fulcrum of managing the consequences. The act immediately triggered a series of practical and, then, existential questions about what a post-Soviet Russia would look like. The practical questions had to do with Soviet institutions of governance, debt, and property. These were resolved in fairly short order as most Soviet institutions became Russian ones. Debt was largely absorbed by Russia, while negotiations finally resolved the status of Soviet property and assets abroad. Soviet bases became Russian military bases, and the question of future withdrawals was a matter to be negotiated bilaterally.

The existential questions were harder to answer. Since the establishment of the Soviet Union, its Russian identity was sublimated and deemphasized. Russians were not supposed to be the state-bearing nation for the USSR; it was a multinational anti-imperial federation. Nevertheless, Russia was the center and Russianness—in differently accented formations as *russkii* and *rossiyanye*—the core identity.[32] The new Russian state now had to determine what made someone Russian. How should the Russian nation be imagined: in state-territorial terms (those born within the borders of the Russian Federation), in ethno-ascriptive terms (those who considered themselves Russian), in linguistic terms (Russian-language speakers), or in historical and aspirational terms (those residing on "historic Russian lands")?

Questions of identity and territory are always deeply entangled. The territorial implications of the collapse of the USSR for Russians were profound. Like Germany, Austria, and Hungary after World War I or Serbia as Yugoslavia collapsed, Russia became a state on a territorial palimpsest considerably smaller than the canvas of its history. Certain well-known places of memory were now foreign, whereas others appeared foreign but were domestic. "It is a strange Russia that includes Chechnya but excludes Crimea," remarked the journalist Vitaly Tretyakov.[33] For intellectuals like Solzyhenitsyn, the rebuilding of Russia inevitably raised the question of the state's borders. Is Russia merely what the Soviet Union has left it? Should it not also include concentrations of ethnic Russians in Belarus, Northern Kazakhstan, Crimea, and southeast Ukraine or, more broadly, former tsarist lands like most of Ukraine, Moldova, and the Baltic states? Is Russia an empire or a nation-state, a great civilization destined to dominate the heart of Eurasia, an imperial motherland that has responsibilities to all Russian-speaking peoples, or a state that needs to adjust to a postimperial age?

Russian geopolitical culture formed around these spatial and existential questions. Inevitably this debate encroached upon the concerns of neighboring states, for how Russia defined its spatial identity and diaspora had significant implications for their territorial integrity and security. Were Russians beyond Russia marooned co-ethnics, separated from their homeland nation by the accident of administrative borders becoming international borders, a nation fragmented and victimized? Were they stranded agents of a modernizing and civilizing mission, now about to be unfairly maligned as imperial agents and occupiers by nationalizing elites within successor states? Or should they be viewed as national minorities just like any other, ordinary residents of newly emergent states that ought to locally integrate? What about those who strongly identified with Russian culture but now found themselves beyond the borders of the state that was still their cultural hearth? The "Russian question" touched all states in post-Soviet space and left all with security dilemmas that were inevitably intertwined.

The first decade of the Russian Federation saw the development of an intense debate over the direction of the country among intellectuals with ties to various networks of power. President Yeltsin even designated a team of scholars to work together to find a new, acceptable definition of the *Russkaya ideya* ("Russian idea"). One notable feature of the debate was the prominence of classic geopolitical concepts and themes, a somewhat surprising development in that geopolitics was a taboo subject associated with Nazi expansionism during the Soviet period. But the circumstances under which Russian elites groped for redefinitions of the state were extreme, a "wild east" of collapsed state institutions, crumbled power verticals, upended norms, rampant inflation and thuggish capitalism. A "Weimar Russia" proved fertile ground for a Weimar-like geopolitics as an answer to the spatial fragmentation and territorial confusion of the moment. There are now many studies of the various debates in Russian geopolitical culture from this time to the present.[34] In abridged form, three distinctive visions of Russia characterize post-Soviet Russian geopolitical culture: a liberal European Russia, a revived imperial Russia, and an independent great power Russia (Table 2.1). These would provide the triangulating template for Russian geopolitical culture as it entered the twenty-first century.

TABLE 2.1 Competing Visions of Russia

Geopolitical Vision	Westernizing Russia	Imperial Russia	Strong Russia
Vision of state	Russia as a modern multiethnic federation with stable borders.	Russia as the center of a great imperial civilization in Eurasia.	Russia as the Great Power (*Derzhava*) in Eurasia.
Cultural identity	Pluralist.	Russians first among equals. Orthodoxy and Russkii central.	Nominally pluralist. Islamic fundamentalism a nascent threat.
Master narrative of the national interest	Integration of Russia into the global economy and liberal hegemonic order.	Former imperial lands settled by ethnic Russians and compatriots should "return" to Russia.	Reestablishing state power vertical: territorial integrity and state strengthening.
View of Soviet collapse	Generally positive. Opportunity to create a new modern democratic Russia.	Conspiratorial. Traumatic. Loss of empire, greatness.	Mixed. Recognition of Soviet weaknesses but also of regrettable loss of great power status.
Attitude toward near Russian diaspora	Encourage them to become citizens of country of residence.	Fragmented nation. Russian people now artificially divided.	Encourage dual citizenship and special relationships within common geopolitical space.
Attitude toward post-Soviet states	Acceptance but verbal protest at their nationalizing policies.	Hostility toward their independent policies and orientations.	Acceptance but effort to create a common geopolitical space (Russian sphere of influence).
Attitude toward post-Soviet borders	Acceptance of new legal borders.	Revisionist. These borders are artificial and unjust.	Pragmatic acceptance. Need to accept world as it is.
Attitude toward post-Soviet breakaway territories	Official rejection of separatism.	Support for these "stranded" imperial territories.	Contingent support.
Attitude toward Ukraine	Ukraine is an independent sovereign state.	Ukraine a patched-together artificial country. Mostly a little brother Slavic nation within historic Russian space.	Ukraine a fellow Slavic state that is of vital national security interest to Russia.

Attitude toward Western states	Friendly. Openness to cooperation on multiple fronts.	Hostile toward the United States but less so toward European powers.	Pragmatic but suspicious. Cooperation on some issues of common concern possible.
Threat discourse	Neo-Communism. Anti-liberal nationalism. Extremism. Modernization failures.	Corrupt Westernization. Rival empires. Conspiracies against Russia's restoration of imperial strength and glory.	Separatism, Islamic terrorism, unpatriotic oligarchs, great powers rivals, sponsored subversion.
Internal differences	Neoliberal internationalists. Social Democrats. European developmentalists.	White: Orthodox Tsarist Red: Neo-Soviet Brown: Neo-Fascist Multicolored: Great Russian	Defensive realist narrow nationalism. Offensive realist power projection. Modernizing Eurasianist.
Intellectual proponents	Gennadi Burbulis Valery Tishkov Alexei Arbatov Dmitri Trenin	Alexander Panarin Eduard Limonov Alexander Dugin Alexander Solzhenitsyn	Andranik Migranyan Yevgeny Ambartsumov
Political proponents	Andrei Kozyrev (partial) Boris Nemtsov Vladimir Ryzhkov Gary Kasparov	Yuri Luzhkov Konstantin Zatulin Aleksei Mitrofanov Vladimir Zhirinovsky Gennadi Zyuganov Dmitri Rogozin Konstantin Malofeev	Yevgeny Primakov Vladimir Putin

Liberal European Russia

The initially ascendant geopolitical vision in Russia's emergent geopolitical culture was that of a liberal European Russia. This was Russia as a responsible successor state to the Soviet Union, an international law–abiding status quo state that recognizes the territorial integrity and inviolable frontiers of its newly independent neighbors. This was also a Russia as a responsible, stability-oriented European power that adheres to the principles of the Charter of Paris and its aspirations to create a Europe free of imperial spheres of influence. Many influential figures around Yeltsin held liberal European values and sought to guide Russia's foreign policy conduct in alignment with them. Though relations with successor states would have complications from the outset, the Russian state consistently signed treaties and declarations that reaffirmed its adherence to the principles of territorial integrity and the inviolability of frontiers.[35] The preamble to the Alma Ata Declaration by eleven Soviet Republic leaders enlarging the Commonwealth of Independent States on December 21, 1991, for example, rearticulates Helsinki Final Act principles including "recognizing and respecting each other's territorial integrity and the inviolability of the existing borders."[36] In unconditionally affirming successor borders and condemning separatism, Russia was asserting that the geopolitical rules of post-Soviet space would be mainstream liberal ones based on agreed principles of international law. The Yeltsin administration position was very much part of the mainstream consensus of states on this issue.

Co-existent with the Yeltsin's administration's liberal European vision of Russia was its definition of citizenship within the new Russian Federation in nonethnic terms. Yeltsin adviser Gennadi Burbulis championed an inclusivist multiethnic, multicultural, and multiconfessional vision, one very much in keeping with official Soviet ideology, in the crucial first two years, as did other then-influential figures like the academic Valery Tishkov. Yet from the outset there were ambivalences in the Yeltsin administration's attitude toward those looking to Russia in its neighboring states. The term of choice the administration developed to describe them was "compatriots abroad." In April 1992, just four months after the Belavezha Accords, Yeltsin spoke about how "[t]wenty-five million of our compatriots in these countries must not and will not be forgotten by Russia." As nationalizing projects in the newly independent states got underway, driven by sometimes chauvinistic nationalist visions, concern about discrimination against ethnic Russians and nontitular peoples grew within Russia. The initial policy response of the Yeltsin administration to the twenty-five million was to champion the notion of dual citizenship. Promoted by Foreign Minister Andrei Kozyrev and a Presidential Commission on Citizenship, the policy drive, however, yielded little: most neighboring countries declined to sign treaties allowing dual citizenship. In 1995 Yeltsin launched a government program for building relations with compatriots living abroad. The program conceptualized compatriots abroad as made up of three categories of people: Russian citizens living in the near abroad, former Soviet citizens who have not obtained, or been allowed to obtain, new citizenship (as occurred when Estonia and Latvia classified Soviet-era settlers there as illegal migrants), and those in the newly independent states who wished to maintain cultural ties to Russia. With Russia gradually liberalizing its citizenship criteria as nationalizing states restricted theirs, the upshot was a modest growth of de facto dual citizenship without the consent of the newly independent states.[37] This issue would later become deeply politicized and securitized during Putin's rule.

Imperial Russia

The preeminence of a liberal European vision of Russia was always contested, but it took the results of the December 1993 Duma elections to publicly dislodge it in political discourse. The big surprise in those elections was that Vladimir Zhirinovsky's Liberal Democratic Party of Russia (LDPR) won 22.92 percent of votes in the federal electoral district, becoming the largest party in Russia's first post-Soviet multiparty parliamentary elections. Frequently derided as a clown, Zhirinovsky became a fixture of the Russian political scene thereafter, albeit one that never subsequently gained so high a vote share. Many factors accounted for his relative success in this election, but undoubtedly one was his vivid articulation of Russian imperial fantasy amid crisis and collapse. In his 1993 autobiography *Last Thrust to the South*, Zhirinovsky locates the main source of Russia's troubles with its southern neighbors, arguing that Russia needs to expand outward to annex not only the territories of the former Soviet Union but also regional powers like Turkey, Iran, and Afghanistan.[38] His vision gives Russia's search for warm water ports an image of continent-conquering swagger, that of Russian troops washing their boots in the waters of the Indian Ocean.

Zhirinovsky was only one of a series of geopolitical entrepreneurs at the time. Indeed, territorially revisionist geopolitical fantasy became a distinctive genre within Russian intellectual life, with intellectuals, artists, writers, and politicians peddling messianic spatial fantasies to each other. Two others that would go on to forms of influence and infamy like Zhirinovsky were Alexander Dugin and Dmitry Rogozin. The son of a high official in the Soviet military—some sources allege his father was a general in the *Glavnoye razvedyvatel'noye upravleniye* (GRU), the Main Intelligence Directorate of the General Staff of the Soviet and subsequently Russian Armed Forces—Dugin was perhaps the most entrepreneurial of all in promoting his views without ever securing a stable, prominent place in official structures of power. An autodidact intellectual with an interest in the occult and far-right European thought, Dugin was briefly involved with the fascist network *Pamyat* (Memory) before founding his own publishing house and think tank in the early nineties.[39] During this time, he became an adviser to Gennady Zyuganov, leader of the Communist Party of the Russian Federation, as well as a writer with the "national-patriotic" newspaper *Den* (Day), later renamed *Zavtra* (Tomorrow), edited by Alexander Prokhanov.[40] *Zavtra* was at the fulcrum of an effort by Soviet imperialists, with Orthodox coloration, like Prokhanov and fascist mystics like Dugin, to create a "red-brown coalition" that would promote a "new order" of space, identity, and power in Russia. Dugin found an effective geopolitical imaginary for this project in the writings of early-twentieth-century Russian émigré scholars: Eurasianism. A mega-geographic signifier that could be bent to the purposes at hand, Eurasianism was, above all else, a vehicle for the construction of imperial Russian nationalist visions of post-Soviet space. Defined in opposition to Westernizing visions of Russia, Eurasianism was founded around the conceit that Russia was geographically exceptional by virtue of its situatedness between Europe and Asia. Russia's destiny was to be an imperial power bridging continents. The irony was that the conceptual raw material for Eurasianism was borrowed from European sources, especially fascist literature; its interest in Eastern or Asian regions was purely theoretical. Two were lifted from early-twentieth-century geopolitical thought, itself based on recycled classical-age ideas. The first was Mackinder's use of the classical world's conception of an eternal, elemental conflict between continental landpower and maritime seapower. Dugin cast this as a spiritual clash between tellurocracy (earth) and thalassocracy (water). The second was the German *geopolitik* vision of a world divided into distinct *grossraum*, large autarkic pan-regions with a great imperial power at their center.

Dugin proved to be a prolific writer and activist. In 1993 he helped found the National Bolshevik Party with Eduard Limonov, a charismatic polemicist and iconoclastic writer. The party was only one of many relatively minuscule fascist groupings at the time. Others included the Black Hundred and Russian National Unity, a breakaway faction from *Pamyat* led by Aleksander Barkashov. Most championed forms of supremacist ethnonationalism married to an expansive imperialist state, though Russian National Unity's form was so imitative of German Nazism that it was "largely outside the bounds of the nationalist traditions of Russia."[41] These movements were strongly opposed to the Kremlin, with many organizing the defense of the Russian parliament from Yeltsin's assault against it in September–October 1993. Limonov was there as was Aleksander Borodai, a writer and editor of *Zavtra* who would later turn up as a Russian leader of the separatist revolt in eastern Ukraine in April 2014. Both he and his friend, Igor Girkin, had fought in Moldova for Transnistria's pro-Moscow rebels. So also was Barkashov. Despite developing considerable ties to countercultural youth networks, Limonov and Dugin's National Bolshevik Party never had much electoral success. Dugin left the party in 1998 after mounting disagreements with the mercurial Limonov.

Thereafter Dugin transitioned from the role of political prophet to that of geopolitical expert. The patronage and support of Russian military structures facilitated the career change. In 1997, with the aid of General Nikolai Klokotov and support of the commander of the General Staff Academy and later Russian minister of defense Igor Rodionov (1996–1997), he published *Osnovy Geopolitiki: Geopoliticheskoe Budushchee*

Rossii (*The Foundations of Geopolitics: Russia's Geopolitical Future*). The book was designed as a new handbook of geopolitical analysis for students at the Academy of the General Staff of the Russian Armed Forces, but it also articulated an agenda for revisionist geopolitics across Eurasia. The work was a considerable success, going through numerous revised and expanded editions in subsequent years. The book served as cultural capital for Dugin to transition into formal structures of power. In 1998 he became an adviser on geopolitics to Gennady Seleznyov, then speaker of the Russian Duma. The following year he founded the Center for Geopolitical Expertise in Moscow. Dugin was now a Kissinger-like guru with a product: Eurasianism as the new "national idea."

A second geopolitical entrepreneur who got his start in the nineties is Dmitry Rogozin. A journalism graduate and senior Komsomol activist, Rogozin was initially active in efforts to preserve the Soviet Union before its collapse. Thereafter he helped found a group called "Union for the Rebirth of Russia" before establishing in 1993 the organization that would be a vehicle into the country's political power structures, the Congress of Russian Communities (*Kongress russkikh obshchin*, or KRO). The KRO was an organization focused on the rights of ethnic Russian communities in the successor states of the Soviet Union. Its raison d'être was the conviction that the Belavezha Accords had left Russians, in Rogozin's words, "foreigners on native land" and that they were now a "nation split into fragments." The KRO was thus no coordinating lobbying group; rather, it was a Russian nationalist party that sought to broaden the borders of Russia by mobilizing Russians within and outside the state so that all Russians could live in one state. In a manner akin to Serb "renewal" nationalism at the time (with whom Rogozin developed ties), the manifesto of the party explained that the "creation of an ideology of unification of Russian people [*russkie*] is the only way to secure the survival of the Russian nation [*russkaya natsiya*], Russian culture [*russkaya kul'tura*] and the Russian state [*Rossiiskoe gosudarstvo*]." As the English political geographer Alan Ingram explains, the KRO engaged in some conceptual entrepreneurship with the notion of *russkaya natsiya* as the party sought a moderate ethnonationalist appeal. It was defined as being virtually equivalent to a Russian nationalist interpretation of *russkii narod* as three ethnoses, Great Russians, Little Russians (Ukrainians), and White Russians (Belarusians), a deeply problematic conceit that denied the latter identities separate and equal status. In addition, it also encompassed a series of other communities, Cossacks, people from the Urals, people from Siberia, and various other smaller indigenous communities who consider themselves *russkie*. The KRO's vision of "Russian communities," in other words, was a mélange of reworked ethnonationalist and Soviet conceptions that rested on a paternal imperial conception of Russianness. This was marketed as tolerant of other nationalities—the KRO sought to distinguish itself as moderate and respectable relative to the "red-brown coalition"—only so long as the paternalistic privilege of *russkii* nationalism was not challenged. In effect, as Ingram notes, this meant "the negation of all other nationalisms within the geopolitical space claimed for Russia."[42] In a 1994 press conference Rogozin stated he was sure "that Russia will be much broader than its current borders—with Belarus, naturally, Crimea, Ukraine and Kazakhstan." A surprising absence was Moldova, for Rogozin was outspoken in his defense of the PMR (Transnistria). When the Russian interventionist "hero" of Transnistria, Commander of the 14th Army General Aleksandr Lebed, joined the KRO in 1995, the party got a significant boost but still failed to achieve the 5 percent of the national vote necessary to get federal list representation in the Duma.[43]

Dugin and Rogozin were only two of many entrepreneurs in Russian geopolitical culture in the nineties. Another figure associated with the KRO was the Ukrainian-born economist Sergey Glazyev, later a prominent figure behind the separatist project of Novorossiya (which encompassed his home city of Zaporizhia). Prominent at the time also was the anti-Communist dissident Alexander Solzhenitsyn. He had returned to Russia in 1994 and proceeded to agitate for a civilizational conception of Russia that was aspirationally anti-imperialist but ineluctably caught in imperial conceits of Orthodoxy and Russianness.[44] Moscow mayor Yuri Luzhkov (in office from June 1992 to September 2010) also played to revisionist sentiments during his long tenure, agitating over Crimea and building aid networks for Abkhazia and South Ossetia in later years. So too did Saint Petersburg's mayor Anatoly Sobchak (in office from June 1991 to June 1996) in whose administration Putin first learned the game of politics. After passage of the December 1, 1991, referendum on independence in Ukraine, for example, Sobchak declared that Russia had historically handed over to Ukraine, "a whole series of Russian provinces, the so-called Novorossiya, whose population is for the most part Russian" and that the Russian minority in Ukraine was threatened with "forcible Ukrainianisation."[45] The depth of Sobchak's concern for Russians beyond Russia is questionable. It is most likely he, as a leading politician at the time, saw the issue as one that could be used for political gain.

But it was the leader of the Russian Communist Party (KPRF), Gennady Zyuganov, who became the standard-bearer for the "national-patriotic" network of opposition to the Yeltsin administration. He championed the restoration of the Soviet Union, mixing Eurasianism and Communist orthodoxy as justification. After Lebed's first-round elimination and his second-round defeat to Yeltsin in the 1996 presidential elections, Zyuganov continued in Russian politics, writing a series of texts promoting neo-Soviet visions of Russia's role in post-Soviet space.[46] Though those promoting imperial visions of Russia were a heterogeneous group, they all shared a common position: they were territorial revisionists. The borders had to change.

Great Power Russia

While they may have failed to take the Kremlin, those pushing revisionist geopolitics profoundly shaped Russian geopolitical culture in the nineties. As early as April 1992, Yeltsin presidential adviser Sergei Stankevich identified two poles developing in Russian geopolitical culture: Atlanticism and Eurasianism.[47] He sought to outline a third, one that would "look for a new balance of Western and Eastern orientations." A policy based on the pursuit of pragmatic interests alone, he argued, is inadequate. "[W]e need a mission . . . one that does not degenerate into messianism." Over its two terms the Yeltsin administration had considerable difficulty formulating that sense of mission in positive terms, for itself, for elites in the new neighboring states, and for Western powers. Nevertheless, it emerged from the very practical needs of the time—from both crisis events, demanding responses in post-Soviet space and in the Balkans, and longstanding habits within a bureaucratic apparatus previously Soviet and now Russian. That sense of mission was the (re)building of Russia as a great power.

Boris Yeltsin was the first democratically elected leader in Russian history. His power struggle with Soviet president Gorbachev saw him deliberately undermine the power vertical holding Russia and the Soviet Union together. While his declaration during a visit to Tatarstan for autonomous republican leaders to "take as much sovereignty as you can swallow" worked in 1990, it created significant difficulties for him in subsequent years as he sought to reestablish Kremlin authority over rebellious and independent-minded regions within Russia. The local leadership of Tatarstan and Chechnya, who both proclaimed full independence from Moscow, proved particularly difficult to manage, and their freelancing exacerbated a sense of territorial crisis. Fatefully, Yeltsin eventually turned toward military force to subdue the rebellious leader of Chechnya, Dzhokhar Dudayev. The move backfired and Chechnya became a very public site of failure and humiliation for the Russian state. The restoration of Russian control over Chechnya would become deeply symbolic of the reassertion of Russian power more generally under Vladimir Putin.

The nationalizing policies of the newly independent states also created significant difficulties for Yeltsin, most especially Estonia and Latvia's move toward ethnic-based citizenship and disenfranchisement of their Russian populations.[48] There was outrage across the political spectrum in Russia in August 1992 at the exclusion of 600,000 "noncitizens" (overwhelmingly Russians) from Estonian presidential and parliamentary elections. Stankevich termed Estonia's laws an "apartheid system," while another Yeltsin adviser threatened later that year to use Russian enclave populations as "geopolitical levers" against nationalizing states.[49] With instability and wars ranging in Azerbaijan, Armenia, Moldova, Georgia, and Tajikistan, Russia's newly emergent near abroad was a major security and humanitarian concern. The plight of Russian-speaking populations in Transnistria (where fighting erupted from March to July 1992) and Abkhazia (where war raged from August 1992 to September 1993) received particular media attention. Critics rounded on Russia's support in the UN Security Council for sanctions against Yugoslavia (May 30, 1992) yet insufficient support for Transnistria, as well as Yeltsin's overtures to Japan on possible resolution of their dispute over the Kurile Islands. Strong statist factions (*derzhavniks*, from the Russian word *derzhava* or great power) within the Supreme Soviet, the Kremlin, the Russian military, and groups representing military-industrial interests (the Civic Union forum and Council for Foreign and Defense Policy) all pushed for more forceful articulation of Russia's national security interests within these conditions of crisis.

The result was dubbed by some "Russia's Monroe Doctrine" but was less one public declaration than a series of policy documents, findings, and public declarations over an eighteen-month period from March 1992 to December 1993 (see Table 2.2). Fyodor Shelov-Kovedayev led the Foreign Ministry's effort to develop a new "foreign policy concept" for Russia's relations with the near abroad. Evgenii Ambartsumov, chair of the Committee on International Affairs in the Supreme Soviet, commissioned a report recommending that Russia take on the role of "Eurasia's gendarme."[50] One of its authors was the academic Andranik Migranyan. He argued Russia should emulate the United States' Monroe Doctrine and "declare to the world community that the entire geopolitical space of the former USSR is a sphere of its vital interests." Only Russia should be the factor determining the fate of the geopolitical space of the former USSR.[51] A published draft of the Russian military concept described the violation of rights of Russians "abroad" as a serious casus belli.[52] Sergei Karaganov, a leading figure in the Council for Foreign and Defense Policy and subsequent Yeltsin adviser, argued that Russia was forced to play an active imperial role in its backyard, a role he described as winning over local princes, dispatching forces, and rescuing people. "This is a thankless task but one that history has set before us and one that we have partly brought on ourselves."[53]

Kozyrev's team circulated its draft foreign policy concept in December 1992. A revision formed the basis of a National Security Directive in April outlining the key tenets of the Russian "foreign policy concept." The document listed nine "vitally important interests" of the Russian state. Only three of them dealt with the world beyond the former Soviet border, indicating how much the Russian state was consumed at the time with issues close to home. The first and overriding priority was "securing state sovereignty and territorial integrity." Three dealt with the near abroad, stressing the need to deepen ties, protect the rights of the members of the Russian Federation's ethnic groups in the near abroad, and protect the "rights and interests of citizens and organizations of the Russian Federation abroad."[54] This

TABLE 2.2 Evolution of Russia's "Monroe Doctrine," March 1992–December 1993

Date	Actor	Source Document	Rhetoric
March–Sept. 1992	Foreign Ministry (Shelov-Kovedyaev)	*Russia in the Near Abroad* and *Strategy and Tactics of Russian Foreign Policy in the New Abroad*	Russia should seek international recognition as a "leader [in terms] of stability and military security on the entire territory of the former USSR," and be acknowledged "as having quite special interests in the region."
May 1992	Russian Military	Draft Russian Military Doctrine	Violation of rights of Russians "abroad" a serious casus belli.
August 4, 1992	Prof. Andranik Migranyan, adviser Duma Joint Committee on Foreign Affairs	Recommendations on Russian Foreign Policy Concept (interview *Rossiiskaya Gazeta*)	"Russia should declare entire geopolitical space of the former USSR a sphere of its vital interests (like US's Monroe Doctrine)."
August 19, 1992	Sergei Karaganov	Council for Foreign & Defense Policy	Russia should return to its traditional (imperial burden) role.
December 1992	Russian Foreign Ministry	Foreign Policy Concept draft	Use force where necessary "to achieve firm good neighborliness."
February 28, 1993	President Boris Yeltsin	Speech to Civic Union Congress	"Stopping all armed conflicts on the territory of former USSR is Russia's vital interest. The world community sees more and more clearly Russia's special responsibility in this difficult undertaking."

(*continued*)

TABLE 2.2 Continued

Date	Actor	Source Document	Rhetoric
April 28, 1993	President Boris Yeltsin	National Security Directive: Key Tenets of Concept of Foreign Policy	"Protection of rights of the members of the Russian Federation's ethnic groups in the Near Abroad" (No. 7)
September 28, 1993	Foreign Minister Kozyrev	Speech to UN General Assembly	Russia had a "special role and influence over the former Soviet republics, including the Muslim countries in the south."
October 8, 1993	Foreign Minister Kozyrev	Interview with *Izvestiya*	If Russia did not intervene in conflicts in the "Near Abroad," it would be in danger of "losing geographical positions that took centuries to conquer."
October 10, 1993	Foreign Minister Kozyrev	Opinion editorial *Washington Post*	"Protection of legitimate rights of the millions of Russian-speaking minorities in the former Soviet Republics, the economic reintegration of the republics and peace-making activities in conflict areas: All of these are an *objective necessity*."
November 2, 1993	Russian Military	New Military Doctrine	Suppression of Russians in Near Abroad is a military threat to Russia.

agenda of reintegration and protection was not ostensibly a revisionist one, but it was the foundation for asserting a traditional great power "sphere of influence" over the near abroad. In a February 1993 speech before the Civic Union—a group mobilized against radical economic reforms in which Vice President Alexander Rutskoi was prominent—Yeltsin called for the UNSC to make Russia the "guarantor of peace and stability in regions of the former U.S.S.R." The reaction was positive domestically but negative on the part of the United States and governments in successor Soviet states. Leslie Gelb dubbed Yeltsin as Monroe in the *New York Times*.[55] U.S. officials were concerned about a rising imperial nostalgia within Russia. During a visit in April 1993, Representative Tom Lantos asked Rutskoi why his wall still displayed a large map of the Soviet Union. Rutskoi reportedly replied that the country was in a transitional state and the map may eventually be accurate again.[56] Initially U.S. officials hoped to broker diplomatic efforts at conflict resolution in those successor states wracked by violence, pointedly refusing to give the Russian military a free hand in the region.[57] While the United States eventually conceded to Russian peacekeeping missions in Georgia and Moldova, it adhered firmly to the legal principle of *uti possidetis*. Russia, for its part, consolidated a doctrine that viewed the near abroad as the zone where Russia's crucial vital interests were concentrated, where threats were gathering, and where Russia needed to preserve a military presence and capacity to act.[58] The new Russian military concept adopted in December 1993 affirmed this understanding. Thus, from the very outset, there were considerable tensions between Russia's securitized vision of the near abroad and the U.S. and Western view of its ostensibly separate and independent legal sovereignty.

Mugged by the realities of geopolitical instability, economic crisis, great power competition, grandstanding by national-patriots, and resistance by powerful institutional bureaucracies, outspoken Westernizers like Kozyrev and Yeltsin shifted toward the *derzhavnik* camp. Initiative on making policy, especially in the conflict zones of the near abroad, tended to lie with the Russian armed forces rather than the Ministry of Foreign Affairs or the Kremlin.[59] U.S. plans to expand NATO to the former Warsaw Pact countries of Hungary, Poland, and the Czech Republic, violating tacit understandings from the Cold War settlement as Moscow saw it, further consolidated sentiment around *derzhavnik* factions though Yeltsin was initially mollified.[60] By the mid-1990s there was an emergent consensus in the Kremlin that Russian foreign policy should be organized around fostering multipolarity in the international system. Russia did not have a natural partnership with Western powers. It needed to be sober about the realities of geopolitics and the competitive policies of rival great powers. Kozyrev's departure in 1995 and replacement by Yevgeni Primakov marked only a manifestly public triumph of the *derzhavniks*: within the state security bureaucracies they had already won. Primakov pursued a pragmatic realist policy as more and more elite sentiment flirted with revisionist positions. By 1999, after the Kosovo War, Russia was disenchanted with its Western partners. The dash by Russian troops from Bosnia to Pristina airport symbolically expressed a desire to break free of Euro-Atlantic strictures and undertake unilateral actions.[61] The humiliating outcome revealed that Russia still had some way to go to rebuild its capacity to effectively use its military forces in the international arena.

The three visions described here are ideal types; in practice discourse and practitioners drifted across lines that blurred. Formed in an era of crisis and collapse, they represent very different responses to the shock of the collapse of the USSR. For a Westernizing elite at the outset, the crisis was a moment when Russia could reinvent itself anew as part of a liberal Western "civilized world." But this discourse soon had little traction and support within the former Soviet ministries, now charged with running the splintering territorial assemblage that was the Russian Federation. Economic crisis and dramatically collapsing standards of living induced by muddled privatization schemes wreaked havoc on everyday life. This created a moment of opportunity for "national patriots" for whom the collapse of the USSR was felt as a moment of painful national dismemberment and loss. This fostered a not always conscious desire for a forceful reassertion of Russian power and glory before the world. Russia required remasculinization.

For those managing the crises, the imperatives for the security and power of the Russian state were state strengthening at home and power projection in the near abroad. Ostensibly now the separate domains of domestic and foreign policy, the crisis of Russia's inner abroad and its near abroad were structurally linked because their source was the same catastrophe: the disintegration of the USSR. It followed that any effort to reverse this disintegration, to reassemble the power vertical and restore power to the state, would extend into the near abroad as well. State strengthening did not stop at the borders of the Russian Federation for the pathways of the state, its military footprint and logistical lines, extended well beyond the legal Russia recognized as a separate state in 1992.

Did the elite debates about the "Russian national idea" really matter to ordinary Russians? Public opinion research from the time indicates that there was widespread disengagement. Russians were preoccupied with making ends meet and had a limited appetite for elite geopolitical fantasies.[62] But they did care about their everyday security and prosperity. By the end of the nineties, after years of difficult adjustments, these were under threat from terrorist outrages that the authorities seemed powerless to stop.

Putin's Revanchist Agenda

Vladimir Putin became the leader of Russia at a moment of great fear and territorial crisis. Nominated and confirmed as prime minister in early August 1999, he assumed office as militant networks centered in Chechnya sparked an uprising in the highlands of Dagestan. Russian troops responded by entering Chechnya, thus igniting a second war over the breakaway region. The following month Russia was rocked by a series of explosions that hit four apartment blocks in the cities of Buynaksk, Moscow, and Volgodonsk. In this atmosphere of crisis and terror, Vladimir Putin projected himself as a new kind of leader, a sober and tough national security professional who could speak plainly and order harsh measures. Putin's vow to "wipe them out in the outhouse" on September 23, 1999, made him instantly popular across Russia.[63] At that moment, Putin became Russia's territorial tough, the man who would reverse its status as a territorial victim.

Putin came to power with what he saw as a philosophy, not an ideology. The Millennium Message, a vision document released by Putin's team two days before Yeltsin resigned and that made him the president, outlines an agenda of what was to be done to revive Russia.[64] Few states, it noted, have faced so many trials as Russia in the twentieth century.

Gross domestic product (GDP) dropped by almost half in the nineties and was now ten times smaller than GDP in the United States and five times smaller than GDP in China. Russia lagged the world in productivity, technical standards, capital investments, defense capacity, health standards, and average lifespan. Russia's future depended on the lessons it learned from the Soviet past and the costly experiments ("taken from foreign text-books") of the nineties. As regards the former, Putin's attitude is clear: "It would be a mistake not to see and, even more so, to deny the unquestionable achievements of those times. But it would be an even bigger mistake not to realize the outrageous price our country, and its people had to pay for that Bolshevik experiment. What is more, it would be a mistake not to understand its historic futility." In this and many later statements on the Soviet legacy, Putin's attitude rests on a conservative statist philosophy, one that has pragmatic respect for what makes the state strong and has an aversion to perceived ideological "experiments." For Putin, the task before Russia is to consolidate a new "Russian Idea" around stability, "fundamental political rights and human liberties," traditional Russian values, belief in the greatness of Russia, social solidarity and statism. The latter is expressed as an organic historical determinism and national exceptionalism:

> It will not happen soon, if it ever happens at all, that Russia will become the second edition of, say, the US or Britain in which liberal values have deep historic traditions. Our state and its institutes and structures have always played an exceptionally important role in the life of the country and its people. For Russians a strong state is not an anomaly to fight against. Quite the contrary, they see it as a source and guarantor of order and the initiator and main driving force of any change.... The public looks forward to the restoration of the guiding and regulating role of the state to a degree that is necessary, proceeding from the traditions and present state of the country.

This disposition was in sharp contrast to Yeltsin's inaugural declaration that "the strength of the state lies in the well-being of its citizens."[65] It foreshadowed a resolute reestablishment of a power vertical within the Russian state that returned fundamental power to the *siloviki* and concentrated it in the hands of the Kremlin. Nominally committed to the practices of capitalist enterprise, democratic governance, media freedom, and respect for human rights, the Putin years saw the establishment of a system (*systema*) that thoroughly compromised and subordinated all to the imperative of state control dressed in the language of sovereignty and nationalism.

Though not often described as "revanchism," the term is an accurate description of this agenda.[66] The French word *revanche* is not only associated with the reclamation of lost territories. Instead, it is associated with a desire in a competitive game to recover past position, power, and status. It describes, for example, the desire of French state elites, in the wake of defeat in the Franco-Prussian War of 1870–1871, to recover the state's power, honor, and respect. While this restorationist agenda found particular expression in their desire to recover the "lost territories" of Alsace and parts of Lorraine, it was about much more. The territory became a symbolic object for the realization of strength, dignity, and esteem. Revanchism, thus, is a form of affective geopolitics, one that is not inevitably about territorial aggrandizement. Putin wanted to make Russia great again. Russia must "rise from its knees." His often explicit use of the term "geopolitical" signified an understanding of international relations as a potentially Darwinian game of power politics where only strong states survive.

The North Caucasus was the geopolitical theater where Putin launched his heroic starring role as the defender of Russia's territorial integrity, status, and power. In the bloody years that followed in Chechnya, Ingushetia, and Dagestan, Putin struck a Faustian bargain at odds with his revanchism. The sponsorship of the former Chechen rebel leader Akhmad Kadyrov, and subsequently his son, Ramzan Kadyrov, created a space of exception within the Russian Federation: rule of law based on concession of a free hand to an internal warlord.[67] The personalized terms of the Putin-Kadyrov link was redolent of a mafia relationship, and it underscored how part of the *systema* established by Putin involved pragmatic relationships with local strongmen whose backgrounds and practices were unseemly if not outright criminal.

Putin proved to be a pragmatist in his relationships with Western powers. Absent from his rhetoric for the first decade was the deep-seated anti-Westernism that characterized the Soviet Union. Rather, Putin wanted Russia to belong, and to use Western investments and technological know-how to restore Russia to great-power status. The terrorist attacks on New York and Washington of September 11, 2001, provided an occasion for Putin to demonstrate solidarity with the United States.[68] As noted earlier, Putin saw the attacks as an opportunity for "civilized powers" to unite in a common alliance against terrorist forces, particularly terrorism sponsored by Islamic states.[69] His tacit approval to U.S. bases in Kyrgyzstan and Uzbekistan did not sit well with Russia's national-patriots or with some Putin supporters. Public opinion was suspicious.[70] Yet Putin made his "risky Western move" out of respect for Bush's response to the attacks and genuine belief in the commonality of ends. The move was not meaningfully reciprocated by the Bush administration, which went on to take a series of unilateral decisions that alienated Russia and many of its Western allies. Putin's disenchantment with the possibility of an alliance with the United States set in after the Bush administration's withdrawal from the 1972 Anti-Ballistic Missile Treaty. Putin suspected Georgia of secretly harboring terrorists just as Britain harbored his sworn enemies.[71] The Iraq invasion deepened the disenchantment. In Putin's eyes, the United States was the superpower tearing up all the established rules and doing what it liked without regard to international law.

A series of Chechnya-related terrorist attacks, culminating in a school siege in the North Ossetian town of Beslan in September 2004, involuntarily returned Putin to the traumas his rule was supposed to overcome. In both an earlier theater siege in Moscow and at the school in Beslan, the Putin administration resorted to force.[72] The Dubrovka Theater siege of October 2002 ended after security forces used poison gas to drug both terrorists and hostages. Due to poor medical response procedures, 133 hostages died. Security forces stormed School Number 1 in Beslan (see Figure 2.1) on the third day of the siege there, triggering a denouement that left 334 hostages dead, 186 of them children. Addressing the nation after the Beslan tragedy, Putin declared that "[w]e showed ourselves to be weak and the weak get beaten." In the event, he detected outside forces seeking to dismember Russia. "Some would like to tear from us a 'juicy piece of pie.' Others help them. They help, reasoning that Russia still remains one of the world's major nuclear powers, and as such still represents a threat to them."[73] The affective geopolitics at work here is explicit. Russia's territorial integrity

FIGURE 2.1 School Number 1, Beslan, North Ossetia. Putin described the September 2004 terrorist attack there as an international plot against the territorial integrity of the Russian Federation. Author photo.

was under attack, with dismemberment and the exposure of impotence in the face of separatist terrorism visceral fears. Only a strong state can survive. After Beslan, Putin abolished the direct election of regional governors. Thereafter, he would directly appoint them (the law was subsequently changed again to allow some elections).

The memory of the Beslan tragedy was still fresh when Putin gave the Federal Address describing the collapse of the Soviet Union as one of the geopolitical catastrophes of the century. In this he was echoing many others, including Solzhenitsyn, who described it as "the Great Russian Catastrophe."[74] The affective appeal of Soviet images of state strength only increased as Russia's weaknesses and struggles became more manifest. The viral spread of a mistranslated quote as a Putin reproach no doubt added to his resentment at Western, particularly U.S., behavior in the international arena. Two years later, Putin's frustrations found expression in a speech at the Munich security conference. "We are seeing a greater and greater disdain for the basic principles of international law," he charged. "One state .... the United States, has overstepped its national borders in every way. This is visible in the economic, political, cultural and educational policies it imposes on other nations. Well, who likes this? Who is happy about this?"[75] Until that time, Putin's revanchist agenda had not significantly altered the frozen conflicts in Moldova and Georgia: its primary geopolitical arena was the inner not the near abroad. That was about to change as the United States and its allies sought an endgame to the status of Kosovo. With no prospect of a settlement, the Bush administration decided they were going to go along with Kosovo unilaterally declaring itself independent from Serbia and pronounce the legal case sui generis, a precedent of no kind. The year 2008 would prove to be a very significant year in the remaking of post-Soviet space.

3

A Cause in the Caucasus

IN DECEMBER 2007, DAMON WILSON returned to the White House to take a position as senior director for Europe in the National Security Council of George W. Bush. Having spent the previous year in Iraq, Wilson was back working on an issue he was passionate about: North Atlantic Treaty Organization (NATO) enlargement. Prior service in the State Department, the NATO secretary general's office, and the White House gave Wilson familiarity with Euro-Atlantic divisions on the subject. Thrust into preparation for the forthcoming NATO summit in Bucharest, he was surprised that no internal policy process had yet generated a formal presidential decision on whether the United States was willing to offer a path to NATO membership for Georgia and Ukraine.[1] Both states underwent "color revolutions" that saw fraudulent election results overturned and new elections sweep dynamic Westernizing leaders into power, events many Russian officials viewed as Western-fomented coups.

Three years later in 2007, things were not looking so positive in either state. In Georgia, Mikheil Saakashvili's government had violently suppressed antigovernment demonstrations a few weeks earlier, while Ukraine's pro-Western leadership had descended into internal factionalism. Wilson, however, knew how strong the president's instincts were on support for fledgling young democracies in post-Soviet space. Bush had announced his commitment at the outset of his presidency in a speech at Warsaw University where he declared: "No more Munichs, no more Yaltas."[2] During Bush's tenure, NATO had admitted seven new member states, including the Baltic Republics, tacitly acknowledged as part of the Soviet Union at Yalta in 1945.[3] Approaching his last NATO summit, Bush had a legacy opportunity to push enlargement farther east and south, to large strategic territories that were part of the original Soviet Union.[4] Secretaries Condoleezza Rice and Robert Gates were skeptical but others such as U.S. ambassador to NATO, Victoria Nuland, were supportive. After a "deep dive" into the question by White House staff, Bush decided in late February that the United States should mobilize all its diplomatic power to offer a Membership Action Plan (MAP), a first step toward NATO membership, to both Georgia and Ukraine at Bucharest.[5] National Security Council staff led a process that formalized Bush's decision as the official U.S. government position and started lining up allies against skeptics within NATO, the most significant of whom was German chancellor Angela Merkel.

The result of the Bush administration's push to expand NATO yet farther into former Soviet space was a diplomatic showdown at the Bucharest summit. American and German diplomats labored hard to draft compromise language that would bridge their differences. To add to the drama, NATO had invited President Putin to the final day of the summit as a gesture of goodwill. Russian government officials had long made it clear to Bush, Merkel, and others that offering membership to Georgia and Ukraine in NATO crossed a "red line" with Russia.[6] In the crucial final negotiations, it was the "new Europe" leadership of Lithuania, Poland, and Romania that crafted the compromise text.[7] The Bucharest Declaration released on April 3, 2008, welcomed Ukraine's and Georgia's "Euro-Atlantic aspirations for membership in NATO." The Declaration then added: "We agreed today that these countries will become members of NATO."[8]

With this simple assertion, the Bucharest Declaration gave voice to a radical policy ambition. A military alliance designed to confront and contain the Soviet Union was now declaring that it was the manifest destiny of two former core territories of the Union of Soviet Socialist Republics (USSR), now complexly divided successor states, to become members of the West's Cold War alliance. The collective security agreement that was the bedrock of NATO would extend deep into former Soviet lands up to the western and southern borders of the Russian Federation. A U.S. president was setting the United States military on a path to becoming the guarantor of the security of states whose borders were decided by Bolsheviks. These countries were located thousands of miles from the United States and, further, were areas of vital national security interest to their neighbor Russia, a country with which they shared a complex history of intimacy and antagonism. The United States and its military allies envisioned NATO becoming the security power-broker in Russia's near abroad. U.S. troops in the future might well find themselves defending the birthplace of Stalin against Russian troops just a few miles away. What made this remarkable combination of geopolitical ambition and geostrategic overreach possible?

This chapter develops an answer to this question through an examination of one of a number of crucial relationships shaping the outcome at Bucharest: the post–Cold War relationship between Georgia and the United States. The Bucharest Declaration cannot be explained solely by reference to George W. Bush's administration, though its role was critical as we shall see. But there were also larger structural factors at work, and a history of U.S.-Georgian friendship that reached back more than a decade. This relationship evolved to become one characterized by affective rather than strategic calculations and understandings. Here I will explain how the state of Georgia became a "cause in the Caucasus" for many in the U.S. foreign policy community.

Regardless of Geography

To explain why the Bush administration lobbied so vigorously at Bucharest to extend NATO, we must begin with U.S. geopolitical culture, which is marked by a longstanding tension between normative visions and pragmatic practices. Normative geopolitics concerns aspirational and ideal visions of world order—how things "ought to be" in the abstract without reference to constraints and necessary compromises. An eighteenth-century form of normative liberal utopianism inspired the establishment of the United States and gave it a high-minded creed that disguised the compromises with human liberty it involved in practice. To believers, the United States was seen not as an ordinary territorial state but as the expression of universal ideals and values, the homeland of freedom, the land of liberty. As the United States became a world power in the twentieth century, U.S. exceptionalism was less about geographic location than this idealization of the state as the geopolitical embodiment of universal aspirations and ideals. To normative geopoliticians, the United States was an "empire of liberty" with a providential mission to spread freedom and democracy.[9] Russia functioned as a "dark double" for the United States, a despotic and enslaving empire that inspired a "free Russia" crusade in the late nineteenth century.[10] The United States' encounter with the Soviet Union adopted much of the same script, casting the relationship overwhelmingly in moral and theological terms in the early decades of the twentieth century. It was not until November 1933 that the United States finally recognized the Soviet Union.

The Cold War was characterized by permanent competition between normative and pragmatic strains in U.S. geopolitical culture. On the one hand, containment was an expression of pragmatic geopolitics, an accommodation to the hard realities of Soviet power on the European continent and elsewhere. It was characterized by recognition that organized containment was preferable to war, even though the American and Soviet blocs were now in a worldwide geopolitical competition for influence and power. Nuclear weapons, and later nuclear parity and the accompanying condition of mutual assured destruction, strongly mandated that competition be conducted through proxy powers and controlled. Both powers unsteadily found a path toward "peaceful coexistence." On the other hand, the doctrine of containment coexisted with an ongoing critique of its supposed passivity and compromise of U.S. ideals. Candidates running for political office in the United States, and some generals also, often gave free expression to fantasies of fighting and defeating the Soviet Empire. Smearing one opponent as "soft on Communism" was standard fare.[11]

This normative critique of containment found expression in two geopolitical conceits that predated the Cold War but became popular during it. The first was the notion of "captive nations." This was the conviction that the Soviet Union was an empire that had kidnapped historic nations in Central and Eastern Europe, and thereafter in lands beyond. This notion played particularly well among immigrant communities in the United States from these areas. The Yalta Conference with the Big Three (Joseph Stalin, Franklin Roosevelt, and Winston Churchill) in February 1945 was seen as a great betrayal, a needless concession to a Soviet "sphere of influence" over Eastern Europe. The claim was pure normative geopolitics, simultaneously refusing engagement with foreign policy complexities while implying that liberation for captive nations was a matter only of a deficiency of will.[12] Writing in *Life* magazine before the 1952 presidential election, John Foster Dulles argued that "liberation from the yoke of Moscow will not occur for a very long time, and courage in neighboring lands will not be sustained, unless the United States makes it publicly known that it wants and expects liberation to occur."[13] Once in office facing the complexities, Dulles modified his views. Nevertheless, his administration had to contend with the geopolitical rhetoric they cultivated and amplified to get elected. Legislation passed by Congress and signed by President Dwight Eisenhower in 1959 established a National Captive Nations Committee and created a Captive Nations Week to spotlight their condition through "ceremonies and activities" each July (it still exists). Twenty-two different "nations" were recognized by the legislation as captive to "the imperialist policies of Communist Russia" (it was inconceivable that Russians themselves might also be captives).[14] Important here is the prevailing conceit that certain nations have been captured by an irredeemably imperialist Russia, and that the United States has a moral responsibility to recognize and try to rescue them. This "rescue fantasy" is not insignificant in understanding the affective force behind NATO expansionism even after the Cold War.

The second conceit was the idea of rollback. This was the belief that the Soviet Union's grip over the "captive nations" of its empire could be challenged and undone. The notion was always popular among certain elements of the U.S. national security state established in the wake of the Cold War.[15] It found expression in NSC-68, the influential hardline policy document written by a committee headed by Paul Nitze, which Truman signed after the outbreak of the Korean War. The later phase of the Truman administration gave lip service to the notion but backed away from newly aggressive polices in practice. Eisenhower promised a more assertive policy than containment, but his administration also recoiled in private from offensive military plans, preferring an emphasis on "psychological warfare" instead. In public, however, the administration's rhetoric was still strident. The tensions between normative and pragmatic strains of U.S. geopolitical culture had tragic implications in Hungary in 1956.¹⁶ Cold War propaganda operations like Radio Free Europe/Radio Liberty, which were institutional expressions of normative geopolitics, broadcast encouragement to Hungarians rebelling against the Communist regime and its Soviet backers. Yet the United States chose not to respond militarily to the Soviet crackdown, acknowledging in practice that Hungary lay within a Soviet sphere of influence.

U.S. geopolitical culture largely processed the end of the Cold War as a victory for the policy of containment and for NATO's steadfastness in the face of the Soviet threat. Initially, there was some discussion about the redundancy of NATO in the transformed security environment. The violent dissolution of Yugoslavia, however, underscored how order and stability in post–Cold War Europe was fragile and uncertain. A dangerous security vacuum seemed apparent. Within the Clinton administration, there was a conscious effort to find a successor doctrine to "containment," an interagency competitive process some jokingly described as the "Kennan sweepstakes."¹⁷ Clinton's national security advisor, Anthony Lake, give a name to the doctrine the administration decided it liked best: enlargement. "The successor to a doctrine of containment must be a strategy of enlargement—enlargement of the world's free community of market democracies." The doctrine rested on a longstanding Wilsonian conceit in U.S. geopolitical culture, namely that "to the extent democracy and market economics hold sway in other nations, our own nation will be more secure, prosperous and influential, while the broader world will be more humane and peaceful."¹⁸ Enlargement justified the Clinton administration's decision to expand NATO. This decision was taken in 1993 and gradually introduced to the United States' NATO allies and to Boris Yeltsin.¹⁹ It was one of many Western polices that encroached on Russia's historic interests that Yeltsin had to accept with resignation. A second one that was deeply galling was the NATO war against Serbia in 1999. As the war raged, the NATO Council met in Washington, DC, and celebrated its fiftieth anniversary by launching a new legitimating "strategic concept" for the alliance. The summit's Washington Declaration announced that NATO "remains open to all European democracies, regardless of geography, willing and able to meet the responsibilities of membership, and whose inclusion would enhance overall security and stability in Europe." The "regardless" clause implicitly clashed with the "enhance" clause, but this was obscured by universal pronouncements. Rhetorically NATO was no longer a geostrategically grounded alliance but "an essential pillar of a wider community of shared values and shared responsibility."[20]

Within the U.S. Republican Party, more aggressive versions of enlargement were being discussed. One influential voice was that of Senator John McCain, a Republican from Arizona. McCain was a prisoner of war in Vietnam for six years, during which time he was tortured. In the early nineties McCain was close to the leadership in the Pentagon and became the spokesperson for their reluctance to use U.S. force to bring the Bosnian civil war to an end. In this cause, he debated and opposed his good personal friend Senator Joe Lieberman, a Democrat from Connecticut.[21] While a reluctant hawk on Bosnia, McCain was, at the same time, an outspoken proponent of advancing American values across the globe. In January 1993, he assumed the chairmanship of the International Republican Institute (IRI), an ostensibly nonpartisan quasi-governmental organization, funded through the National Endowment for Democracy (NED) established by President Ronald Reagan in 1983 to advance the spread of democratic institutions and practices across the world. The IRI's stated goal is to "expand freedom throughout the world," and to this end it created a series of programs, in partnership with local democracy promotion groups and other international organizations, to advance the cause of democracy. One program IRI developed identified emergent young leaders in the post-Soviet space and brought them to Washington for conferences and training programs. It was at one IRI-hosted event in Washington in 1995 that Senator McCain met a talented young student from Georgia by the name of Mikheil Saakashvili. An intensely energetic figure, Saakashvili had a cosmopolitan education, first in Tbilisi, then Kyiv, and thereafter in the United States at George Washington University and at Columbia University.

The idea of rollback did not disappear after the end of the Cold War. Running for his party's nomination for president in 2000, Senator McCain declared that if he became president he would institute a doctrine of "rogue state rollback": "I would arm, train, equip, both from without and from within, forces that would eventually overthrow the governments and install free and democratically-elected governments."[22]

The chief state in McCain's sights at this time was Saddam Hussein's Iraq. As long as Hussein was in power, McCain held, Iraq would be a threat to U.S. security. A neoconservative group strongly committed to the state of Israel, the Project for the New American Century (PNAC), and the Washington think tank, the American Enterprise Institute, came together to advance this agenda in 2002 by establishing a pressure group, modeled on earlier lobby work by many of the same figures on NATO enlargement, called the Committee for the Liberation of Iraq. Among others serving as board members were Senator McCain; Senator Lieberman; and a director at PNAC and IRI, previously an aid to the Republican Senate Majority Leader, Randy Scheunemann. The pressure group championed a rich Iraqi exile, Ahmed Chalabi, as their desired replacement for Saddam Hussein. Later the same figures would rally to the cause of Mikheil Saakashvili. The success of networks of influence in Washington that specialize in promoting foreign politicians as pro-American leaders depends, in part, on influential political figures in the U.S. power structure believing strongly in the expansion of liberty as a uniquely American cause. This vision of American exceptionalism was bipartisan and one most U.S. politicians endorsed. Senator McCain was open about his attraction to romantic causes and heroic rebels—independent men of action fighting for freedom. During his 2000 presidential campaign McCain told an interviewer that his favorite literary character was Robert Jordan from Ernest Hemingway's *For Whom the Bell Tolls*: "I am an incurable idealist and romantic. Robert Jordan is everything I ever wanted to be."[23] McCain updated his action hero identification in the 2000 Republican presidential primaries to Luke Skywalker from the *Star Wars* franchise films.[24]

David and Goliath

From the very outset, affective ties and symbolic debt shaped U.S. foreign policy toward Georgia. In the late eighties, Soviet foreign minister Eduard Shevardnadze built up a strong working relationship with U.S. secretary of state James Baker and German chancellor Helmut Kohl. Their relationship helped facilitate a peaceful end to the Cold War in Europe and the reunification of Germany. When he returned to Georgia in January 1992, Shevardnadze became parliamentary speaker and later president. The United States and Germany sought to repay him with financial assistance in consolidating Georgian independence. Georgia and the United States did not establish formal diplomatic relations until April 1992. A month later, Baker visited Tbilisi to see his friend, signaling by his very presence that the United States was interested in the fate of Georgia.[25]

The U.S. commitment to Shevardnadze from the outset led it to overlook contentious aspects of his rule. A few months after meeting with Baker, Georgian military forces sought to violently suppress Abkhaz separatist moves, triggering a bloody civil war. As already noted, after protracted battles and intervention by militias from the North Caucasus on the side of the Abkhaz, the Georgian army, with Shevardnadze as its commander, was forced into a humiliating withdrawal from Sukhum(i). At least two hundred thousand ethnic Georgians were forcefully displaced from Abkhazia thereafter (some later returned). Military defeat in Abkhazia and active rebellion in western Georgia forced Shevardnadze to turn to Moscow for military assistance. In a meeting with Russian president Boris Yeltsin, Shevardnadze agreed to have Georgia join the Commonwealth of Independent States and its aspirant military alliance the Collective Security Treaty Organization (CSTO). As part of the arrangement, Georgia ceded to Russia control of four Soviet bases on its territory and access to its Black Sea ports. In return, Yeltsin's government supplied weapons to Shevardnadze's army, fuel to Georgia as a whole, and Russian protection for its railways and ports. With the Russian support, Shevardnadze's army thereafter crushed a rebellion by supporters of the former Georgian president, the messianic Zviad Gamsakhurdia.

The October 1993 deal enabled Shevardnadze to survive the debacle in Abkhazia. It was nevertheless a second humiliation following on from the first, both of which became fused in the minds of Georgian nationalists. Whereas the Abkhaz read the war as victory over Georgian imperialism, the Georgians read it as defeat by Russian neo-imperialism, with the Abkhaz as mere pawns. Fearing Russian dominance in Abkhazia after the Georgian army's defeat there in September 1993, Shevardnadze tried to get the United Nations (UN), the OSCE, and the United States to send an international peacekeeping force to the region. While the first U.S. ambassador to Georgia, Kent Brown, privately advocated that a large contingent of U.S. troops be sent as peacekeepers, the idea was a nonstarter in Washington, DC.[26] The United States lost eighteen soldiers in Mogadishu in early October 1993, with hundreds of Somalis killed. Calls for U.S. intervention in Bosnia-Herzegovina were also being strongly resisted within the Clinton administration. In analogizing U.S. interest in Haiti to Russian interest in Georgia in January 1994, Clinton implicitly acknowledged the importance of geographic location in peacekeeping. To some this smacked of the United States tacitly endorsing spheres of influence. To others it was mere acknowledgment of the realities of geography and power politics.[27]

The realities facing Shevardnadze at the end of 1993 were grim: territorial fragmentation, lawlessness, pervasive state corruption, compromised sovereignty, and personal insecurity. In Shevardnadze, however, Georgia had an experienced leader highly skilled in the art of political messaging. Nicknamed the "silver fox," he had a reputation for cunning. As first secretary of Soviet Georgia, Shevardnadze had famously flattered Russians in Moscow by declaring: "[F]or Georgians, the sun rises not in the east but in the north, in Russia."[28] But he also knew how to play Georgian nationalism, and he managed to retain the Georgian language's constitutional status as the sole state language despite plans to reform this in 1977–1978. As Soviet foreign minister, Shevardnadze was charged with implementing Gorbachev's policy of "new thinking" in foreign policy. Gorbachev made common security challenges central to his vision and spoke of a "common European home" shared by the Cold War blocs. Shevardnadze envisioned a day when "East and West would recover their original geographical meaning, taken from them by postwar politics."[29] In 1988 and 1989 he approved what some termed the "Sinatra doctrine," the policy of letting Warsaw Pact countries choose their own way forward. Accumulating enemies within the power ministries in Moscow, he resigned in December 1990 while warning of a coming dictatorship.[30] Faced with a post–Cold War world now gravitating around a U.S.-led liberal order, Shevardnadze adapted. Together with his advisers, he developed what would become Georgia's marketing pitch to the international community. This comprised a set of geopolitical storylines that situated Georgia's fragmented territorial order and geostrategic situation within broader U.S.-sponsored geopolitical discourses.

Beyond keeping Shevardnadze alive (he survived three assassination attempts), the United States had no country-specific geopolitical interests in Georgia at the time. Rather, the United States had a commitment to the norms and principles of the liberal hegemonic order it sought to create in post–Cold War Europe and beyond. Three principles were longstanding and open for appropriation and local rewriting by Shevardnadze: an anti-imperialist commitment to the self-determination of small states, a commitment to competitive rather than monopolistic economic systems, and an interest in promoting democratization and liberal values across the world. Georgia's new geopolitical story drew upon self-images of linguistic, religious, civilizational, and geographic exceptionalism. Georgia is an ancient culture with its own unique written scripts that date from the conversion of local nobles to Christianity in the fourth century. Religion fostered messianic antemurale myths of Georgia as a Christian bastion in the Caucasus and facilitated later alliance with Russia. Georgia is a country with a rich literary tradition and widely recognized high cultural achievement. Finally, Georgia is located in a compelling physical environment that features scenic mountains and beaches, temperate climate, hospitable people, and fine cuisine.[31]

But according to the storyline, exceptional Georgia has long been a victim abused by surrounding great powers. The contemporary condition of Georgia provided more than sufficient evidence of its suffering and victimhood: thousands of impoverished displaced persons, territories lost to separatists, and Russia as the seeming imperial manipulator behind it all. This storyline worked well within a U.S. geopolitical culture predisposed to see the world in terms of Russian imperialists and captive nations. But it also benefited from the adherence by the United States to more abstract principles of international law as a response to the Soviet collapse. As we already noted, in automatically following the legal principle of *uti possidetis* after the USSR's dissolution, the United States was taking the side of the central Georgian government against the claims of separatist forces in Tskhinval(i) and Sukhum(i). This made it difficult for the United States to be viewed as a neutral party in peace-resolution efforts over these conflicts. Further, in routinely articulating its respect for the "territorial integrity" of Georgia, the United States was implicitly endorsing Tbilisi's vision of Georgia. U.S. officials had to determine whether

Georgia's condition of territorial fragmentation was evidence of its geopolitical victimization at the hands of Russia. Many concluded that it was, especially figures in the U.S. Congress who were predisposed by Cold War habit to anti-Russian sentiments. This disposition took the form of a popular understanding of Georgia as a plucky David battling the Goliath to the North. Georgia was an underdog nation fighting for freedom that deserved the support of the United States.

The second storyline was the presentation of Georgia as a geostrategic location. In making this argument, Shevardnadze and his advisers returned to a trope of connectivity he had used as Soviet foreign minister, the idea of a Great Silk Way uniting Europe and Asia.[32] This notion was refashioned to envision Georgia as an emergent transit space connecting European markets to the Caspian Sea and Central Asian oil and gas, an important transit country on a Great Silk Road.[33] Shevardnadze invested so much in this vision that he wrote a book about it while in office.

The story acquired resonance because of broader regional geopolitical developments and emergent U.S. strategy in the mid- to late nineties. The so-called contract of the century, which saw a consortium of oil companies from six different countries reach an agreement with the government of Azerbaijan in 1994 to develop three Caspian Sea oil fields, inaugurated a decade-long geopolitical game over how to develop Caspian and Central Asian energy reserves. As a rule, the "Russia first" policy of the first Clinton administration subordinated regional issues in the Caucasus and Central Asia to stability in U.S.-Russian relations. That began to change as U.S.-Russian relations soured and the United States adopted more aggressive policy positions against Iran.[34] Rhetorically, the United States explained its energy policy in the region as driven by concern for diversity of supply as well as support for the efforts of newly independent states to reduce their dependence on Russian markets. There was more to it than this, for the Caucasus and Central Asia were returning as locales of great power rivalry. In geostrategic terms, the United States sought to prevent any significant north-south pipeline projects and to ensure that Caspian and Central Asian energy flowed to markets via a western corridor of pipelines across states it considered friendly. Turkey was to be rewarded, Russia's dominant power position curtailed, and Iran isolated. The Karabakh conflict eliminated Armenia as a potential transit country and left Georgia as the transit country on a route to NATO member Turkey. The National Security Council in the White House worked with major oil companies to advocate for the multiple pipelines west policy, with the former driving the process as the economic feasibility waxed and waned.[35] Its first success was an agreement in March 1996 between Shevardnadze and Heydar Aliyev, president of Azerbaijan, to reconstruct an oil pipeline following an established pipeline route from Baku to the Georgian Black Sea port of Supsa, a project balanced by Aliyev's agreement the month prior to a repurposing of an existing Baku-Novorossiysk pipeline to transport Caspian oil to Russia. Three years later the Baku-Supsa pipeline opened.

An even more ambitious plan to construct a new Baku-Tbilisi-Ceyhan oil pipeline followed, a project linking the Caspian to the Mediterranean. Visiting Washington, DC, in July 1997, Shevardnadze joined the already considerable lobbying effort for the project. Former secretary of state Baker helped out by publishing a *New York Times* opinion editorial touting the vision.[36] In October 1998, the leaders of Georgia, Turkey, and Azerbaijan, as well as Uzbekistan and Kazakhstan, publicly committed to it. Construction began in 2002, and the thousand-mile-long project was finally completed in 2005. All of these projects helped Georgia present itself to the United States and other patrons as a geostrategic location, though this "objective fact" was a function of considerable effort and investment to make it such.

The third storyline was the vision of Georgia as a country on a path of reform and westernization. Shevardnadze's image in the West helped considerably at the outset, for he enjoyed trust and respect at the highest levels in the United States and Germany. This was a function of the afterglow of his time as Soviet foreign minister and not the more relevant history of his time as first secretary of the Georgian Communist Party. But Shevardnadze's habits were not democratic, though he was not an authoritarian like Aliyev. He was a skillful bureaucratic player and patronal figure accustomed to using ideological blandishments as an instrument of power. He grasped the importance of bringing in new modernizing and reformist forces into the political process in Georgia, most notably Zurab Zhvania, who was cochair of Georgia's Green Party. Zhvania, in turn, recruited a group of young reformers, including Mikheil Saakashvili. Both won seats in the December 1995 elections as members of Shevardnadze's party, the Citizens Union of Georgia. These reformers gave the country a young face and helped it considerably when they visited Brussels and Washington to solicit funds for various capacity-building projects. With reformers increasingly marginalized in Russia and elsewhere, Georgia was the next best hope for democratic transformation. Senator John McCain greatly admired Shevardnadze and developed a strong interest in Georgia. After Shevardnadze's visit to Washington in July 1997, McCain led a congressional delegation on an extensive "Silk Road" travel itinerary that began in Turkey; took in Georgia, Azerbaijan, and four states of Central Asia; and then ended in China. Visiting Georgia for the first time, McCain met Shevardnadze as well as Zhvania and Saakashvili. Also on the trip was Randy Scheunemann, who met Georgia's new reformers for the first time, including Saakashvili who was a minor figure at the time. The U.S. ambassador, William Courtney, summarized the prevailing conception of Georgia among U.S. officials at the time: it was a Caucasian Baltic state, a country stuck in an unfavorable location that wanted to be European.[37]

Shevardnadze's vision of Georgia as part of a New Silk Road was a geoeconomic rather than a geostrategic envisioning of space, a dream of open commercial networks rather than closed military spheres. In his book Shevardnadze took care to argue that a New Silk Road was a means to strengthen East-West cooperation, and that the project must include Russia. Most of the young reformers around Shevardnadze, however, did not see Georgia as a "bridge" or "crossroad" between East and West. Instead, they aspired to a radical break with the Soviet past and with a Russian imperialism they despised. This desire found an alternative geopolitical vision in the writings of an early-twentieth-century group of Georgian literary figures, the *tsisperkhantselni*, who were largely educated in Western Europe.[38] Building upon the notion that Georgia was an ancient Christian civilization, they argued that Georgia was, thus, primordially European. The destiny of the Georgian national movement was to reject Asian influences and "reunite" with European cultural space. The conceit of Georgia as an ancient European country whose destiny was to return to European civilization was the alternative geopolitical vision Georgia's young reformers mobilized around.[39] It was compelling and politically convenient, an appealing story for potential

Western patrons that obscured the messy territorial legacies of the different modes of imperial incorporation of Georgia and its subregions over the prior two centuries. As time would reveal, it also obscured the uncertain commitment of some of its ardent proponents to European values and democratic practices.

As he forged the geopolitical culture of an independent Georgia, Shevardnadze cautiously tilted Georgian foreign policy toward the West. In 1994 he enlisted Georgia in NATO's Partnership for Peace program. The following year he backed the U.S.-led diplomacy that helped end the Bosnian war and later, breaking with Russia, supported NATO's war against Serbia over Kosovo. In 1999, the country became a member of the Council of Europe while it left the CSTO. The same year it secured agreement from Russia in Istanbul to withdraw from two of four military bases in Georgia (Vaziani outside Tbilisi and Gudauta in Abkhazia). The outbreak of a second Chechen war complicated Georgia's relations with Russia considerably. Russian forces wanted to use Vaziani and other bases on Georgian territory for their offensive. They also proposed to attack the Pankisi Gorge in northern Georgia adjacent to Chechnya, viewing it as a sanctuary for "Chechen terrorists." (Fighters were indeed in the area and later the warlord Ruslan Gelayev, "defense minister" of the unrecognized Chechen Republic of Ichkeria, would use it as a sanctuary after defeats in 2000.) Shevardnadze refused, much to the displeasure of Putin and the Russian military.[40] It was within the context of his presidential reelection campaign and this perceived challenge to Georgian territorial sovereignty that Shevardnadze stated in a *Financial Times* interview in October 1999 that Georgia intended to "knock very hard on the door" of NATO for admission by 2005.[41] A statement of independence and geopolitical aspiration that played well at home, the remark became a sore point in his dealings with the Russians thereafter.[42]

Shevardnadze was remarkably successful in attracting U.S. aid to the country. By 2001, Georgia was the third-largest recipient of U.S. aid per-capita, an astounding achievement for a country run by a former Communist that was without strategic resources, a powerful diaspora, or a record of democratic achievement. Considerable portions of the aid provided by the United States were stolen. Meanwhile, reform programs never touched the core power ministries of the state. Corruption, nepotism, and espionage flourished in the ministries of Defense and Internal Affairs while key proposed reforms were never adopted. The country's media, however, remained independent and diverse. Saakashvili broke with Shevardnadze in September 2001 after accusing top officials of corruption. Zhvania, considered by many the political heir of Shevardnadze, quit as parliamentary speaker that November. (His replacement, Nino Burjanadze, also eventually broke with Shevardnadze.) The same year, a visiting U.S. academic penned a damning exposé of Georgia, dubbing it a "Potemkin democracy."[43]

"This Is Our Neighborhood Now"

On November 26, 2001, a formerly retired U.S. colonel by the name of Otar Joseph Shalikashvili and the U.S. defense attaché to Georgia and Azerbaijan met with the Turkish General Staff to discuss the implications of the September 11 terrorist attacks for the Caucasus region. Shalikashvili had recently been deputized by Donald Rumsfeld to be his senior defense consultant on Georgian defense matters. Otar's father Dimitri was a Georgian noble who served in the tsarist army and then the army of the Democratic Republic of Georgia before being forced into exile as the Red Army captured Georgia. Dimitri ended up in Poland where he married a countess, had three sons, and in 1941 enlisted in the Georgian Legion, a force of ethnic Georgians recruited by the Wehrmacht to fight against the Soviet Union. The legion was later incorporated as the *SS-Waffengruppe Georgien* and transferred to Normandy, where he was captured by the British army. In the early 1950s, Dimitri and his family moved to the United States. Otar joined the U.S. Army and trained for covert operations in the mid-1950s, eventually rising to become commander of the 10th Special Forces Group.[44] His brother John also joined the U.S. Army and rose dramatically to become the first foreign-born chairman of the Joint Chiefs of Staff from 1993 to 1997. A Special Forces enthusiast, Rumsfeld wanted to communicate to the Turkish military what 9/11 meant in the Caucasus and Central Asia.[45] Shalikashvili conveyed the message directly. "This is our neighborhood now," he announced before a stunned Turkish General Staff.

Shalikashvili's message was delivered at a moment of dramatic developments in global geopolitics. The United States had just launched an air war (Operation Enduring Freedom) against the Taliban in Afghanistan. Russia had provided consent to the United States to use former Soviet airfields in Central Asia as staging posts for its war in Afghanistan. Years earlier, in the wake of the Soviet collapse, the Turkish state had visions of an extended Turkic sphere of influence through the Caucasus to Central Asia. Though this had not flourished as hoped, various oil and gas pipeline projects from the region to Turkey were in the works. Further, the Turkish military had military cooperation agreements with Georgia and Azerbaijan that involved provision of military equipment, training of officers and soldiers, and the modernization of defense academies and military bases. The Turkish military, for example, had spent over a million dollars refurbishing the Vaziani military base south-east of Tbilisi after Russian forces had vacated it in June 2001, following an agreement reached in Istanbul. Now the United States was telling Turkey to move aside. The message did not go down well.[46]

The U.S. war against the Taliban government in Afghanistan launched on October 7, 2001, propelled the U.S. military into the Caucasus and Central Asia in a significant way. In fairly short order, the United States codified military-basing agreements with Uzbekistan and Afghanistan, negotiated overflight rights with Kazakhstan, and acquired contingent use of the airport at Dushanbe in Tajikistan. Responding to Shevardnadze's request for help with Pankisi, the United States established a "train and equip" (GTEP) mission between February 2002 and March 2004 that saw U.S. military trainers deploy to Georgia for the first time. The program sought to build the capacity of the Georgian state to assert control over Pankisi. The United States had to secure congressional authorization to train Georgian Ministry of Interior (MIA) Special Forces as well as regular military units (the MIA would later play a significant role in Georgia's fraught relationship with South Ossetia and Abkhazia). The larger U.S. concern was to shore up the state capacity of a friendly country in the Caucasus and signal support for its sovereignty and territorial integrity. In an October 2003 visit to Georgia, Senator McCain sought to see Pankisi for himself, but for security reasons had to settle for a low-altitude flight over the area and a visit to Georgia's mountain border with Dagestan.

Predictably the sight of U.S. soldiers (originally 10th Special Forces Group but later U.S. Marines tasked through European Command) training in Georgia caused an outcry in Russia among great power and revisionist factions.[47] Vaziani, the military base the Russians had retreated from in June 2001 and the Turks had refurbished, for example, was now hosting U.S. soldiers and aircraft. In October 2001, exile Chechen rebels and local Georgian guerilla fighters killed around forty people, including five UN personnel, in mountainous Abkhazia (discussed further in chapter 4). Russian foreign minister Ivanov conveyed Russia's concerns to U.S. secretary of state Powell in a meeting at the end of February 2002. Under pressure to take a tougher line against the growing U.S. presence in Russia's "backyard," Putin instead gave the United States and Georgia the benefit of the doubt. On March 1, 2002, Putin met with Georgian president Eduard Shevardnadze in Kazakhstan. There he publicly pledged his support for the U.S. military initiative: "Russia supports the international anti-terrorist efforts, and could only be glad to witness counter-terrorism in the Pankisi Gorge."[48] This was to be the high-water mark of cooperation in U.S.-Russian-Georgian relations. In August 2002, Russian fighter aircraft, taking inspiration from the new U.S. doctrine of preemptive attacks on "terrorists," bombed Pankisi a number of times, killing an elderly woman. "The international community has just crushed the nest of international terrorism in Afghanistan," Russia's defense minister, Sergei B. Ivanov, declared. "We must not forget about Georgia nearby, where a similar nest has recently begun to emerge."[49] Another Kremlin source described Georgia as having "an enclave of international terrorism."[50]

The United States' global war on terror (GWOT) brought temporary respite from the external and internal pressures building upon Shevardnadze to reform. The instructions to the new U.S. ambassador to Georgia, Richard Miles, in 2002 was to "practice tough love" with Shevardnadze. This proved challenging, for Shevardnadze was unfailingly courteous with the ambassador and receptive to his suggestions but there was little follow-through.[51] One thing that Shevardnadze did do was publicly move Georgia further in the direction of NATO. In November 2002 at a NATO conference in Prague, he formally proclaimed Georgia's desire for membership and explained his rationale: "NATO membership means security for Georgia. It means that we will have final security guarantees. Throughout our history, we have seen a lot of hardship, and I think that today the only right decision is to become a member of NATO."[52] NATO secretary George Robertson visited Georgia in May 2003, declaring: "[W]e welcome Georgia's ambition to further integrate in the Euro-Atlantic structure."[53] In July Georgia approved a one-hour test flight of NATO's AWACs surveillance plane within Georgian airspace, provoking a sharp Russian government protest because the plane "could provide surveillance over a large portion of Russian territory without ever entering the country."[54]

Georgian membership in NATO was not likely to happen quickly. In the meantime the Bush administration's contentious invasion of Iraq in March 2003 created a new opportunity for Georgia to build goodwill with NATO's most powerful member. Shevardnadze signed up Georgia to the relatively small "coalition of the willing" to stabilize Iraq, and that August over one hundred and fifty Georgian soldiers deployed to Tikrit as part of Operation Iraqi Freedom. Before the month was out, a number were injured while conducting mine-clearing operations. Subsequent injuries and deaths suffered by Georgian forces on U.S.-led missions in Iraq and Afghanistan established what some U.S. military commanders and politicians would come to characterize as a "blood bond" between the countries.[55]

Shevardnadze is the grandfather of all the high-profile policies later identified with his flamboyant successor, Mikheil Saakashvili. Under his rule, a narrative was developed that artfully presented Georgia to Western donors and patrons within the terms of broad liberal principles and goals. A skillful geopolitical entrepreneur in his dealings with the United States, Shevardnadze sold a particular geopoliticization of Georgia to U.S. constituencies that reflected their own aspirations and hopes. All of this served to attract foreign aid and perpetuate his political survival. Though their political styles were very different, there are important continuities in how Shevardnadze and Saakashvili responded to the structural challenges of Georgia's geopolitical location and territorial fragmentation. Both aspired to have Georgia escape its geography by building a series of network relationships with Western companies, institutions, and powers. It was an open question how far this strategy could be pursued without a strong counterreaction from Russia. Saakashvili would test these limits.

Beacon of Liberty

What was to become known worldwide as the Rose Revolution was the successful ouster of a sitting president by a relatively small urban opposition movement protesting fraudulent election results. It was, as Lincoln Mitchell has shown in depth, an accidental revolution yet one that was instantaneously mythologized by participants and international observers in different ways.[56] Funded by the Soros Foundation and other international nongovernmental organizations, Georgian civil society groups demonstrated through exit polls and parallel vote tabulation that the 2003 Georgian parliamentary elections were marked by fraud and manipulation on the part of Shevardnadze's coalition and its allies. Leading street protests against the falsified results, Mikheil Saakashvili crystalized the opposition's demand (without consulting them): Shevardnadze must resign.[57] The dramatic denouement came as Saakashvili, a single rose in hand at the head of a protesting crowd, burst into an inaugural gathering for the new parliament and drove Shevardnadze from the podium, symbolically taking his place and even finishing his cup of tea. After brief negotiations the following day, Shevardnadze unexpectedly resigned and handed power over temporarily to Burjanadze, who presided over new parliamentary and presidential elections that saw Saakashvili elected president with almost 96 percent of the vote.

The compelling visuals of an aging Soviet autocrat being ousted peacefully by an idealistic young cosmopolitan gave the moment more mythic qualities than it deserved. Shevardnadze, after all, was a longstanding U.S. ally while it was his own former government ministers who led the coup against him. But the image of peaceful regime change to a new generation of Georgians overrode the details. "It was a moment where the people spoke. It was a moment where a government changed because the people peacefully exercised their voice and raised their voice. And Georgia transitioned to a new government in an inspiring way."[58] This was the Rose Revolution in the words of President George W. Bush. And his government moved quickly to back the new leadership, establishing an interagency group to assess their needs. When Zhvania called the U.S. Embassy in late November to report that he found the Georgian state treasury empty, the U.S. government stepped in and, in an unusual move, approved a direct fiscal grant to pay pensions and salaries.[59] Funds supporting new parliamentary and presidential elections in January were also released. Further, in an important geopolitical signal, U.S. secretary of defense Donald Rumsfeld visited Georgia in early December to, in his words, "underscore America's very strong support for stability and security and the territorial integrity here in Georgia."[60] The United Nations Development Program and the Soros Foundation also came through with crucial early funding, establishing a joint-capacity building fund to pay Georgian state officials a sufficient salary to live and work well without recourse to bribes.

Saakashvili had proven himself a charismatic and bold leader, but he also had evident impetuous and monomaniac qualities. Russian and U.S. diplomats had worked together to resolve the Georgian election crisis with Foreign Minister Ivanov, whose mother was ethnic Georgian, playing a brokering role.[61] Saakashvili promised the Russians that if he became president, counterterrorist cooperation from Georgia would improve significantly. Both Ivanov and his U.S. counterpart Secretary Powell attended Saakashvili's inauguration, but they were accorded a very different reception. Powell had a front-row seat, whereas Ivanov was ushered to the back. The American anthem was played; the Russian was not. Both encountered a demonstratively rebranded Georgia, with large draped flags from Saakashvili's United National Movement now as the new official Georgian state flag, and a European Union flag also prominently on display.[62] Saakashvili explained this in his inauguration speech (official English translation):

> Georgia should be formed as the state assuming international responsibility, as the dignified member of international community, as the state, which regardless the highly complicated geopolitical situation and location, has equally benign relations with all its neighbors, and at the same time does not forget to take its own place in European family, in European civilization, the place lost several centuries ago. As an ancient Christian state, we should take this place again. Our direction is towards European integration. It is time for Europe finally to see and appreciate Georgia and undertake steps towards us. And first signs of these are already evident. Today, we have not raised the European flag by accident—this flag is Georgian flag as well, as

far as it embodies our civilization, our culture, essence of our history and perspective, and vision of our future.[63]

The address announced not only a geopolitical aspiration but also a self-aggrandizing rhetorical style that was to mark Saakashvili's rule. Its enduring features were essentialist claims about Georgian identity as European, and inflationary claims about its geostrategic significance. The hyperbolic geopolitical rhetoric was organically connected to Saakashvili's passionate preoccupation: restoring Georgia's territorial integrity. Days before his inauguration, Saakashvili had taken what was billed as a "spiritual oath" at the grave of David the Builder, a celebrated medieval king who wielded together fissiparous fiefdoms into a single kingdom. "Georgia's territorial integrity is the goal of my life," Saakashvili declared before thousands gathered outside the Gelati Monastery in Kutaisi. "At the grave of King David we must all say: Georgia will be united, strong, will restore its wholeness and become a united, strong state."[64] Adopting the five-cross Tamar flag as the new state flag was an act filled with symbolic meaning. Georgia was reborn as the successor state to a golden age medieval state, its territory marked by Christian crosses, simultaneously sacred and suffering. Theatrically presenting his presidency within a pantheon of Orthodox heroes, he cast himself as a new savior, someone who would resurrect the nation and heal its dismembered geo-body. To the worshippers, he vowed: "I would better die rather than to disappoint you."

This heady mix of Georgian spiritual nationalism and geopolitical embellishment was inevitably going to cause strains in Georgia's relations with Russia, not to mention among communities with different traditions, some of whom were concentrated in its breakaway territories. Shevardnadze's ouster caused policy splits within the Kremlin between great power and revisionist camps. The leadership of Abkhazia and South Ossetia, along with Aslan Abashidze, the kleptocratic ruler of Adjaria, flew to Moscow for meetings.[65] Saakashvili himself headed to Moscow after his inauguration for a prescheduled meeting of the leaders of the CIS countries in a confident mood. There he promised tougher actions against "Wahabbism" in the Pankisi Gorge.[66] Putin showed his guests Stalin's dacha and praised how simply the dictator lived. Saakashvili was positive about the meeting and about Putin as a role model more generally, a position he and Zhvania shared.[67] Ivanov later recalled: "Saakashvili greeted Putin very emotionally, very joyfully, and said that he greatly respected him as a politician and had always dreamed of being like him." Later Saakashvili claimed that Putin expressed concerns about U.S. influence behind the Rose Revolution, a view Saakashvili strenuously denied. (Putin came to hold a view that was the opposite of U.S. Rose Revolution mythology, namely that it was a U.S.-instigated coup.)[68] While it took some time for a rupture to play itself out publicly between the two men, the origin of their deeply personal antagonism lay in Putin's perception that Saakashvili had double-crossed him. Putin, the physically diminutive leader of a former imperial power, was a Soviet KGB veteran, whereas Saakashvili, the physically tall leader of a small, proud country intent on displaying its independence, was an anti-Soviet revolutionary.

Saakashvili's next major trip was to Georgia's preferred patron, located thousands of miles from the Caucasus. Waiting for a meeting with George W. Bush at the White House, Saakashvili hastily read the president's State of the Union Address delivered a month prior and then quoted it verbatim to Bush when they met. Bush was charmed and took an instant liking to Saakashvili.[69] In front of the press after their meeting, Bush was effusive with praise: "I'm impressed by this leader. I'm impressed by his vision, I'm impressed by his courage. I am heartened by the fact that we have such a strong friend, a friend with whom we share values." Turning to Saakashvili he said: "I'm proud to call you friend."[70] And so began another consequential geopolitical relationship in strongly personalized terms, Bush the admiring paternal figure, Saakashvili the eager-to-impress young leader.[71] To Bush and key members of his White House team, Saakashvili was "our guy" in the Caucasus. Friendship and attendant quasi-familial emotions (loyalty, commitment, pride, and protection) thus came to define the U.S.-Georgian relationship at the highest levels in ways that greatly complicated deliberative calculations of strategic interest. Significantly intensifying a dynamic of fealty-for-support established under Shevardnadze, Georgia would make great efforts to demonstrate its loyalty to the White House, while many U.S. political leaders would come to feel they "owed" Saakashvili and Georgians for their courageous efforts.

Saakashvili already had friends, as we noted, in high places in Washington, DC, before coming to power. One friendship that stood out was with Matthew Bryza, a Polish American diplomat close to his own age and temperament. A Stanford graduate, Bryza worked in the White House under Condoleezza Rice when Saakashvili came to power. He had first visited Georgia in 1998 and again in 2003 when former secretary of state Baker had sought to broker a preelection deal between Shevardnadze and the opposition. Bryza had found Saakashvili intense and confrontational when they first met but he was also impressed by his vision and energy. Zeyno Baran, a Turkish-born analyst who ran a Caucasus-focused program at the Center for Strategic and International Studies in Washington, became a major Saakashvili supporter (she and Bryza later married). After a long dinner with Saakashvili in 2002, Bryza too became a Saakashvili enthusiast: "I remember being dazzled. Oh my God, this guy actually has vision, and he is committed to doing exactly what we would like." But it was about more than the man:

> What got us all so enthralled by the Rose Revolution was that the people who ran it were people like [Saakashvili] who had been educated here. They were veterans of our exchange programs, and our reform promotion programs; who else would you want to be in charge of a country when our whole foreign policy, that of George Bush at the time, was about democratization. ... Bush believed it, and this is what we were supposed to be fighting for as American officials. Don't we believe in democratization? And suddenly we've got this group of twentysomethings, full of energy, full of idealism, no professional experience but they've been formed by our academic institutions, and they are in charge. Wow! What an opportunity. Can't squander that.[72]

Bryza became the primary figure in charge of Georgia within the Bush administration, first within the White House and later at the State Department as a deputy to the assistant secretary of state in the Bureau of European and Eurasian Affairs, Daniel Fried. While some veteran diplomats criticized his relationship with Saakashvili as overly personalized, Bryza saw Georgia, as did all Georgia enthusiasts within the administration, within the terms of President

Bush's Freedom Agenda. Bush had pledged the United States to the ultimate goal of ending tyranny in the world. He had promoted the spread of freedom as the great alternative to what he characterized as a terrorist ideology of hatred. (We should recall that Bush is a sincere evangelical Christian.) Expanding liberty and democracy, Bush reasoned, will help defeat extremism and protect the American people. Saakashvili did more than rhetorically frame Georgian interests within the terms of Bush administration discourse. He absorbed it almost verbatim and, with the enthusiasm of a convert, rearticulated an amplified version right back to Bush and others. The dynamic that took hold was a relationship of mutual admiration and affirmation, with top Bush administration officials projecting onto Saakashvili their own ideological aspirations, while Saakashvili mirrored these and, in turn, projected his own aspirations onto the Bush administration.

While Saakashvili was selling a rebranded image of Georgia on the international stage, U.S. officials and other internationals in Georgia were dealing with its sharp contradictions. Saakashvili seemed to rule like a Bolshevik, an impatient revolutionary at the head of a vanguard charging to a new utopia. His overriding interest was to "change the mentality" of the country while strengthening the capacity of the state to govern effectively and become militarily strong enough to unify its legally recognized territory. Youthfulness and hard-charging zeal characterized the revolution: no longer was Georgia a country for old men. In February 2004, the Georgian parliament passed a series of measures that greatly strengthened executive presidential power and enabled the confiscation of assets of those suspected of corruption. Saakashvili quickly fired 80,000 state employees, including most of the state's KGB-trained security personnel as well as its entire traffic police. Pressure on the media and prominent businesses began almost immediately. Georgia's standing in a worldwide index of press freedom got worse, not better, after the Rose Revolution.[73] The judicial system was used to break the existing political economy in Georgia, with businessmen arrested and threatened if they did not pay exorbitant fines. The personalized nature of governance in Georgia was all too clear in a generally positive portrait of Saakashvili in the *New York Times* magazine that nevertheless featured his shouting "To jail" after his prosecutor general calls to ask for instructions about the fate of a businessman with ties to the president.[74]

Georgia analysts spoke of a "post-revolutionary syndrome," of "government by adrenalin" and "Bonapartism."[75] In the fevered atmosphere, critics became "traitors" and "Russian agents." Over time, Saakashvili's authoritarianism deepened and all the key opposition figures behind the Rose Revolution broke with him. In early 2005 the Parliamentary Assembly of the Council on Europe passed a resolution expressing concerns about developments in Georgia: "the post-revolutionary situation should not become an alibi for hasty decisions and neglect for democratic and human rights standards."[76] But it was. Criticism would grow sharper as it became apparent that a government espousing freedom and democracy was intimidating citizens through judicial blackmail, gutting the independence of the media and governing through a culture of fear. Despite these manifest contradictions, which only deepened in the subsequent years and were unavoidable when Saakashvili violently suppressed antigovernment demonstrations in November 2007, the Bush administration remained positively disposed toward Saakashvili. Experienced national security officials were amazed at the access he enjoyed to the White House. Georgia was the recipient of increased levels of U.S. development aid. The Millennium Challenge Corporation provided a series of rounds of funding for infrastructure across Georgia.

Four major factors account for the exceptional level of support Saakashvili's government enjoyed in the run-up to the August 2008 war. The first is the personality and leadership style of Saakashvili himself. To the key figures in the Bush administration, and on Capitol Hill, President Saakashvili was a very impressive figure. In fluent English he energetically cultivated members of the U.S. Congress and proved a charismatic and charming interlocutor. Yet, while Saakashvili could be charming, he was also impatient and inclined to hear only what he wanted to hear. One figure that experienced this was U.S. ambassador Richard Miles. Conveying bad news or U.S. views that were critical while also constructive was not welcome. Instead, from the outset, Saakashvili was on the phone daily to the White House and Capitol Hill. "He was like LBJ with the telephone," Miles recalled, cold-calling various figures to chat and lobby. Despite pleas to proceed through the regular diplomatic chain of command, Saakashvili ignored the request.[77]

The authoritarian and centralized form of Saakashvili's governance was also a matter of concern to some U.S. officials. Steven Larabee of the Rand Corporation, for example, led an effort to establish a National Security Council structure for the Georgian government. The effort, however, floundered when it became apparent that Saakashvili's inner circle had no interest in establishing robust democratic procedures of accountability for the security structure.[78] Saakashvili's authoritarianism, however, had supporters within U.S. political and think tank circles. Vice President Cheney and the American Enterprise Institute were strong proponents of the doctrine of unitary executive power, and this was applied to justify Saakashvili's dismissal of state employees and use of coercive methods against perceived corrupt businesses to extract revenue. Saakashvili's emphasis on increasing military spending to create a strong Georgian military was also viewed sympathetically.

Second, Saakashvili took Shevardnadze's geopolitical policies to a new level. He portrayed Georgia as a former captive nation struggling to free itself from its imperialist oppressor. It was David against Goliath, the plucky small nation standing up to the bully. Saakashvili deepened Georgia's support for the U.S.-led GWOT. In March 2004, Georgia made its air space, road, and rail infrastructure available to NATO to send supplies necessary for the sustainment of NATO forces in Afghanistan. In August, Georgia announced it would send troops to support NATO operations in Afghanistan. In November, Georgia announced an increase in its deployment of troops to Iraq to 850, while the United States announced a new military assistance program for Georgian soldiers. This later increased to 2,000, making Georgia the third-largest troop contributor to Operation Iraqi Freedom. All of these deployments earned Georgia military assistance, an Individual Partnership Action Plan with NATO (October 2004), and considerable levels of positive will from the leadership of the U.S. military. Some U.S. military attaches became enthusiastic advocates for the country, while the U.S. Marine Corps became increasingly invested in helping it train and equip its soldiers.

Saakashvili also built a new strategic relationship between Georgia and Israel. Georgia has long had a small, thriving Jewish community and two of its members, both fluent in Hebrew, became influential advisers in Saakashvili's government.[79] Davit Kezerashvili was born in

Tbilisi but educated for a time in Israel before returning to Georgia. He was a founding member of Saakashvili's United National Movement party, and Georgian minister of defense in 2008. A decade older, Timuri Yakobashvili worked for Georgia's ministry of foreign affairs before becoming the minister in charge of reintegrating the breakaway territories. He was a key figure in August 2008 and later Georgian ambassador to the United States. Israel long had an interest in cultivating Georgia given its location north of Iran. Georgia wanted to enhance its military capabilities and had already purchased some weapons from Israel. Saakashvili first visited Israel in July 2004 and worked to create a strategic military relationship between the countries that met with the approval of the Bush administration. Georgia bought a range of weapon systems from Israel, including unmanned aerial vehicles. Private Israeli military contractors (Defense Shield and Global CTS) provided training for elite Georgian military units alongside Americans. Israel allegedly secured access to two airfields in the south of Georgia should it decide on preemptive military action against suspected Iranian nuclear facilities.[80] Rhetorically, Saakashvili saw Israel as an inspirational model for Georgia, the "original" David against Goliath.[81] He saw Ben-Gurion as a role model and admired Israel's aggressive military posture.[82] Georgia would be an "Israel in the Caucasus." Aligning publicly with Israel also helped win friends and consolidate alliances on Capitol Hill. For example, Democratic congressman Eliot Engel, an influential figure on the House Foreign Affairs Committee and stalwart supporter of Israel, also became a major supporter of Georgia.

A further example of affective alignment was Saakashvili's enthusiastic embrace of neoliberal ideology and reform schemes. Saakashvili persuaded Kakha Bendukidze, an ethnic Georgian who had made a fortune as a businessman in Russia, to become minister of the economy in June 2004. A committed libertarian, Bendukidze sought to privatize as much of the state as possible and deregulate economic activity. Under Bendukidze, Georgia adopted a series of neoliberal reforms. Saakashvili's government also set about improving the ranking of the state on international ranking indexes, like the World Bank's "Ease of Doing Business" index.[83] Bendukidze sought to institutionalize the teaching of libertarianism in Georgia.[84] His friend, the Russian libertarian Andrei Illarionov, also left Russia at the same time, moving to the Cato Institute in Washington, DC. There he became a major supporter of Saakashvili's Georgia. Saakashvili's relationship to the Cato Institute, however, was complicated by the fact that it advocated a minimalist foreign policy that did not look favorably upon U.S. support for client states like Georgia. Instead, Saakashvili found supporters within the right-wing American Enterprise Institute, the conservative Heritage Foundation, and among some in the Center for Strategic and International Studies.

A third factor propelling Georgia into a special relationship with the United States was the desire on the part of the Bush administration for a Freedom Agenda success story. The purest statement of this creed was Bush's second inaugural address on January 20, 2005. In it he argued "the survival of liberty" in the United States "increasingly depends on the success of liberty in other lands. The best hope for peace in our world is the expansion of freedom in all the world. America's vital interests and our deepest beliefs are now one."[85] In keeping with the most expansive traditions in U.S. geopolitical culture, Bush's program was universalist, intended for application regardless of culture and location. It was to be "the policy of the United States to seek and support the growth of democratic movements and institutions in every nation and culture, with the ultimate goal of ending tyranny in our world." With Iraq and Afghanistan wracked by civil war, however, Bush had few examples of success to tout. One seeming point of light, however, was Georgia. Already scheduled for a Victory Day commemoration in Moscow, Bush's team decided to honor Georgia with a presidential visit. On May 10 he addressed the central square in Tbilisi (renamed Freedom Square), declaring that because of the Rose Revolution "Georgia is today both sovereign and free, and a beacon of liberty for this region and the world." Georgia, he explained, was making many "important contributions to freedom's cause" by its troop deployments to Afghanistan and Iraq. Its most important contribution, however, was as a model to other countries:

> But before there was a Purple Revolution in Iraq, or an Orange Revolution in Ukraine, or a Cedar Revolution in Lebanon, there was the Rose Revolution in Georgia. Your courage is inspiring democratic reformers and sending a message that echoes across the

world: Freedom will be the future of every nation and every people on Earth.... Now, across the Caucasus, in Central Asia and the broader Middle East, we see the same desire for liberty burning in the hearts of young people. They are demanding their freedom—and they will have it. As free nations, the United States and Georgia have great responsibilities and together we will do our duty. Free societies are peaceful societies. And by extending liberty to millions who have not known it, we will advance the cause of freedom and we will advance the cause of peace.[86]

This was the gospel of democracy and thoroughly in keeping with the messianic liberal tradition of American geopolitical culture. In a press conference Bush explicitly linked Georgia to Iraq, suggesting, "[I]t's this democratic movement that took place here in Georgia that is going to help transform the greater Middle East. And that's important for people in Georgia and around the world to understand, that democracies in the greater Middle East will make the world a more peaceful place. A democracy in Iraq will send such a strong and vivid example to others about what is possible."[87]

While his visit was a public endorsement of Saakashvili, Bush privately cautioned Saakashvili on two separate occasions against any further military adventurism in South Ossetia and Abkhazia. The U.S. president made it clear that the United States believed such acts would draw a Russian response and that the United States would not come riding like the cavalry to Georgia's rescue.[88] The disjuncture between the lofty public rhetoric and the private practical caution perhaps caused some cognitive dissonance among Saakashvili's team. More likely, however, was that they chose to believe what they already were inclined to believe, namely that the United States viewed Georgia as a special friend and partner in the fight to advance the cause of freedom.

A final factor shaping the U.S.-Georgian relationship prior to the August War was the Georgian lobby that had emerged in Washington, DC, from 2004 onward. Prior to Saakashvili, Georgia had a very limited lobbying operation in the U.S. capital. After the Rose Revolution, however, Saakashvili decided to invest heavily in promoting Georgia in Washington, DC, tapping former supporters and boosters to do so. The key lobbyist he decided to employ was Randy Scheunemann, the IRI board member who previously worked on campaigns supporting NATO enlargement and the Iraq war. Scheunemann ran a small firm called Orion Strategies that specialized in promoting newly democratizing states. He and Saakashvili met in February 2004 and signed an agreement to promote Georgia in Washington, DC. That fall Scheunemann organized a retreat on a small island on the Potomac for Georgia's new leaders and a group of U.S. supporters. Together they worked out a strategy to secure a range of sources of U.S. support for the country.[89] While Scheunemann's primary networks were among Republican groups and sympathetic media outlets (the *Wall Street Journal* editorial page, the *Washington Times, Fox News,* the *Weekly Standard*), the cause of Georgia was bipartisan from the outset. Prominent Democrats like Richard Holbrooke and Ron Asmus became major supporters. Asmus visited the country for the first time in the summer of 2004 and was, he later wrote, "soon captivated. . . . How could we not support them."[90] He agreed to look into the question of gaining NATO membership. Scheunemann worked with the staffs of Senators Joe Lieberman, Joe Biden, and John McCain to craft legislation that favored Georgia. One early victory was qualifying Georgia for Millennium Challenge funding despite its past record of corruption: the new government, it was argued, cannot be held responsible for the failures of the old. Asmus began to work on revitalizing Georgia's case for NATO membership. A culmination of this effort was Senate Resolution 439, twinned with a similar House of Representatives resolution, which passed the U.S. Senate by unanimous consent two months before the NATO Bucharest Summit. It declared that the United States should take the lead in supporting the awarding of a Membership Action Plan to Georgia and Ukraine. The resolution was cosponsored by Barack Obama.[91]

Scheunemann helped the Georgians refine their pitch to congressional members and to the U.S. policy community. Initially this was a low-information geopolitical pitch. Saakashvili described it thus: "We would have a big map and tell them, 'This is America. This is Russia. This is Georgia. This is the pipeline. This is why Georgia matters.'"[92] The Rose Revolution made matters easier, and Saakashvili began to attract admirers across the political spectrum in the United States. Numerous think tanks and academics were enlisted to promote the cause of Georgia within the Beltway, while those in the more specialized field of Caucasian policy were invited on study trips and to lavish conferences in Georgia.[93] Reflecting on the U.S.-Georgian relationship before 2008, former deputy national security advisor in the Bush administration (2007–2008) Ambassador James Jeffrey was scathing in his condemnation of the role of lobbyists in making causes out of ambitious foreign politicians. Their strategy was to wrap their clients in the U.S. flag and tell people in Congress that they are "one of us." "There's a game plan for this, there's a template and its Congress, think-tanks, rich people, some academics . . . make us a cause. We are awash in this shit as a country."[94] Boosterism of this type, Jeffrey believed, undermined executive branch foreign policy making and sent mixed messages overseas to U.S. allies and adversaries.

The considerable efforts made by the Bush administration to have Georgia granted a membership action plan yielded a pledge, not a pathway. British prime minister Gordon Brown underscored the confused compromise of the Bucharest Declaration's text—Georgia and Ukraine "will become members of NATO"—when he commented to Bush: "I am not sure what we did here. I know we did not extend MAP. But I'm not sure we didn't just make them members of NATO."[95] Saakashvili was bitterly disappointed there was not a formal MAP invitation but grateful to Bush. "He really fought for us at this NATO summit in Bucharest. When I went into the room, he looked like he was just back from the OK Corral—red-faced, very tired, exhausted."[96] The strange compromise—a future promise but no practical path—laid bare the internal divisions within NATO over Georgia and Ukraine.

President Vladimir Putin arrived in Bucharest on the second day of the summit. Prior to the summit, Bush had phoned Putin and asked that he not embarrass him with a hostile broadside. Putin was gracious at the evening reception. The next day, after Putin and his team had time to digest the summit's declaration, Putin gave a speech in which he reportedly passed over the toughest remarks in the prepared draft, though he was still characteristically forthright.[97] Russia, he declared, was "always open for cooperation on the basis of equality and taking into account each other's interests." But NATO, a Cold War alliance, was relentlessly expanding and now operating only a few hundred kilometers from Saint Petersburg. Further, recognition of Kosovo's unilateral declaration of independence had created considerable difficulties in Russia's neighborhood, for it had similar contentious places that required careful policies: Abkhazia, Transnistria, South Ossetia, and Nagorny Karabakh. The ethnic conflicts between Georgians and Abkhazians, and between Georgians and Ossetians, were more than a hundred years old. Georgia wrongly believed NATO can help it restore its territorial integrity. His view was that in order to solve its territorial problems Georgia should not join NATO but instead "be patient and establish a dialogue with these small ethnic groups. And we will try to help, by the way, Georgia to restore its territorial integrity." Putin's most pointed remarks, in the conclusion of his address, concerned not Georgia but Ukraine (which we will take up in chapter 5).

The Bucharest summit of 2008 was a deeply consequential one in the history of post–Cold War Europe. NATO, at the urging of the Bush administration, had declared as manifest destiny that Georgia and Ukraine would join the alliance. Russia's declared "red line" was openly crossed. At his press conference after the summit, Putin reiterated Russia's position once again: "We view the appearance of a powerful military bloc on our borders ... as a direct threat to the security of our country. The claim that this process is not directed against Russia will not suffice. National security is not based on promises."[98] Soon afterward, international observers began to suspect that the rules of geopolitics in the Caucasus were changing. Saakashvili, in response, raised the stakes even further.

4

Territorial Integrity

ON THE EVENING OF August 7, 2008, Inal Pliyev was working late at his office in the center of Tskhinval(i). A former journalist, Pliyev was head of communications for the self-declared South Ossetian Republic. Earlier in the evening, Georgian president Mikheil Saakashvili had declared a unilateral ceasefire after days of skirmishes between Georgian forces and South Ossetian militias. Pliyev, however, was still in the office because of information about increasing Georgian artillery and armor concentrations near the town. "First we heard what sounded like grenade launchers—after the years of conflict everyone here knows what sound is made by which weapon. I did not pay much attention to that."[1] But when he heard the first sounds of Grad missiles, Pliyev turned off his computer and ran for his life. "All parts of the city came under fire simultaneously. It was so intense, that you couldn't even register a fraction of time between explosions, there were multiple explosions every second. The fire was non-stop. Electricity and gas supplies were cut off during the first minute of the shelling, and for the most part phone service was also cut off."[2] One shell fell next to the government building where Pliyev and his colleagues huddled. "The building shook so much that part of the ceiling bent down, and we ran into an underground bunker in a nearby non-government building. Explosions were becoming louder and even more frequent. We could not leave our hideout, and everyone was getting ready to die. Even more we feared being taken prisoner by Georgian soldiers. It was especially terrifying when we heard machine gun fire. Our only thought was to avoid being taken prisoner at any cost. Our only hope was for the Russian air force, we were waiting for it to come, so that Georgians would leave our city. But it wasn't coming."[3]

Pliyev had his mobile phone, and as its battery ran out he spoke to various Russian media outlets pleading for Russian military help. The whole of central Tskhinval(i) seemed to be on fire. The number of dead was unknown, but Pliyev told the media that "I think we can talk about no less than two thousand dead and together with villages there could be as many as five thousand."[4] Dashing to his home many hours later as Georgian tanks and infantry fought the town's Ossetian defenders, Pliyev discovered his family dwellings destroyed, a car of fleeing residents with everyone dead inside, and a Georgian tank operator lifeless in his gutted tank. Russian fighter planes were forcing Georgian soldiers to retreat, while Russian soldiers were on the way to rescue the town and its inhabitants.

The circumstances of the outbreak of the August War of 2008 are deeply contested. For Ossetians like Pliyev, the war was a renewed genocidal campaign against the Ossetians by the Georgian state. For the Georgian government, the war was a Russian invasion of sovereign Georgian territory, and more broadly an "attack on the West."[5] For Moscow, the war was a defensive response grounded in its historic responsibility to protect a vulnerable nation, most of whose members were now Russian passport-holders, from genocide. A report by an international commission organized by the Council of the European Union—officially, the Independent International Fact-Finding Mission on the Conflict in Georgia, or simply the Tagliavini Report after its head, the experienced Swiss diplomat Heidi Tagliavini—concluded that "open hostilities" were initiated by the Georgian government on August 7 but that the Russian government bore some responsibility for the circumstances that led up to the beginning of "large-scale armed conflict in Georgia." Opinions varied within the United States. Republican presidential nominee John McCain condemned the "stark international aggression" against Georgia. The world, he wrote, "has learned at great cost the price of allowing aggression against free nations to go unchecked." In solidarity with the victimized nation he told President Saakashvili: "today we are all Georgians."[6] Others privately saw fault on all sides. U.S. secretary of defense Robert Gates believed that the "Russians had baited a trap, and the impetuous Saakashvili walked right into it."[7] The description recognized the war as co-created, placing proximate blame on Saakashvili but structural blame on the Russian government.

Missing from most interpretations of the August War, particularly in the United States, is any account of the intra–South Ossetian and intra-Georgian dimensions of the territorial conflict over South Ossetia. Instead, debate was dominated by abstract moralized dichotomies—the West versus Russia, freedom versus imperialism, the small nation versus bullying empire, and, tacitly and unconsciously, "us versus them"—that made little or no reference to geographic particulars within Georgia or South Ossetia. But the August War was not only or even chiefly a global event (the predominance of this scalar framing is an aspect of the conflict itself). Instead, it was a complex, multiscalar event, one that requires understanding of how the different parties to the conflict—the Russian-backed South Ossetian regime in Tskhinval(i), the Georgian government and its South Ossetian regime north of Tskhinval(i), and the Russian state, as well as, at some remove, the foreign policies of the United States and the European Union—pursued overlapping geopolitical strategies that had negative interactive and polarizing effects. A deeper account of the conflict is required to grasp its historic emergence, conditioning geographies, and evolving cultural geopolitics. It is incomplete to treat the August War unidimensionally as either "Georgian genocide" or "premeditated Russian aggression," and those who force these frames express rather than analyze the conflict.[8] Instead, the war was a consequence of a concerted effort by the Georgian state under Mikheil Saakashvili to recover its "lost territories" and the reaction this provoked within the de facto states and the Kremlin. Saakashvili's revanchist project was concealed by its articulation within principles that the liberal international order broadly accepted (including Russia), namely that states should enjoy sovereignty and territorial integrity. But the pursuit of territorial integrity by Saakashvili's government was a form of geopolitical revisionism that threatened the frozen conflict lines within Georgia, a condition Georgian nationalists experienced as an intolerable fragmentation of their nation-state. This determined Georgian revanchist drive encountered a reassertionist Russia and stimulated it toward a more expansive revanchism of its own. Radical nationalist ambition to "take Georgia back" in Tbilisi—more focused on Abkhazia than South Ossetia because of its greater number of displaced former residents—empowered radical countervisions in Russia to shore up the de facto regimes in Sukhum(i) and Tskhinval(i). Tragically, this emergent conflict acquired a deeply personalized form and culminated in a violent paroxysm that revealed aspects of the personalities of the two leaders—recklessness from Saakashvili and punitive reaction from Putin—that made for an ugly little war.

This chapter provides an account of the geopolitical circumstances that led to the August 2008 assault on Tskhinval(i) and South Ossetia. Divided into four parts, it provides a brief background summary of South Ossetia as a contested space, the revanchist commitments of Mikheil Saakashvili, and the proactive policies his government adopted in pursuit of territorial integrity. It ends with a brief discussion of who started the August 2008 war.

A Contested Space

The relationship of South Ossetia as a distinct entity to the state of Georgia has always been contentious. For hundreds of years Ossetians, a Caucasian nationality numbering just half a million people who speak an Eastern Iranian language and Russian, have lived on both sides of the Caucasus Mountains. Ossetians trace their ancestry to the Alans, rulers of a kingdom in the North Caucasus before the Mongol invasions.[9] Most reside on the northern slopes, in the internal Russian ethnic republic of North Ossetia, but there has long been a community on the southern slopes also, a fact recognized by the creation of South Ossetia as an autonomous region within Soviet Georgia. Over time more Ossetians lived beyond this entity than within it, in Tbilisi and many other Georgian cities. Most spoke Georgian and were well acculturated to Georgian mores. The second husband of legendary Georgian Queen Tamar (1184–1213) was an Alan prince, so Ossetians had a place in the Georgian mytho-dynastic pantheon. While relations with local Georgians were largely peaceful for almost seventy years, the collapse of Communist rule provoked a clash between competing nationalisms, Georgian and Ossetian, over control of the territory of South

Ossetia. To Georgian nationalists at the time, the region was an imposition of Soviet occupation, its ethnic Ossetian residents recent invaders of primordially Georgian land. In Georgia, the north-central region is named Shida Kartli (lit., "Inner Kartli"), and the area around the capital of South Ossetia (Tskhinval to Ossetians and Tskhinvali to Georgians) is described as Samachablo or the Tskhinvali region. The names situate South Ossetia within golden age imaginings of Georgia. Shida Kartli signified the heartland, while Samachablo means "fief of the Machabeli clan," a Georgian aristocratic family who once controlled the area. Past dynastic "ownership" implies contemporary Georgian national "ownership," the names marking the territory as the nation's patrimony. "Tskhinvali region" is the replacement name given to the area by the Georgian government after it unilaterally abolished the autonomous status of South Ossetia in 1990.

The local Ossetian story is one of victimization at the hands of an imperialist Georgia practicing a genocidal form of violence.[10] This narrative framed the August 2008 war as the third installment of genocidal Georgian violence against the Ossetian people, building on a victimhood story that originally gained traction in the late 1980s. The first campaign was by the Menshevik-controlled Democratic Republic of Georgia, the independent Georgian state proclaimed in May 1918 and overthrown in February–March 1921 by the Bolshevik Red Army. The tumultuous period between 1918 and 1920 saw three separate revolts by Ossetian peasant groups in what was administratively encompassed by the Tiflis, and to a lesser extent the Kutaisi, *gubernias* (provinces) of the Tsarist Empire, until May 1918. The most consequential of these was a revolt by Bolshevik-led Ossetians in April 1920, which saw them capture Tskhinval(i) before retreating in the face of a counteroffensive by Georgian National Guard soldiers in June. During the course of this campaign, Georgian government forces retaliated by destroying some 40 Ossetian villages; at least 5,000 people perished, mostly displaced persons who died from starvation and illness. An estimated 20,000 to 35,000 Ossetian residents were forced to flee across high mountain passes to North Ossetia. In response to the legacy of distrust created by this violence and displacement, a compromise was forged between Ossetian leadership demands for unification with North Ossetia and Georgian reluctance to create any special Ossetian district. That compromise was the creation of a South Ossetian Autonomous Oblast (SOAO), which was approved by Bolshevik Georgian Central Committee decree on April 20, 1922. Though recognized as a separate ethnoterritorial region, this was a lesser status than Abkhazia and Adjara, both of which became Autonomous Soviet Socialist Republics (ASSRs) within Georgia (though in different ways). The historian Arsène Saparov argues convincingly that this move was an attempted conflict resolution in response to independent Ossetian forces from below and not a "divide and rule" strategy from above.[11]

As an autonomous oblast, South Ossetia had a local parliament that gave it an important forum for nation-building and later for grievance articulation. Russian and Georgian, not Ossetian, were the languages of administration (standard practice with administrative oblasts in the Soviet Union), while ethnic Georgians and Ossetians (whose language repertoire was Ossetian at home, and largely Russian but also Georgian at work and in public) enjoyed mostly positive relations under the Soviet Union. Strains did exist over linguistic script and education, with a policy from 1938 to 1953 of using Georgian script for the Ossetian language, creating a resented split from the Cyrillic-based Ossetian taught in North Ossetia. The moment was anomalous, for at a time of Russification for most Soviet minority nations, Abkhaz and Ossetians were being forcefully acculturated to Georgian, not Russian. Intermarriage was common, and while many Ossetians living in Soviet Georgia became culturally Georgian as a means of social mobility, most also had strong ties with ethnic kin in North Ossetia. Many had family members who moved to Prigorodnyy on the outskirts of Vladikavkaz, the capital of North Ossetia, as part of an organized hiring of workers in 1944–1947 and subsequently in the 1950s. Seasonal migration across the Caucasus for work left the region embedded within broader networks. Plans to create a reliable all-weather communications route between North and South Ossetia were opposed by Georgia's leadership. In the late sixties, however, a highly decorated Ossetian Soviet Army General, General Issa Pliyev (commander of the North Caucasus Military District during 1958–1968), provided the lobbying power necessary to establish such a route.[12] In 1975 construction of the Transcaucasian (TransKam) highway connecting Alagir in North Ossetia to Java in South Ossetia, thus directly linking Vladikavkaz to Tskhinval(i), began. Construction involved the creation of high mountain roads and a tunnel almost four kilometers long. The TransKam highway and Roki tunnel, which was finally opened in 1984, made it much easier for the region's residents to travel and trade with North Ossetia. It also brought Ossetians from north and south of the Caucasus closer together, rekindling aspirations to create a unified Alania within the Soviet Union. In the Soviet census of 1989, the South Ossetian autonomous region had a population of just under 100,000, including 65,232 declared as Ossetians and 28,544 as ethnic Georgians, with Russians, Armenians, and Jews making up the rest. In Soviet Georgia as a whole, there were over 164,000 Ossetians, meaning the majority of Ossetians lived beyond South Ossetia.

The chaotic period between 1988 and 1992, when Soviet authority collapsed and Georgian nationalism became a replacement among most ethnic Georgians, is represented as a second genocidal campaign by the Ossetian nationalist narrative. The culture of Georgian nationalism during this period, most especially after the death of twenty protesters in April 1989, was radical, rejectionist, and romantic. "The peculiarity of Georgia" at this time, the eminent Georgian scholar Ghia Nodia noted, "consisted in the fact that it was a single republic where the 'irreconcilable' mentality dominated the opposition political agenda."[13] Moderate liberal nationalists were eclipsed by radicals who viewed the entire Soviet period after the defeat of the first Georgian republic in 1921 as an "occupation." Longstanding Soviet administrative structures, such as the SOAO, were considered Bolshevik encumbrances upon a spiritual Georgia, a patrimonial geobody of Georgian Orthodox monasteries and churches that was tragically divided and dismembered.[14] Ossetians were conceptualized as "ungrateful guests" on historic Georgian soil, outsiders whose real home was across the Caucasus in North Ossetia.[15] Few articulated these sentiments more forcefully than Zviad Gamsakhurdia, a KGB-persecuted dissident who became the first democratically elected president of Georgia on May 26, 1991.[16] Gamsakhurdia owed his rise, in no small part, to his campaign against Ossetians and the SOAO, leading a protest caravan of between 12,000 and 15,000 Georgian nationalists against the entity on November 23, 1989, which triggered a blockade of Tskhinval(i)and the first ethnicized violence in the area.[17] Among the first group of Ossetians blocking the convoy from entering Tskhinval(i) was the head of the local Komsomol, Eduard Kokoity, a Soviet champion wrestler born in Tskhinval(i). After heated negotiations with Georgian Komsomol members in the convoy, they were able to prevent it from entering the town.

Just as significant as the uncompromising nature of Georgian nationalist rhetoric at this time was the proliferation of nationalist militias, so-called informals (*neformaly*), who took advantage of the collapse of authority and police power to plunder the property of vulnerable groups. Nationalist discourse about ancient ethnogenesis and territorial ownership provided a rationalization for this criminal activity, unleashing fears and fantasies on both sides. Ossetians across Georgia were targeted, and many fled to South and North Ossetia for safety. Armed groups formed around workplaces, social clubs, neighborhoods, and charismatic leaders.[18] Ossetian militia groups, led by various local figures—Soviet army veterans of Afghanistan, black marketeers, Komsomol leaders, sports club members, and others—fought back. Since 1988, the Ossetian popular front organization *Adamon Nykhaz* (Popular Shrine) pushed an agenda of greater autonomy and later outright separation from Georgia.[19] Moves by it and local Georgian nationalists fed off each other, creating a tense environment pervaded by ethnopolitical fears.[20] To marginalize the *Adamon Nykhaz*, the Georgian Soviet passed a law in August 1990 banning regional parties from participating in Georgia's first multiparty elections. The SOAO Soviet responded on September 20, declaring South Ossetia a separate sovereign republic within the USSR. This, in turn, provoked the new nationalist-dominated Georgian parliament on December 11, 1990, to abolish South Ossetia as an autonomous oblast completely, and to declare a state of emergency in the Tskhinval(i) and Dzau/Java districts the following day. To Tbilisi, South Ossetia no longer existed.

USSR president Gorbachev intervened at this point, declaring both Tskhinval(i)'s and Tbilisi's declarations unconstitutional. But Gamsakhurdia, newly empowered as the chair of the Supreme Council of Georgia, denounced Gorbachev's decree as "interference in Georgia's internal affairs and encroachment on its territorial integrity."[21] Tskhinval(i) descended into violence after the murder of some Georgian militia members, prompting several thousand Georgian police, new National Guard troops, and "informals" (*neformaly*) to occupy the town on the night of January 6–7, 1991. Citizens were killed, properties looted, and the town's Ossetian theater and graveyards despoiled. In rural areas, there was also considerable destruction of dwellings and plunder of private property. The violence of January 1991 pushed South Ossetia and the surrounding regions of Georgia into a negative spiral of ethnic polarization, violence, and displacement. Tskhinval(i) was blockaded again, with power, communications, and food lines cut. Soviet Central Television characterized the situation in February 1991 as "worse than Leningrad in 1942. The entire city is without heating and electricity . . . there is no food."[22] Fighting within the town left it divided into largely Ossetian and Georgian zones, with the Georgian forces eventually expelled from the section they held. Ossetians living in the predominantly Georgian communities to the north of the town were themselves driven out and forced into largely monoethnic settlements, with their houses plundered and robbed. Ossetians living beyond the SOAO in other parts of Georgia were also forced from their properties, with thousands fleeing to North Ossetia. The result was the "unmixing" of the population in those settlements across South Ossetia, where ethnic Georgians and Ossetians lived together, and beyond. Because relations were generally good between Georgians and Ossetians in the Soviet period, there were many mixed marriages and families caught in the middle.[23] People were forced to choose their primary ethnic allegiance, even though many had Georgian and Ossetian relatives. Militia groups on both sides acted with impunity.[24] Eduard Kokoity led the youth wing of the South Ossetian defense efforts under the direction of Gregory Kochiev (known as "Grey's Group") and later helped create a body to coordinate command of the local South Ossetian self-defense units. In the wake of the Georgian attack, a rump parliament in Tskhinval(i) moved toward full separation from Georgia. On January 19, 1992, one year after the ransacking of Tskhinval(i), residents in the areas controlled by Ossetian militias held a referendum to approve the proclamation of an independent Republic of South Ossetia. It was not until weeks after a peace agreement signed in Sochi by Shevardnadze (who had replaced Gamsakhurdia after a violent coup in January 1992) and Russian president Boris Yeltsin on June 24, 1992, that an uneasy peace was reestablished in the region through the mechanism of a Joint Control Commission.

The violent clashes between 1989 and 1992 in South Ossetia were disastrous for the region. Population numbers declined while economic activity collapsed. The Soviet-era infrastructure of roads and water, gas, and electric lines that connected the towns and villages in the region to each other were now broken. Violence had produced a South Ossetia polarized into largely monoethnic villages in close yet complex proximity to each other. It was particularly significant that the series of mostly monoethnic Georgian settlements north of Tskhinval(i) in the Greater Liakhvi river valley—south-to-north for eight kilometers, the villages of Tamarasheni, Kheiti, Kvemo-Achabeti, Zemo-Achabeti, Kurta, Kekhvi, and Kemerti—enabled Georgian forces to control movement along the TransKam and flows of water, electricity, and gas into the South Ossetian capital (see Figure 4.1). This left the town permanently vulnerable and forced Ossetians in Tskhinval(i) to travel on an unpaved mountain pass called the "Zar road" to get around any blockade, then back onto the TransKam farther north where ethnic Ossetian settlements predominated. Other parts of South Ossetia had similar enclaves. Residents of Georgian villages had to improvise their own mountain pass roads to connect themselves to the rest of Georgia. A few ethnically mixed villages remained, while ethnically mixed families were found all over.

Wounds from the violence of the 1989–1992 period run deep in South Ossetia. A South Ossetian memorial book claims over one thousand people were killed, three-and-a-half thousand wounded, over twenty thousand South Ossetians forced to flee, and 117 out of 365 Ossetian villages burnt and destroyed during this period.[25] Many incidents from this time are recalled, but one is particularly salient in public memory. On May 20, 1992, a convoy of Ossetians fleeing the region on the Zar road was shelled by Georgian forces from the Liakhvi valley below them. Thirty-three people were killed, nineteen of them women and children, with thirty people left severely wounded. The incident was widely publicized in North Ossetia and provoked a renewed flow of arms and fighters to the Ossetian forces in the region.[26] A tall black statue of a wailing woman, looking skyward next to a Christian cross, memorializes the incident on the now-paved road (see Figure 4.2).

The Dagomys Agreements (also called the Sochi Agreement), and its subsequent elaboration, institutionalized a post-Soviet South Ossetia

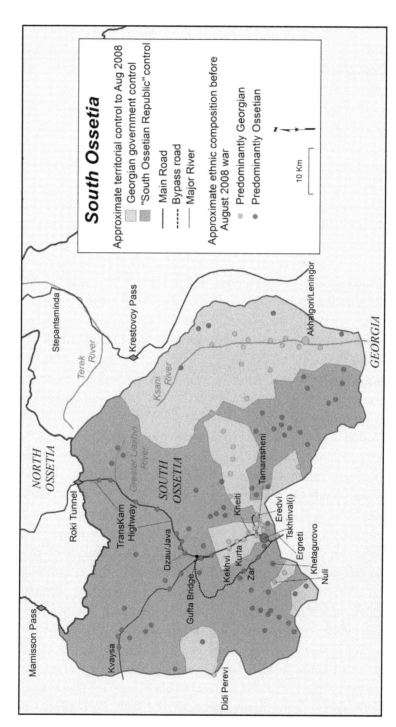

FIGURE 4.1 Map of South Ossetia.

FIGURE 4.2 Memorial to Ossetian refugees killed by Georgian irregulars on the Zar Road, May 29, 1992. Author photo.

that was an in-between space between Russia and Georgia proper, with important dependencies and ties to both. Firstly it established a Joint Control Commission (JCC) consisting of Georgian, Russian, and North and South Ossetian representatives plus those from the Organization for Security and Cooperation in Europe (OSCE) to supervise observance of the agreement and craft conflict-resolution measures as needed. Second, it established Joint Peacekeeping Forces (JPKF), made up of Georgian, Russian, and Ossetian units, which operated under a mandate from the JCC and the accumulation of agreements it developed. Protocols following the June 1992 agreement defined a "zone of conflict"—a circle of fifteen-kilometer radius measured from the center of Tskhinval(i). It also established a "security corridor" that designated an area inside the former autonomous oblast and beyond it in Georgia proper, where peacekeepers could establish checkpoints.

The security corridor traversed the administrative border and was not divided into separate distinct spaces or areas of control. Russian peacekeeping forces, therefore, could legally establish checkpoints within uncontested Georgian territory, provided they were within the agreed security corridor (although, in practice, they tended not to cross into Georgia).[27] With a geographically circumscribed mandate, small numbers, local composition, and a limited international dimension, these forces were never equivalent to United Nations peacekeepers in other conflict zones, like the Implementation Force (IFOR) in Bosnia-Herzegovina deployed in the wake of the Dayton Peace Accords. The Dagomys Agreements acknowledge that Russia was the de facto keeper of the peace in the area. From Georgia's perspective, the agreement effectively legitimated Russian proxies and the de facto Russian occupation of Georgian territory.

The Dagomys Agreements were limited and imperfect instruments for handling the South Ossetian conflict. Nevertheless they created a forum for dialogue, and in the subsequent decade progress was made on certain issues. All sides recognized the importance of restoring freedom of movement and providing regular and reliable electricity and water flows. After 1996, peacekeeping checkpoints were scaled back and trade between the different parts of South Ossetia, and with Georgia proper, took off. As a more open space located on an important trans-Caucasus highway, South Ossetia became an attractive land entrepôt for the movement of untaxed commodities. Moving commonplace goods, like oil, flour, and cigarettes, as well as contraband products, like drugs, guns, and counterfeit goods, through the Roki tunnel and down the TransKam became a highly lucrative business, with the Russian authorities guarding the entrance receiving payments to look the other way. Commercial flows created intraethnic interdependencies, but these rested on the leveraging of South Ossetia's extralegal status for capital accumulation. In a location on a hill outside Tskhinval(i) near the village of Ergneti, there developed a large outdoor market where traders small and large from all over the Caucasus met to make deals, purchase goods, and exchange commodities. The market provided a livelihood for many South Ossetians (ethnic Ossetians, Georgians, and others) who had few other options as well as for networks of traders in Shida Kartli province and beyond across the rest of Georgia.

The period from 1996 to 2001 saw the return of some optimism in South Ossetia, though life was still difficult. Ethnic tension subsided as all sides sought ways to make money. The Shevardnadze government adopted a relatively tolerant attitude, and the polarization wrought by violence began to unwind. The Russian Federation provided salaries and pensions to South Ossetians under the control of the local de facto government—grandiosely the Republic of South Ossetia but in practice nothing more than a small municipal government—while gas and electricity lines from Georgia provided intermittent heat and power (a condition it shared with the rest of Georgia). A sizable number of Ossetians, especially those regularly moving back and forth between Vladikavkaz and Tskhinval(i), acquired Russian passports under the terms of the prevailing Russian citizenship law.[28] A drinkable water supply pipeline ran from Edisi in the north to Tskhinval(i), rehabilitated by European Union peace-building funds. Women traders from the Georgian enclave north of Tskhinval(i) visited the town to sell the produce they had grown in their gardens and processed themselves. In South Ossetian Republic elections, Lyudvig Chibirov, an Ossetian professor, came to power first as chairman of the parliament and then in 1996 as the first "president of South Ossetia." Under the auspices of the Russian government, Chibirov held a number of rounds of talks with Shevardnadze that yielded some agreements and came close to the terms of a comprehensive settlement.[29] Stalemate and local discontent, however, hurt his popularity in South Ossetia.

A significant turning point in relations between Georgia, its breakaway regions, and Russia was the eruption of fighting in October 2001 in the Upper Kodori Gorge, a sparsely populated mountain valley within Abkhazia not fully controlled by the de facto authorities in Sukhum(i) (see Figure 4.1). Ruslan Gelayev, an infamous Chechen warlord who had escaped capture by Russian forces in March 2000 by fleeing across the Caucasus Mountains to Georgia, led a group of his rebels and joined local Georgian fighters known as the Forest Brothers in an attack on villages in the upper valley. Scores were killed, including nine in a UN helicopter shot down by the rebels.[30] The incident underscored the entanglement of Russia's war against separatists in the mountains of Chechnya and Dagestan with mountain regions (Pankisi, South Ossetia, and Kodori) on the southern slopes of the Caucasus range in Georgia. The attack occurred just as Russia was aiding the United States in removing the Taliban from power in Afghanistan and while Shevardnadze was visiting the United States. Conspiracies abounded as to how the Chechen fighters were able to reach the region and what their alliance with Georgian fighters meant.[31] To Russian security officials, this was an alliance of terrorists tacitly supported by elements in the Georgian state, who were only too happy to aid Russia's enemies. Their distrust of Shevardnadze, perceived as undermining Kremlin power from Eastern Europe to the Caucasus, ran deep.

In late 2001, Eduard Kokoity returned from Moscow to his hometown and ran for president, defeating Chibirov in the first round and another candidate in the runoff.[32] Tensions ran high because the Tedeyev group, a prominent Ossetian family jostling to control transit on the TransKam, had allegedly funded Kokoity's campaign.[33] Jambulat Tedeyev was a world wrestling champion and Russia's wrestling coach. Thereafter, the group was able to take charge of the republic's border and customs service. Kokoity's ascent was also enabled by his connections to networks within the Russia security services and, allegedly, organized crime. He was a member of the Eurasia Movement, an organization established by Alexander Dugin in May 2001 to promote his revisionist imperial geopolitical agenda. Kokoity claimed that South Ossetia was historically part of the Russian Empire and had never left it (Figure 4.3). Its historic and geopolitical destiny, he claimed, was unity with Russia.

The official Russian position remained that both South Ossetia and Abkhazia were unresolved conflicts within the territory of the state of Georgia. A new Russian citizenship law in 2002, however, complicated matters by simplifying procedures for residents in the de facto entities to acquire Russian nationality. The administrative change led to a significant increase in the number of Russian passport-holders in both de facto states.[34] Critics dubbed the policy "passportization" and viewed it as a strategy by the Kremlin to reimperialize the near abroad.[35] Russian nationality, however, was chosen by and not imposed upon residents of the de facto states, whose local "citizenship" granted them no international right to travel. The policy change proved significant because it gave bureaucratic state form to imagined communities of identity and belonging—*russkii* (ethnic Russians), *rossiyanye* ("Russians"), *sootechestvenniki*

FIGURE 4.3 Eduard Kokoity, Tskhinval(i), dedicating a statue to Pushkin, March 2010. Author photo.

("compatriots"), and those within *Russki Mir* (the "Russian world")—beyond Russia's borders. Geopolitics was more transparently expressing itself as biopolitics (the management of populations). A Russian leader could now claim a right to intervene in the near abroad to "protect Russian citizens."

Getting Georgia Back

As Kokoity and his entourage consolidated power in South Ossetia, political tensions with Georgia continued to rise. First, the appearance of U.S. troops in Georgia in March 2002 made Kokoity and others in South Ossetia nervous, fearing an alliance between the United States and Georgia would embolden Tbilisi to move against Tskhinval(i). There was a direct threat as about 1,500 Ossetians lived in the Pankisi Gorge, a hundred kilometers to the east, and were likely to be impacted by military action there. The South Ossetian assembly appealed to the Russian State Duma to send more peacekeepers since "Russia is the only guarantor of peace and tranquility in the Transcaucasus."[36]

Second, Kokoity moved to take over the lucrative TransKam freight trade himself in 2003, ousting the Tedeyev group. He also formed his own political party, *Edinstvo* (Unity), modeled after Putin's *Yedinaya Rossiya* (United Russia), and with the slogan "Unity—Our Road to Russia," all in a bid to tighten his grip on South Ossetia's parliament.[37] Third, the Rose Revolution in Georgia brought Mikheil Saakashvili to power in Tbilisi, by far the biggest threat the Tskhinval(i) power structure had heretofore faced.

The Georgian nationalism Saakashvili articulated was Janus-faced.[38] On the one hand, he was political heir to the irreconcilable mentality of Zviad Gamsakhurdia and the idea of a spiritual Georgia. Like Gamsakhurdia, Saakashvili had a messianic vision of his role in Georgian political life. Also, he brought a radical ethos to the challenges of governance, one that disdained compromise, favored bold authoritarian moves, and expressed itself in populist "heroic-aesthetic gestures."[39] An emotive speaker and charismatic presence, Saakashvili breathed youthful energy into classic tropes of Georgian exceptionalism—its ancient origins, historic glories, sacred mission, and victimization at the hands of neighboring empires. Spiritual nationalists viewed Georgia as an incomplete achievement, an Orthodox Christian outpost whose historic patrimony was denied to it by Soviet imperialism and thereafter by puppet regimes disguising Russian occupation.[40] Spiritual nationalists sought the "unification of Georgian lands" and the return of those forcefully displaced from their homes in Abkhazia and South Ossetia. Saakashvili personalized this desire by citing Abkhazia as a place where he had family roots: "the fact that I often dream Abkhazia means that this region is a constituent part of the soul of the Georgian people."[41] Spiritual nationalists spoke of Georgian lands as dismembered, as under occupation, and as annexed.[42] Saakashvili used this language both before and after he came to power.[43] His first day as president saw him pardon a group of imprisoned Gamsakhurdia supporters.[44] In March 2007, his government organized the reburial (the third) of the remains of Zviad Gamsakhurdia with full state honors. "Unity," Saakashvili told the ceremony, "is most important for Georgia, for our identity. Our force lies in our unity."[45]

On the other hand, Saakashvili was a modernizing state builder who rhetorically articulated Western liberal values of freedom, tolerance, and multicultural identity. This was a nationalism of civil inclusion and respect for the identities of the many nonethnic Georgians who live within Georgia. The language Saakashvili used to articulate such civic nationalism was a reworking of the rhetoric of U.S. civil nationalism, a reflection of both his own American experience and the influence of U.S. political advisers like Daniel Kunin.[46] Articulated in Saakashvili's fluent English to Western audiences, it presented a positive and encouraging image of Saakashvili's beliefs. Like many civic nationalisms, however, this liberal rhetoric had implicit limits. Georgia was multiconfessional but Orthodox Christianity had a privileged place (as the new flag made visible). Georgia was home to different nations, but all had to learn Georgian and demonstrate loyalty to an ethnic Georgian nation-state.[47] The two-sided nature of Saakashvili's rhetoric was evident from the outset. Saakashvili's spiritual oath at Gelati Monastery gave voice to heroic ethnonationalism, while his official state inauguration before an international audience articulated liberal themes. While Saakashvili had a high tolerance for contradictions, the audiences he addressed did not. Saakashvili made it even more difficult for himself by raising their expectations, telling, for example, those gathered at Gelati that the next presidential inauguration would be in Sukhumi. Saakashvili made a surprise visit to Tskhinval(i) on January 3, 2004, just before his election as president, greeting people on the street before driving north to Tamarasheni where he promised polling stations for the next Georgian presidential election.[48]

The contradictions in Saakashvili's rhetoric were obscured at the outset by a common language about priorities and principles. Georgia was practically a failed state. Radical state-building reforms were an absolute necessity, and corruption the agreed enemy. For Saakashvili "each corrupted official is a traitor to state interests."[49] Fighting corruption required establishing a rule-of-law state and this, according to the young reformers, required strong executive rule. This was a Chicago-school form of liberalism that accepted authoritarianism as an indispensable stage in a transition from Communism. Chile under Pinochet was the implicit model. Despite earlier commitments to balanced power, Zhvania went along with constitutional changes that strengthened Saakashvili's executive powers at the expense of parliament.

Fighting corruption also required confronting the breakaway territories, regions that had become, in liberal international characterization, "geopolitical black holes." Here the war against corruption made common cause with another shared rhetorical commonplace: territorial integrity. No democracy, Saakashvili declared, "can allow black holes to exist on its territory."[50] Saakashvili characterized his goals colloquially as "getting Georgia back." "It is the responsibility of every citizen of this country to get Georgia back. It is the essence of life and a goal for each of us. Getting Georgia back means its unification."[51] To modernizing nationalists, and Western liberal powers, territorial integrity was an international legal right, a prerequisite for a modern sovereign state. All states need to be able to control their own borders to establish tax regimes and governance. While self-evident and rational to outside governance experts, international support for "territorial integrity" as a principle in the Georgian context translated into tacit support for revanchist geopolitical aims. To the de facto regimes in Sukhum(i) and Tskhinval(i), Saakashvili's rhetoric about territorial integrity was an existential threat.

Saakashvili's initial approach to the challenge of Georgia's breakaway territories reflected an incoherence emergent in his contradictory nationalism. One the one hand, he saw the problem as a challenge to rebuild the coercive capacity of the Georgian state so it was strong enough to take back the breakaway regions. This was the heroic stance, the hard power option. On the other hand, he recognized that it was important for Georgia to try to appeal to ordinary people in the de facto states. This was liberal idealism, the soft power path. The two approaches sat uneasily together and, over time, his attraction for the use of the coercive instruments of the state undermined his efforts at building Georgian soft power. At the outset, however, his discourse was full of wishful thinking and platitudes. There was no coherent and detailed strategy. For example, in a meeting with business leaders at the end of 2003, he expressed reluctance to use force to restore Georgia's territorial integrity, noting that the only alternative is "economic leverage." The hard power dimension of this is softened by his expression of it: "I am sure, that if the Abkhaz and South Ossetian sides will see that the economy is growing in Georgia they will come to us. We should attract them with economic opportunities."[52]

Saakashvili, however, was impatient, and readily admitted so.[53] He wanted quick results and calculated that the fastest way to achieve them was by building the coercive capacity of the state.[54] Boldness had served him well in the past.[55] Thus, from the outset Saakashvili's government set about rebuilding the Georgian military. Defense spending rose dramatically, from $93 million in 2003 to $1.2 billion in 2007, a more than tenfold increase in four years. Military spending in 2007 accounted for 9.2 percent of Georgia's rising gross domestic product (GDP), the second-highest in the world (after Oman) and higher than Israel's figure of 6.8 percent that year.[56] Service within U.S.-led "coalitions of the willing" in Iraq and Afghanistan was a direct way for Georgia to have its military expertly trained and the drive for NATO membership an immediate way for it to tap NATO expertise to construct new military bases facing both South Ossetia (in Gori) and Abkhazia (in Senaki). Military service was reorganized and training camps for reservists created. "[E]very Georgian citizen should be able to take a weapon in their hand and if necessary offer resistance to aggression against this country."[57] Saakashvili spent days living with troops and reservists.[58] Separate patriotic camps for kids and teenagers from 2005 taught an authoritarian form of patriotism, including the principle that "the leader is always right."[59] Military parades returned as moments of pride and display. Georgia purchased Soviet-era tanks and military vehicles from Ukraine and an array of newer weapon systems from Israel. It also employed private U.S. mercenary companies, including Military Professional Resources Incorporated (MPRI), to game-plan military offensive operations against its separatist regions. MPRI had previously worked on planning Operation Storm, a successful three-day military offensive in August 2005 by the Croatian military against Republika Srpska Krajina, a separatist statelet supported by the Milošević regime in Serbia in 1991. Croatian president Stjepan Mesić visited Georgia in May 2006 and he and Saakashvili discussed what Georgia could learn from Croatia's experience with "sponsored separatism."

Saakashvili did try to reach out to ordinary people in the breakaway regions. Concurrent with the launch of antismuggling operations against South Ossetia, his government initiated a program to provide free fertilizer and humanitarian aid to residents of the breakaway region.[60] He used his first Independence Day speech to address

Ossetians and Abkhaz in their own language.⁶¹ Alania TV, a television station aimed specifically at Ossetians, was established and began broadcasting. Saakashvili also criticized past actions by Georgia, such as the abolition of South Ossetian autonomy in December 1990, and held out the prospect of a new "wide autonomy" within Georgia.⁶² The strategy behind these moves was to drive a wedge between elites in the breakaway regions, presumed to be hated kleptocrats, and ordinary citizens. These assumptions, however, were naïve for they failed to appreciate the distinctive identity and aspirations of most Ossetians in these regions, not to mention their enduring memories of Georgian state violence. In the end Saakashvili's government had to confront the reality that negotiations with de facto state elites were required, no matter how much they despised them. His government did offer proposals on special constitutional status for Abkhazia and South Ossetia within Georgia, but these came amid hard power flexing and proved unattractive to the de facto regimes (whose interest was locking in Russia as their ultimate security backstop).

Saakashvili's government thus faced entrenched elites, populations alienated by past Georgian state violence against them, and a Russian state that was soon openly hostile to his geopolitical aspirations. Whether there ever was a possibility of him peacefully achieving territorial integrity for Georgia is questionable. What is apparent is that his leadership style and his policy choices—a function of the fact that power was concentrated in his hands—made that possibility all the more remote. Rather than playing a long, patient game with his adversaries, he displayed a fatal attraction for aggressive military actions. Some of his moves appeared successful but sowed the seeds for future defeat; others backfired immediately. Borders hardened in the de facto states while Georgia began a risky game of geopolitical escalation with Russia that proved disastrous.

Upping the Ante

The period prior to the August 2008 war is a chronicle of untrammeled ambition and aggressive actions leading to escalating competition, polarized populations, and simmering levels of violence that finally boiled over with Georgia's military offensive on the night of August 7. It is a complex story but can be broken down into four discrete phases, each featuring an intensification of the prior phase of relations.

There is little doubt that the Rose Revolution was perceived skeptically by many in the Kremlin, and with genuine fear by the leadership in Sukhum(i), Tskhinval(i), and Batumi (capital of Adjara). Despite numerous trips to Moscow by these local leaders, Putin adopted a pragmatic position as Saakashvili's government confronted the corrupt fiefdom created by Aslan Abashidze during his thirteen-year rule in Adjara. A Black Sea region with a Russian military base and considerable strategic value as an energy distribution port and key border crossing with Turkey, Adjara was an ideal place for Russia to make difficulties for Saakashvili. Instead, Foreign Minister Ivanov and Putin personally brokered Abashidze's exile to Moscow, handing Saakashvili a great victory.[63] Saakashvili characterized it as a "second bloodless revolution" and portentously as "the beginning of the territorial integrity of Georgia."[64] Adjara, however, was a very different place than Abkhazia and South Ossetia. Its population was largely ethnic Georgian and had never fought the central government in a war that produced significant casualties and forced displacement. Instead Abashidze had taken advantage of Tbilisi's weakness to carve out a highly personalized form of autonomy, one that allowed his family to accumulate significant wealth and to act as a power broker within Georgian political life. Saakashvili's government, with Russia's help, had ended that.

Few knew it at the time, but the fall of Adjara proved to be a pivotal turning point in Georgian-Russian relations. High on its success, Saakashvili's government made a series of aggressive moves against the de factos that brought Georgia directly into military conflict with Russia. He should have known better, for Putin allegedly warned him in a phone conservation after Abashidze's ouster: "OK, Mikheil Nikolozovich, we helped you on this one, but remember very well, there will be no more free gifts offered to you, on South Ossetia and Abkhazia."[65] Saakashvili, however, was a man on a mission. South Ossetia was targeted first. Already in the spring of 2004 Georgian MIA Special Forces had begun to disrupt the TransKam contraband trade through seizures of smuggled goods, maintaining checkpoints on Georgian-controlled parts of the highway, and blowing up unauthorized roads. The practices were demonstrative acts of sovereignty and brought large numbers of armed

Georgian forces into parts of South Ossetia they had not ventured into since 1992.⁶⁶ Established JPKF rules were violated, and the JCC negotiation mechanism broke down. The actions sparked fear in Tskhinval(i) and rising tension in Russian-Georgian relations. The Ergneti market was shut down, and this directly impacted the livelihood of most residents of South Ossetia who depended upon the region's "competitive advantage" as an unregulated trade zone. Humanitarian outreach efforts by Georgian nongovernmental organizations, and a visit by Saakashvili's wife, did little to ease suspicion of Tbilisi's moves.⁶⁷ In fact Georgia's actions played into Kokoity's hands, as he was able to frame Saakashvili's actions as an attack on South Ossetia, not on criminality. In local elections that May his *Edinstvo* (Unity) party won two-thirds of the seats in the South Ossetian assembly. On June 5 this new assembly appealed to the Russian State Duma to incorporate South Ossetia into Russia. Establishing a pattern that would repeat, Russia decried the Georgian troop buildup and positioned itself as the protector of the vulnerable Ossetians, while Georgia denounced what it claimed were illegal military convoys coming from Russia. One signal of Russia's attitude toward South Ossetia that was different from Adjara was the appointment of two Russian FSB officers, one an ethnic Ossetian, to top posts in the South Ossetian equivalent, which had kept the Soviet name "the KGB."⁶⁸ Fear brought an end not only to interethnic trade in Ergneti but also to trips by ethnic Georgians living north of Tskhinval(i) to the town to sell market produce. Roadblocks and checkpoints hardened. In a sign of deepening spatial separation, Tbilisi worked to consolidate a bypass road from the Georgian enclave villages into Georgia proper. No longer able to rely on the TransKam immediately north of the town, South Ossetians began to do the same with the Zar road, the lifeline they had used in 1992.

Ossetian militias, which drew members from both North and South Ossetia, prepared for war. Tensions eventually spilled over into kidnapping and sporadic firefights. Gun battles and mortar exchanges broke out between Tskhinval(i) and Tamarasheni in late July and early August 2004. Saakashvili made a hasty trip to the United States, where U.S. secretary of state Colin Powell privately upbraided him for his actions in South Ossetia. Violent incidents there nevertheless continued to metastasize. On the night of August

18–19, Georgian Internal Affairs troops seized three strategic heights to secure a bypass road with the loss of seventeen Georgian and five Ossetian lives. Georgian forces later withdrew, handing control to the JPKF. Saakashvili portrayed the retreat as a "last chance for peace."[69] To Ossetians burying their dead in Tskhinval(i) and beyond, the words "peace" and "Saakashvili" were now incompatible.

In a report soon after the violence, the International Crisis Group wrote that the "remilitarization of the zone of conflict reversed a decade of progress" in South Ossetia.[70] In hastily pursuing territorial integrity by forcefully blocking South Ossetia's economic lifeline, Saakashvili's government made the goal all the more difficult if not impossible to achieve. Furthermore, he had moved Russia from pragmatic collaborator to antagonistic power. Putin's already evident exasperation with Saakashvili deepened.[71] Russian-Georgian relations thereafter entered a new phase of escalating competition. Saakashvili sought to make the best of what was a defeat for his quick reunification strategy. But he did not alter his objective of territorial unity or revise his conceptualization of the problems involved. In December 2004, he appointed his hardline state security minister Ivane (Vano) Merabishvili, who led the antismuggling campaign, as his minister of Internal Affairs. Merabishvili would go on to be the longest-serving of Saakashvili's ministers and one of his closest confidantes as other prominent Rose Revolution Georgians broke with him or died. Together with Merabishvili, Saakashvili would plot a way for Georgia to win militarily in South Ossetia, drawing inspiration from Finland (the 1939–1940 Winter War) to Israel (the Six-Day War) to Croatia's Operation Storm.

The period from August 2004 to July 2006 saw considerable deterioration in Georgian-Russian relations. The Beslan terrorist siege in September 2004 created turmoil in North Ossetia, an island of relative stability and loyalty to the Kremlin in the North Caucasus. Ignored by many analysts was that most of the terrorists were ethnic Ingush not Chechens, and that the horrific attack reopened wounds from a brief war between ethnic Ingush and Ossetians in 1992 over a disputed area on the outskirts of Vladikavkaz called Prigorodnyy.[72] Fighting in South Ossetia contributed to that war and to instability in the North Caucasus more generally. It was thus all the more important for the Kremlin that it have control in Tskhinval(i) so North

Ossetia and, by extension, the rest of the North Caucasus remain as pacified and stable as possible.[73] Russia sealed its border with Georgia immediately after Beslan and accused the country of harboring terrorists.[74] Saakashvili and Putin had a heated exchange at a CIS meeting in Astana on September 16 over Russia's decision to help refurbish its railway connections with Sukhum(i).[75] The move had already been agreed, and Saakashvili's objection underscored his determination to pursue a policy of isolation toward the breakaway territories. At the United Nations later that month, Saakashvili offered a three-stage settlement plan for the territories involving confidence building, demilitarization, and the "fullest and broadest form of autonomy." He did so, however, only after describing the breakaway territories as geopolitical black holes that were safe havens for criminality and terrorism. He appealed to Russia to see them as a common threat "[b]ecause all forms of violent separatism—be they in Tskhinvali, Grozny or Sokhumi, represent destabilizing factors for Russia and Georgia alike."[76] Moscow, however, viewed Saakashvili's government as the destabilizing force in the South Caucasus. In an effort to dampen tensions in South Ossetia, Zhvania and Kokoity met in Sochi in November and signed an agreement on demilitarization. Less than a week later, however, Georgia opened a training camp for reservists in the village of Dzevera, within Georgia proper but also within the JCC security corridor around South Ossetia (it was later closed). Violent incidents such as the shooting of Ossetian soldiers and landmine explosions in Kurta were the reality on the ground. Kokoity, who met with Russian Rodina Party Duma deputies in Moscow in December, predicted a new "Georgian incursion" into South Ossetia and reiterated that South Ossetia seeks to join Russia.[77]

The death of Georgian prime minister Zurab Zhvania from gas poisoning in early February was another fatal turning point in Georgia's relations with Russia. While sharing many of Saakashvili's commitments and attitudes, Zhvania was a respected and skilled negotiator who was cool to the geopolitical amplification strategies of Saakashvili. With his restraining presence gone, Saakashvili came to dominate the political scene in Georgia. Georgia's fate became tethered to the self-aggrandizing quality of his geopolitical entrepreneurship. In this he had willing accomplices. Saakashvili viewed the Orange Revolution in Ukraine as inspired by the Rose Revolution, and built close personal ties with the new Ukrainian president Viktor Yushchenko. Together both sought to revitalize the anti-Russian grouping of Georgia, Ukraine, Azerbaijan, and Moldova (GUAM). In August 2005, they both signed the Borjomi Declaration, a document envisioning a community of states dedicated to "removing the remaining divisions in the Baltic–Black Sea region, human rights violations, and any type of confrontation, or frozen conflict."[78] On the second anniversary of the Rose Revolution, the presidents of Estonia, Georgia, Romania, and Ukraine gathered in Tbilisi for a forum on "Europe's New Wave of Liberation."[79] Later they joined with the leaders of Lithuania, Latvia, Moldova, Macedonia, and Slovenia in Kyiv to proclaim a new nine-state Community of Democratic Choice.[80] Their adversary was clear.

In May 2005, Saakashvili hosted U.S. president Bush, interpreting his visit as a superlative endorsement of Georgia's territorial integrity. It is "the final confirmation that Georgia is an independent country whose borders and territory are inviolable. The red line lies on the Caucasus Range and no one should cross it to this side. Everything that is temporarily on this side should go back."[81] Later he described the United States as supporting the "total integrity of Georgia within its internationally recognized borders."[82] Georgia's pursuit of NATO membership intensified, with Saakashvili describing this as making possible a peaceful settlement of its territorial conflicts because it made Georgia strong.[83] "[O]ur joining NATO is not aimed against anyone. It is a well-thought-out strategy and, I repeat, peaceful step towards making Georgia strong and settling the conflicts in Georgia by peaceful means." Neither Russia nor the de factos were likely to be reassured by this. As Saakashvili's government pushed its three-part peace plan to the international community, the Georgian parliament voted unanimously for the government "to take measures for the withdrawal of Russian peacekeeping forces from the South Ossetian and Abkhazian conflict zones if their performance does not improve within four and eight months."[84] Georgia had legitimate complaints about the peacekeeping system in South Ossetia; actively repudiating Russia's role, however, was tempting fate. Amid these developments there were some positives in Georgian-Russian relations. The Russian State Duma in March 2005 rejected a draft law sponsored by Rodina deputies that would have made it easier for breakaway territories in post-Soviet space to join the Russian Federation. In May, Russia signed an agreement of withdrawal from the Russian military bases of Batumi and Akhalkalaki. The concession, however, may have been a recognition of the need to consolidate Russian forces should Russia and Georgia drift toward war. In December 2005 Kokoity proposed a three-point plan that sounded like Georgia's plan.[85] JCC talks in Moscow, however, floundered as Saakashvili refused to meet with Kokoity.[86] Relations took a turn for the worse when two gas explosions in North Ossetia in mid-January plunged Georgia into darkness and cold. Saakashvili accused Russia of "heavy sabotage" against Georgia and of being an "unprincipled blackmailer."[87] On February 15, 2006, despite U.S. calls for caution, the Georgian parliament unanimously passed a resolution instructing the government to secure a replacement for the Russian-led peacekeeping operation in South Ossetia "with an effective international peacekeeping operation."[88] At the end of March 2006, Russia imposed an import ban against all Georgian wine for "sanitary violations."[89] It later banned Georgian mineral water.

The attack on the existing peacekeeping arrangements in the disputed regions was a second phase of Georgia upping the ante (with Russia retaliating economically). A new phase began in the summer of 2006. The precipitating event was the Georgian government's seizure of the Upper Kodori Gorge in Abkhazia, a vulnerable site of previous violent clashes that was largely under the control of a local ethnic Georgian Svan warlord, Emzar Kvitsiani (now a Georgian parliamentarian). Saakashvili's government maneuvered to displace him from the area.[90] With him gone, Georgia began to implement a policy of standing up parallel governments within the territories of the de facto states. This policy, with echoes of Stalin's proxy actor tactics, Reagan's sponsorship of the anti-Communist Contras in Nicaragua, and the Bush administration's sponsorship of the Iraqi National Congress before the Iraq war, gave localized political form to Georgia's aspiration for "regime change" in the de facto state-controlled areas. It proved to be a particularly divisive move. Violent events in the breakaway regions increased. The same day a new headquarters for Abkhazia's government in exile was inaugurated in the Kodori Gorge, Georgian-Russian relations plunged to new depths as Georgia arrested four Russian military officers and charged them with espionage. The Kremlin reacted furiously, suspending all transportation links with Georgia. It also began an ugly campaign of targeting and deporting Georgian migrants working in Russia. On October 13, 2006, the UN Security Council adopted a resolution on Abkhazia that, while reaffirming the commitment of member states to Georgia's territorial integrity and reauthorizing the United Nations Observer Mission in Georgia (UNOMIG) for six months, urged "the Georgian side to address seriously legitimate Abkhaz security concerns, to avoid steps which could be seen as threatening and to refrain from militant rhetoric and provocative actions, especially in Upper Kodori Valley."[91]

The UNSC rebuke made little difference to the Georgian policy of establishing new forces within claimed de facto state territory. Less than two weeks later a newly established front organization for Tbilisi, the Salvation Union of Ossetia, announced plans to hold an alternative presidential election and referendum in the Georgian-controlled villages of South Ossetia, parallel to those organized by the de facto authorities.[92] The subsequent elections on November 12, 2006, saw a rival theater of elections sponsored by respective patrons Russia and Georgia. In areas controlled by the Republic of South Ossetia, Kokoity was overwhelmingly reelected, whereas in areas controlled by Georgia, Dmitri Sanakoev, a former South Ossetian official now loyal to Tbilisi, was overwhelmingly elected to lead a so-called alternative government.[93] An inauguration ceremony for Sanakoev, attended by the Georgian minister responsible for South Ossetia, was held in the village of Kurta in the middle of the Georgian enclave in the Didi Liakhvi valley.[94] This village was designed as the "capital" of the rival South Ossetian entity.[95]

Thereafter, throughout 2007, the Georgian government poured money—at least $3.5 million—into what became informally known as the Sanakoev Project.[96] Sanakoev was being stood up as an alternative president of the Republic of South Ossetia, yet Tbilisi did not want to describe him as such. The workaround was to fudge the name of the region and Sanakoev's title. Using a law rushed through parliament, Saakashvili proclaimed a "provisional administrative entity" in May with its own budget, Special Forces, and ministers, naming Sanakoev as its head.[97] Sanakoev addressed the Georgian parliament that month, in Ossetian, and later the Parliamentary Assembly of the Council of Europe. Predictably, the Sanakoev Project was strongly condemned by Moscow.⁹⁸ Funds flowing from the Kremlin to the de facto states increased as the geopolitical game became more intense. Violent incidents, including shootouts and landmine explosions, continued throughout the year.⁹⁹ Tskhinval(i) residents suffered water shortages that were blamed on the Georgian enclave to its north sabotaging the pipeline they both depended upon.¹⁰⁰ A war by other means was also unfolding in the skies as Georgia deployed its Israeli drones and Russian aircraft violated Georgian airspace.

The competitive rivalry between Tbilisi and Moscow entered its final phase in February 2008 with the choreographed unilateral declaration of independence by the de facto authorities of Kosovo, the breakaway province of Serbia. The United States and many other states immediately recognized Kosovo as an independent state, opening up a serious divide in the international community over rules governing recognition of secession. Vladimir Putin had long signaled that Russia rejected the U.S. effort to define the Kosovo case as sui generis. How could the United States fully support territorial integrity for Georgia but not for Serbia, self-determination for Kosovars but not for Abkhazians or Ossetians? In late January 2006, he told the Russian cabinet that any solution to Kosovo must be "universally applicable."¹⁰¹ He explained what he meant the next day: "If someone thinks that Kosovo can be granted full independence as a state, then why should the Abkhaz or the South-Ossetian peoples not also have the right to statehood?"¹⁰² Generally accepted universal principles, he argued, are required for resolving these problems. He took up the theme again in June 2007:

> There is nothing to suggest that the case of Kosovo is any different to that of South Ossetia, Abkhazia or Trans-Dniester. The Yugoslav communist empire collapsed in one case and the Soviet communist empire collapsed in the second. Both cases had their litany of war, victims, criminals and the victims of crimes. South Ossetia, Abkhazia and Trans-Dniester have been living essentially as independent states for 15 years now and have elected parliaments and presidents and adopted constitutions. There is no difference. We do not understand why we should support one principle in one part of Europe and follow other principles in other parts of Europe, denying peoples in the Caucasus, say, the right to self-determination.¹⁰³

After Kosovo's declaration and recognition by the United States and others, Putin met Saakashvili on the margins of a CIS conference in Moscow on February 17. According to the Georgian record, Putin explained to Saakashvili: "You know we have to answer the West on Kosovo. And we are very sorry but you are going to be part of that answer. Your geography is what it is."[104] He remarked, with some indignation, a few days later that Kosovo was a terrible precedent that "blew up the whole system of international relations and might provoke a whole chain of unpredictable consequences. Those who are doing this," he warned, "relying exclusively on force and having their satellites submit to their will, are not calculating the results of what they are doing. Ultimately this is a stick with two ends, and one day the other end of this stick will hit them on their heads."[105] Putin had already authorized a detailed contingency military campaign against Georgia to protect Abkhazia and South Ossetia in 2007.[106] Part of the plan required greater integration between the Russian military and Ossetian militias. Georgia's unilateral withdrawal from the JCC, in a bid to force a new process that would include Sanakoev, upped the ante further.

In February 2008, reacting to U.S. recognition of Kosovo, the Russian State Duma passed a motion calling on the Kremlin to recognize Abkhazia and South Ossetia as independent states. On March 6, 2008, Moscow announced it was formally withdrawing from a 1996 pact that imposed trade and other sanctions on Abkhazia. Already ignored in practice, the formal announcement was a signal that Russia was hardening its position. On March 13 the Russian Duma held hearings on possible recognition of Abkhazia and South Ossetia as independent states. A few days later Saakashvili visited the United States, where he was granted an Oval Office press conference with President Bush. Bush announced that his administration would back Georgia for a MAP at the forthcoming NATO Bucharest Summit.[107] In a public talk later that day (March 19) at the Atlantic Council, Saakashvili spoke confidently about Georgia and NATO. He indicated fears about possible Russian recognition of Abkhazia's and South Ossetia's independence in response to Georgia moving toward NATO were exaggerated. He also suggested that the Russian military was stretched thin in the North Caucasus and could not effectively enforce what he termed "Georgia's partition."[108] The following day he met with Secretaries Rice and Gates.

Rumors in Georgian circles, and not-so-oblique statements Saakashvili made at a meeting with think tank analysts, with Deputy Assistant Secretary Bryza at his side, suggested he was preparing an offensive in Abkhazia, likely in May 2008 when spring floodwaters on the Inguri River, separating Georgia proper from Abkhazia and the Gali region, would have subsided.[109] The expectation of war in Abkhazia was so prevalent that the International Crisis Group started work on a report on the prospect (it was released on June 5, 2008).[110] The mixed message sent by the Bucharest Declaration did little to check Saakashvili. On April 19 Putin authorized for the first time formal diplomatic relations between the Russia Federation, Abkhazia, and South Ossetia. The move underscored Russia's deepening commitment to preserving the de facto states. Saakashvili's government brought the issue to the UN Security Council. Georgia's use of UAVs over Abkhazia were perceived by the Abkhaz as reconnaissance in preparation for war.[111] On April 20 a Russian fighter aircraft shot down a Georgian unmanned aerial vehicle over Abkhazia. The following day Saakashvili called Putin to protest Russia's actions. The messages not to attempt any military offensive against Abkhazia should have been clear.[112]

On May 7, 2008, Dmitri Medvedev became the president of Russia. His ascension provided the possibility of a fresh start in Russian-Georgian relations. In Saakashvili's account, Putin was still calling the shots and whenever he called Moscow to speak to the new president he was put through to Putin. Russia sent in troops to upgrade railway infrastructure in Abkhazia, an action Georgia portrayed as menacing, and it was condemned in the West. In a call to Moscow in late May to protest the presence of the railway troops, Saakashvili's conversation with Putin was blunt and crude.[113] Medvedev and Saakashvili finally met on June 6 in Saint Petersburg. Medvedev offered to help Saakashvili broker compromise agreements that would eventually lead to the breakaway territories being reintegrated into Georgia.[114] However, Saakashvili would have to negotiate with the de facto state leadership. He would also have to pledge never to use force in his dealings with the disputed territories. The United States and Germany were supportive of a "no-use-of-force" pledge for all parties but Saakashvili's government, which had invested heavily in its military options, refused. As fire fights and military incursions increased, Condoleezza Rice flew to Tbilisi to pressure Saakashvili. He refused to make any public commitment not to use force. Though strongly discouraging Saakashvili from using force in private, Rice's public remarks did not help de-escalate matters; in fact, they may have encouraged Saakashvili. After reaffirming the United States' commitment to the territorial integrity of Georgia, she concluded her response to a question about Israel and Iran with the words: "we take very strongly our obligation to defend our allies, and nobody should be confused about that."[115] Given Saakashvili's disposition to hear what he wanted, not to mention his conflation of Georgia with Israel, Rice's rhetoric sent a less than clear message at a crucial juncture. Russian president Medvedev claimed that after Rice's visit Saakashvili "simply dropped all communications with us."[116] Some Russian analysts believe that Saakashvili saw an opportunity because he perceived Medvedev as a weak leader.[117]

A few days later, both Russia and Georgia began military exercises. Russia's Kavkaz 2008 exercise across the North Caucasus Military District was ostensibly antiterrorist training, but the 10,000 service personnel, hundreds of combat vehicles, and numerous fighter aircraft, not to mention the accompanying scenario literature, suggested training for conflict with Georgia, with an emphasis on Abkhazia. Afterward, a small Russian force of 1,500 soldiers, as well as tanks and artillery, remained deployed on the Russian border with South Ossetia in a field camp at the North Caucasus Military District's training range near the Mamison pass, approximately thirty kilometers from the Roki tunnel.[118] Meanwhile in Georgia, 2,000 U.S. troops were flown in for "Immediate Response 2008," a joint exercise with Georgian forces. On July 3 a bomb planted in an Ossetian village killed Nodar Bibilov, the leader of a South Ossetian militia. Immediately thereafter, there was an assassination attempt against Sanakoev as he traveled on a bypass road between Georgian enclaves. He survived, but three Georgian special forces with him were injured. Georgian forces subsequently seized the Sarabuki heights, a strategic mountain location northeast of Tskhinval(i). The next day, July 4, 2008, Georgian forces fired several mortar rounds into the town, killing three and wounding eleven.[119] In response South Ossetia declared a general mobilization. An effort by Ossetian militias to retake the heights triggered intensive firefights and shelling. Fighting subsequently spread to the high ground near the

Georgian village of Nuli. On August 1, a Georgian police vehicle on the Eredvi-Kheiti bypass ran over an improvised explosive, wounding six. That evening Georgian snipers targeted South Ossetian policemen in Tskhinval(i) and surrounding villages, killing six and wounding twelve with long-range shots.[120] In response to continued sniper fire against the town, and a cut in its water supply, South Ossetia began evacuating Tskhinval(i) women, children, and the elderly on buses via the treacherous Zar road and onward through the Roki tunnel and down the steep mountainous road to North Ossetia. South Ossetia interior minister Mikhail Mindzayev charged that the Georgian snipers attacking Tskhinval(i) were trained by U.S. and Ukrainian specialists. Both states, according to de facto president Kokoity, were responsible for aggravating the situation.[121] The U.S. State Department, at this time, was actually trying to calm matters with the Russians, with U.S. assistant secretary of state Daniel Fried calling Russian deputy foreign minister Grigory Karasin. After relative calm on August 2, fighting flared up again on August 6, with many civilians wounded in Tskhinval(i) and surrounding villages. That night Georgian infantry brigades moved toward South Ossetia. With war looking imminent, journalists flocked to the area and rumors swirled. Mortar exchanges resumed early on August 7 and that afternoon a Georgian infantry vehicle was hit. Two Georgian soldiers, on peacekeeping duty at the time, were killed and five wounded. The deaths radicalized Georgia's leadership and inclined it toward war.[122] After previously announcing a ceasefire on Georgian TV, President Saakashvili reversed himself and late on August 7 authorized a full-scale Georgian military offensive against South Ossetia.[123]

Who Started the August 2008 War?

In his subsequent justification of why he launched a military offensive against South Ossetia, President Saakashvili claimed that he was responding to a Russian military invasion through the Roki tunnel that was already underway when he acted. A tank column was headed to Tskhinval(i). He could not stand aside and let this happen. No politician in Georgia could hope to survive if he did nothing in response.[124] Thus he launched his attack as a defensive move against the Russian invasion, and in the hope that the international community would "wake up" and force its reversal.[125] Ron Asmus's book *A Little War That Shook the World* is a fulsome articulation of this point of view.

There are a number of reasons to be skeptical of this account and to consider the possibility it was a post-hoc invention to justify a failed military operation to seize South Ossetia. First, the initial announcements of the Georgian government about their attack on Tskhinvali make no mention of an ongoing Russian military invasion launched through the Roki tunnel. The Georgian commander on the scene, General Kurashvili, explained the offensive euphemistically by saying that "the Georgian power-wielding bodies decided to restore constitutional order in the entire region."[126] Georgian officials later argued his statement was unauthorized and emotional, even though it repeated alleged prior language by the Georgian foreign minister (see below).[127] Addressing the Georgian public at 12:20 p.m. the next day, Saakashvili stated that "we initiated military operations after separatist rebels in South Ossetia bombed Tamarsheni and other villages under our control. Most of the territory of South Ossetia has been liberated and is now under the control of Georgian law enforcement agencies." He went on to list a series of villages that had been "liberated," adding that Georgian forces "have surrounded Tskhinvali, most of which has been liberated." Saakashvili made no mention of a Russian invasion, instead charging that the "fighting was initiated by the separatist regime."[128] By the time of his address, the Russian military response to his attack on South Ossetia was already underway. This he represented as "classic international aggression." The international community should know that "Georgia was not the aggressor, and Georgia will not give up its territories."[129] The first official Georgian citation of Russian forces in the Roki tunnel is on August 8, when there were indeed Russian military formations moving through it but in response to Saakashvili's attack. It was three days into the war before the movement of Russian troops through the Roki tunnel was directly linked to Georgia's attack on Tskhinval(i): the timing of the troop movement was revised from 5:30 a.m. on August 8 to 11:30 p.m. on August 7, that is, five minutes before Saakashvili ordered his assault on Tskhinval(i).[130] It was subsequently revised further back again to 3:30 a.m. on the morning of August 7.[131] It is worth noting that Saakashvili and his advisers, who were severely sleep-deprived, already believed that Russia was engaged in a slow-motion "invasion" and "annexation" of Georgian territory in the weeks and months prior. Indeed, rhetorically this was a long-standing ideological claim. Preconvictions likely overrode the precise empirics of the evening of August 7, for the Georgian leadership feared war was imminent.[132] Later, disparate scraps of intelligence would be marshaled to bolster the Georgian claim that a "Russian invasion" was indeed underway before Saakashvili acted.[133] To the Russians, there was no "invasion" but merely the normal rotation and replenishment of Russia's peacekeeping forces in the area.[134] Georgia's submission to the EU Fact Finding Commission, led by Heidi Tagliavini, argued that the buildup of Russian forces in South Ossetia began in early July 2008, and continued thereafter. The Tagliavini Report concluded it was "not in a position to consider as sufficiently substantiated the Georgian claim concerning a large-scale Russian military incursion into South Ossetia before 8 August 2008."[135] The report, however, is definite on Saakashvili's charge of "Russian aggression": "There was no ongoing armed attack by Russia before the start of the Georgian operation."[136]

Second, there is evidence of hubris inside the Georgian government, induced by high levels of military spending, new weapon system acquisitions, and close military cooperation with powerful military states. Private Israeli Special Forces veterans had helped establish and train equivalent Georgian units.[137] Georgian troops had just completed a series of exercises with U.S. troops, and there were as many as 130 U.S. advisers in the Georgian Ministry of Defense at the outbreak of war.[138] Some Ukrainian Special Forces were also on the ground. Georgia had paid military contractors to develop a plan to encircle Tskhinval(i) and take the southern part of South Ossetia before Russia could respond. Asmus notes Vano Merabishvili's preoccupation with Croatia's Operation Storm and Swedish foreign minister Carl Bildt's worry, after dining with Saakashvili at the Bucharest Summit, that the Georgian president believed a bold military move would reap its own rewards.[139] U.S. deputy assistant secretary of state Matt Bryza described the Georgian attitude thus: "They bought enough kit, were well trained enough, had a strong enough fighting spirit that they could stop the Russian advance and negotiate from there and then we would all be involved."[140] Saakashvili reportedly talked with Bildt, Polish president Lech Kaczyński, Lithuanian president Adamkus, and the NATO secretary general de Hoop Scheffer on the day of August 7, briefing them on the escalating violence and likely warning them Georgia would act.[141] Oddly he reportedly did not talk to any senior U.S. government official.[142] In order to launch a surprise assault, Georgian forces needed to be surged to the area. The Georgian Ministry of Defense ordered a mobilization of its reserves at 14:30 on August 7. Approximately thirty yellow city buses of uniformed personnel moved toward Gori and South Ossetia that afternoon and were in place by 18:00.[143] Saakashvili, at the very least, had decided to set up his military contingency plan to seize Tskhinval(i) and consolidate it with the Georgian enclave to its north, in the hope of creating a fait accompli before the Russians could react.[144] Then Polish foreign minister Radek Sikorski, no friend of Russia, stated in a BBC documentary that Georgian foreign minister Eka Tkeshelashvili called him before Saakashvili's order to indicate that Georgia was going to "establish constitutional authority" over South Ossetia, which he interpreted as a statement that they were "moving in" (he wisely replied that they should be careful not to lose the perception that they were the victims).[145] Tkeshelashvili denies ever making any reference to the restoration of "constitutional authority" to him or any of the counterparts to whom she spoke. Her aim, as she saw it, was to stop the escalation of the conflict and reverse the spiral of violence on the ground.[146] She talked numerous times with Matt Bryza in Washington. According to Bryza, she told him that fighting was initiated by the South Ossetians but that Russian troops were also shooting.[147] Further, the South Ossetians were increasing the caliber of their ammunition. Yet despite these intolerable provocations, Saakashvili was going to call a ceasefire. Two hours later she called back and told Bryza that the South Ossetians had intensified their firing and that Saakashvili wanted to lift the ceasefire. Georgian villages in South Ossetia were under heavy attack. Bryza called Rice, who was preparing to go on holiday, and then relayed her advice to Tkeshelashvili: Georgia should pull back its forces to a defensive position. According to Bryza, Tkeshelashvili replied: "[W]e anticipated you'd say that. But we can't stand it any more because we have to protect our villages. We have to protect our people. So we are going to return fire, and we are going to increase the caliber of our return fire. We're going to fight back."[148] And then they ended their phone call. Tkeshelashvili remembers the call differently and denies indicating a decision had already been made. To her, "Washington's message delivered by Matt was completely out of touch with the reality." Indeed, she believed more broadly that "the international community's reaction was out of touch with the gravity of the situation and the determination of Russia to test the effectiveness of the use of force for the achievement of its geopolitical goals."[149]

Boldness in the past served Saakashvili well; he likely believed it would continue to do so.[150] After giving the order to attack Tskhinval(i), Saakashvili reportedly turned to Yakobashvili and said: "Do you think we will end up as Israelis or Palestinians?"[151] The comment, and its subsequent repetition, was probably meant to illustrate the moment as an existential one for the Georgians. But the analogy revealed also the comparative vision and aspirant heroic mythology that had a hold over Saakashvili and his inner circle. Was Georgia's David the Builder going to defeat Goliath? Was this Georgia's 1967 war? Hubris and heroics were still burning two days later when Yakobashvili gave an interview to Israeli Army Radio in which he claimed that a small group of Georgian soldiers were able to wipe out an entire Russian military division, thanks to their Israeli training. "We killed 60 Russian soldiers yesterday alone," he said. "The Russians have lost more than 50 tanks, and we have shot down 11 of their planes. They have sustained enormous damage in terms of manpower."[152] The reality on the battlefield, however, was not David and Goliath redux but one of Georgian forces confronting dogged Ossetian resistance and thereafter overwhelming Russian force.

Third, there is evidence that Saakashvili and Georgian military officials really believed that if their forces got into trouble then the United States would be forced to come to their aid. Outwardly this was delusional thinking. There is no confirmation that any senior U.S. official ever said that the United States would militarily support Georgia if it got into a conflict with Russia. As already noted, President Bush himself had made this clear, twice, to Saakashvili during his 2005 visit. All the top U.S. officials dealing with Georgia have stated publicly that they warned Saakashvili against military action in the breakaway regions.[153] The Wikileaks cables from the U.S. Embassy in Tbilisi reveal that prior to the outbreak of the war Ambassador Teft sought to keep

Georgian government officials from "overreacting to Russian provocations as this would only strengthen Russia's hand."[154] While there was no green light, was there, as Strobe Talbot later put it, "a problem with an insufficiently red light"?[155]

It was not irrational for Saakashvili to believe that the United States would be very supportive of his government should it get into a military conflict with Russia. Georgian troops were fighting alongside U.S. troops in Iraq and Afghanistan, and U.S. troops, as well as U.S. and Israeli military contractors, had trained those he was now sending into battle. The U.S. Embassy noted how Georgian military officials were deeply disappointed that the United States and the West were not providing them military support against Russian attack.[156] This suggests they believed their GWOT efforts for the United States created an implicit debt of honor and obligation. Beyond military circles, many senior Bush administration officials were Georgia enthusiasts and saw Saakashvili as a great friend of the United States. Many American officials served as advisers and employees of Georgian government agencies while an American, Daniel Kunin, was part of his inner circle. Furthermore, Saakashvili had very solid ties to prominent U.S. politicians. He was personally close to the Republican nominee for president, Senator McCain, and had many sympathizers in influential U.S. think tanks and the media. During his March 2008 visit to Washington, Saakashvili both openly discounted Russian military capacity and suggested a war was in the offing. Saakashvili may well have believed that his military action would be admired as courageous and bring its own reward even in defeat. Georgia would become a righteous cause in the U.S. presidential election and throughout the West. To a certain extent this indeed proved to be the case. For example, the prominent neoconservative William Kristol wrote in his *New York Times* column that because Georgian troops were fighting in Iraq, "we owe Georgia a serious effort to defend its sovereignty."[157] What this meant he did not specify. The leaders of Poland, Ukraine, and the Baltic States rallied to Saakashvili's side, and they traveled to Tbilisi for a mass protest against Russia on August 12. McCain tried to make support for Georgia an issue in the U.S. presidential campaign. In late September the U.S. Congress provided Georgia with over $1 billion in assistance amid a major financial crisis. Launching an offensive war—though

Saakashvili's defenders did not see it in these terms at the time—did bring some rewards.[158]

Numerous works by scholars and analysts have been dedicated to the proposition that Russia planned and initiated the August 2008 war.[159] That Russia had a contingency plan to invade South Ossetia and Abkhazia to support their client regimes there is obvious. That Russia, therefore, started the war on August 7, however, is not persuasive. No senior U.S. official holds this view.[160] All circumstantial evidence suggests that the Russian leadership was genuinely surprised by Georgia's attack. In some ways, though not in others, it was a gift that served their strategic goals.[161] But in the first few hours of the conflict the Russian leadership, both civilian and military, was disorganized and incoherent. Its subsequent prosecution of the war was not impressive.[162] The best case for the Georgian claim that can be made is that the relationships forged between the Russian security services and the South Ossetian militias empowered the latter to be aggressive in their response to Georgian military actions prior to Georgia's full-out assault late on August 7. Many have argued that the Russians sought to "bait" the Georgians into starting the war, but such a claim is hard to prove.[163] This "trap theory" was popular among politicians and analysts, but there were too many warnings from Russia to the Georgians not to use force for it to be persuasive.[164] It also ignores the diplomatic efforts involving Russia that preceded the outbreak of war. A variant "bait theory" of the war is that South Ossetia's militias provoked it.[165] That these militias had their own interests must not be forgotten, though bringing war upon their heads and homes without full certainty that Russia would intervene, and do so in a timely manner, is unlikely to have been their primary plan.[166] Instrumental rationality, however, was not in the driver's seat. The South Ossetian position is that they responded with proportionate force to Georgian provocations in the summer of 2008 in order to prevent escalation. Explaining the buildup to the war, South Ossetian spokesperson Pliyev claimed that the South Ossetians pleaded with the Russians to send a military signal in order to "cool the hotheads" in the Georgian military and civilian leadership.[167] This they did with the airspace incursion on July 8 as Rice prepared to visit Saakashvili in Tbilisi.[168] But, as far as they were concerned, the Georgian military forces did not cease their provocations against South Ossetian positions and personnel. They feared war, evacuated as many residents as they could to North Ossetia, and organized volunteers from North Ossetia to help them prepare for war. In the end, South Ossetia's defense forces in Tskhinval(i) and elsewhere had to face the "Georgian invaders" on their own until their Russian allies mobilized in force to come rescue them.[169]

5

Rescue Missions

THE GEORGIAN MILITARY ASSAULT on Tskhinval(i) began with an artillery barrage by truck-mounted Grad missiles that rained down in a largely indiscriminate manner on the urban area.[1] OSCE monitors in the city counted rounds exploding at intervals of fifteen to twenty seconds.[2] Then Georgian forces began a ground offensive. Scores were killed, mostly civilians but also combatants and, significantly, Russian soldiers serving as peacekeepers. Those wounded were taken to makeshift basement "hospitals" as fighting raged on the streets above them. Many bled to death.[3] Inal Pliyev's initial claim, made in the fog of war, that two thousand civilians died proved to be more than a tenfold exaggeration, but it was his number that traveled around the world before any verifiable body count got underway. And attached to the false number was the disputatious charge of "genocide" from the script of the conflict already written by the Ossetian authorities.

News of the Georgian attack and Russian response reached Beijing (four hours ahead of Tbilisi), where U.S. president George W. Bush was standing in a line of visiting dignitaries in the Great Hall of the People to greet President Hu Jintao. A few places ahead of him was Vladimir Putin. They briefly talked, and when Bush returned to his hotel he placed a call to President Dmitry Medvedev.[4] Medvedev, according to Bush, was angry and charged that Mikheil Saakashvili was a war criminal responsible for the deaths of more than fifteen hundred civilians as well as Russian peacekeepers. Bush told Medvedev that the United States was concerned about the "disproportionality" of the Russian response, adding, "We are going to be with them." In his memoir Bush recounts how he sought de-escalation while wondering if Russia would have been as aggressive if the North Atlantic Treaty Organization (NATO) had approved Georgia's Membership Action Plan (MAP) application. Later, at the Opening Ceremony of the Olympics, Bush was in the same row as Putin and asked those in between to shift seats so they could speak. Through a translator Putin also described Saakashvili as a war criminal. Bush replied: "I've been warning you Saakashvili is hot-blooded." Putin responded: "I'm hot-blooded, too." Bush describes his riposte thus: "I stared back at him. 'No, Vladimir,' I said. 'You're cold-blooded.'"[5]

Bush's rendering of his translator-mediated conversation with Putin tells us a great deal about how the outbreak of the August 2008 war was experienced and understood by the U.S. and Russian leadership. First, Saakashvili's move provoked the anger and disgust of the Russian leadership. This was affective not strategic geopolitics, and the language that accompanied it—"war criminal" and "genocide"—was not the language of diplomacy or transactional geopolitics. Instead, this was the language of righteous anger. Bush himself had used similar language about Saddam Hussein in the year prior to the U.S.-led invasion of Iraq, though his use was part of a public relations campaign to justify invasion.[6] Unlike Iraq, Georgia had just killed Russian soldiers and citizens (to them, in response to the killing of Georgian soldiers). Second, Saakashvili's actions placed the Bush administration in a difficult position. Their ally, a leader and a state they had publicly embraced, supported, and financed, had started a war by attacking a client regime of the Russian Federation. Despite repeated exhortations not to attack, Saakashvili had done so anyway. James Jeffrey, the U.S. deputy national security advisor who broke news of the invasion to Bush in Beijing, recalled: "It was obvious to us from the start that Saakashvili had started this. There was no doubt, no doubt."[7] On its face, this was an embarrassment if not a failure of U.S. diplomacy. Bush chose to describe Saakashvili to Putin with a phrase that was both paternalistic and empathetic: he was "hot-blooded."[8] At the same time, he felt the need, irrespective of what had transpired, to tell Medvedev that "[w]e are going to be with them." In this, Bush was articulating paternal solidarity with a younger leader he considered a loyal friend of the United States. The dilemma

Saakashvili created for the Bush administration was in directly pitting affective against strategic geopolitics. It was one that would not be resolved quickly. Third, Bush's presentation of his riposte to Putin in his memoir hints at the pervasive heroic masculinity at work during the August War. Like a cowboy squaring off against an outlaw, Bush stares at Putin and pronounces him "cold-blooded" (a masculine coda to his much-mocked soul sensing in Slovenia). Putin, for his part, was angry, particularly about the deaths of Russian peacekeepers. He was also deeply distrustful of the United States' role. He may well have grasped that the United States did not authorize the Georgians to attack South Ossetia. This is what Bush, in effect, told him. However, whether persuaded by a conspiracy theory or finding it convenient, it became clear subsequently that Putin held the United States responsible for Saakashvili's actions. There was a conspiracy to encroach on Russia's interests in its backyard, and the United States and NATO were behind it.

This chapter examines how the August 2008 war played out in the subsequent weeks after its outbreak. It first examines, in a necessarily condensed manner, the different phases of the conflict and how it altered the political geography of Georgia and its breakaway regions. Second, it considers the storyline that the Georgian government presented to the world about the war. Third, it then documents the Russian storyline on the war. Finally, it examines the White House's response to the war and how it evolved. Crucial to grasping how the war was understood by Russia and the United States, I argue, is an appreciation of the operation of a structurally similar myth of victimization and rescue that is saturated with colonial tropes. To Russia, Ossetians were the victims of a resurgent fascist nationalism that was genocidal in nature. To the United States, little Georgia was the victim of a resurgent imperial Russian aggression. Both storylines feature a small victim nation and a heroic role for a protective, paternal great power. Fortunately, the role certain key decision-makers within the Bush administration defined for the United States was as a responsive humanitarian power, not as a military protector for Georgia. Because of this, a potentially dangerous military confrontation between the two great powers was avoided.

Phases of the War

The Georgian attack against South Ossetia began not only with an artillery barrage and ground offensive against Tskhinval(i) but also a ground attack against Ossetian villages in all four *rayoni* (counties) controlled by the South Ossetian authorities. The initial defensive response was by the Ossetian Defense Ministry, the KGB, police and other paramilitary militias, fortified by volunteers who had traveled to the region as war appeared imminent.[9] As noted in the last chapter, at the conclusion of the Kavkaz 2008 military exercise, the 58th Army positioned two battalion tactical groups, with artillery, armored personnel carriers, tanks, and around 1,500 personnel, in a training base with easy access to the Roki tunnel. Because of the intensified shooting and shelling in South Ossetia, they were on high alert to immediately deploy should their fellow soldiers serving as Russian peacekeepers in the region require support. These forces started moving through the Roki tunnel in the early hours of August 8, with the first clear of the tunnel around 1:40 a.m.[10] Support brigades departed Vladikavkaz around 2 a.m.[11] Subsequently Russian military formations were mobilized from around the North Caucasus Military District. Russia's Airborne Assault Regiment, which had just returned to Pskov after *Kavkaz 2008*, returned to Vladikavkaz and from there deployed into South Ossetia. Volunteer militia units, later involved in pillage, were dispatched to the region from the North Caucasus, including Chechnya. North Ossetian president Taimuraz Mamsurov later traveled to South Ossetia with a column of buses to evacuate civilians, meeting with Eduard Kokoity and promising support.

The war that followed can be divided into five phases.[12] The first phase was the Georgian offensive battle for Tskhinval(i) and the surrounding mountain villages and the initial Russian response on August 8. The previous day, Georgia had assembled a force of approximately 12,000 soldiers and 75 T-72 tanks as well as 4,000 Ministry of the Interior troops with 70 armored vehicles.[13] Georgia began offensive operations against Tskhinval(i) and flanking operations to the west and east to occupy the heights around the town. The plan thereafter was for one brigade to capture the Zar road while the second linked with Georgian forces inside the Didi Liakhvi enclave, with all moving northward toward a convergence north of the enclave in the Ossetian village of Stari Gufta (known as Didi Gupta in Georgian) where the strategic Gufta Bridge was located. Tskhinval(i) would then be encircled, and Georgian forces could then push northward up the TransKam. Smaller Georgian forces attacked Ossetian villages farther north from the west. Ossetian mountain villages in the Akhalgori (Leningor) district were also occupied by Georgian troops.

A series of factors thwarted the Georgian plan. Difficult terrain, effective defensive fighting by Ossetians, and the relatively rapid Russian military response immediately complicated matters. Destroying the Gufta Bridge, a large span bridge over the Patsa River as it joins the Greater Liakhvi River south of Dzau/Java, would have severed the Russian route to Tskhinval(i) and the south. However, Georgian bombs failed to hit the bridge while its ground forces encountered the advancing Russians on the bridge around 4:40 a.m.[14] The Russians lost an armored personnel carrier but secured control of the bridge. The two Russian forces then split farther south, one taking the Zar bypass road toward Tskhinval(i) while the second continued down the TransKam toward the barricade north of the Didi Liakhvi enclave. But their progress thereafter was slow and cautious. Their numbers were small and Georgian aviation and artillery targeted them. Political decision making also needed to catch up. Their delay later led to accusations against President Medvedev (see below). According to the head of Russian peacekeepers in South Ossetia, the first company of Russian forces entered Tskhinval(i) at 2 p.m. on August 9, with the main armored forces only arriving the morning of August 10.[15]

The first evidence of the Russian military response was in the air. Russian aircraft—Sukhoi twenty-five close air-support aircraft, Tupolev twenty-two reconnaissance planes, and Mi-24 helicopter gunships—were mobilized and by 10 a.m. began engaging Georgian planes and Georgian forces in South Ossetia. Around noon on August 8, Russian planes attacked the Georgian military base in Gori. There is also evidence that two SS-21 missiles were fired from Russian territory into Georgia, one hitting a police station in Poti.[16] Reporters later found debris from SS-21 and BM-21 rockets, both of which can carry cluster munitions, in Poti, Gori, and the village of Variani.[17] Human Rights Watch researchers later found evidence of the use of cluster munitions by both Russian and Georgian forces around four towns and villages in Georgia's Gori district.¹⁸ While Saakashvili's television broadcast declared that most of Tshkinval(i) was "under our control," the Georgian war plan was already in serious trouble. Fighting was fierce in Tskhinval(i), with the Georgians losing a number of tanks and their crews. Georgian forces withdrew in the early afternoon only to return and try to advance again that evening.

With aggregate numbers on their side and fresh troops arriving, Georgian forces launched a renewed offensive in the early hours of August 9. The city would be bombarded by artillery all night. Russian units by this time were striving to secure the Zar road. A few reportedly made it through to relieve fellow soldiers in the southern peacekeeping compound at 4 a.m. A convoy of wounded peacekeepers had left earlier and made it to Russian lines on the Zar road. The Georgian offensive continued all morning but never managed to return to positions held the previous day. In the early afternoon advancing Georgian tanks ran into a convoy of Russian forces that included the Russian commander of the 58th Army. In a brief exchange of fire, soldiers on both sides were killed while the commander of the 58th Army, Lieutenant General Anatoly Khrulev, was wounded. The wounded Russians managed to make it to the southern peacekeeping compound only to face the brunt of a renewed Georgian offensive. Other Russian units arriving in Tskhinval(i) that afternoon were thrust into battle with Georgian forces. Close-quarter fighting in the streets of Tskhinval(i) with local Ossetian militias, as well as heavy artillery shelling from Russian positions on the Zar road, blunted the August 9 Georgian offensive. This was the turning point. By the early evening Georgian units were in retreat while Russian forces were finally able to lift the blockade on the southern peacekeeping compound, and evacuate the wounded. Fourteen Russian soldiers serving in the compound were killed, the majority from Georgian tank shelling on the morning of August 8.¹⁹

The second phase saw the Russian-Ossetian counterattack gather momentum as targets across Georgia came under sustained attack. Russia began bombing Georgian military facilities outside South Ossetia from the outset, seeking to target Georgia's logistical tail to South Ossetia. On August 9 it intensified its operations. Some bombs missed their targets, causing civilian casualties. Two apartment blocks were hit in Gori, killing at least seven civilians on the morning of August 9. Some villages in the vicinity of Tskhinval(i) were also targeted. Russia lost three combat aircraft that same morning in a few hours (it lost a total of seven, some to friendly fire).[20] On August 10 the Georgian government declared a unilateral ceasefire and said it would withdraw its forces from South Ossetia. By 2 p.m. its last forces had left the area.[21]

The Russian and Ossetian forces used the opportunity to occupy ethnic Georgian villages that were previously beyond the control of the South Ossetian authorities. On August 8, the Georgian villages in the Didi Liakhvi valley had largely been evacuated via the Kheiti-Eredvi bypass road, leaving the enclave's ten thousand residents without homes and a future as internally displaced persons. Many residents later reported only having one hour to pack and leave.[22] The next evening and into the early hours of August 10 Russian special forces units, including the Chechnya-based Vostok Battalion, conducted so-called clean-up operations within the enclave villages.[23] Ossetian sources indicate the enclave, which they estimated had seven to eight hundred Georgian fighters during the offensive, was mostly deserted except for a few elderly.[24] The TransKam was opened to the passage of Russian armor the next morning.

On August 9 and 10 Russian planes continued bombing sorties against Georgian military facilities, airports, and the port of Poti. A Russian naval squadron patrolled the waters off Georgia's Black Sea coast and sank some Georgian missile boats that engaged them. Other Georgian vessels were sunk. The extensive nature of the Russian response, targeting facilities and infrastructure far from the zone of hostilities, as well as launching cyberoperations, drew widespread international charges that it was "disproportionate" relative to the Georgian military offensive to capture South Ossetia.

The third phase was the Russian ground invasion of Georgia proper. In practical terms, this was barely a discrete phase as counterattack became offensive operation. Russian forces pushing the Georgian army back to Gori continued their hot pursuit beyond the administrative lines of the South Ossetian autonomous oblast.[25] Despite the announced unilateral ceasefire, Georgian artillery continued to shell Tskhinval(i). Russian forces sought to suppress the fire. The move beyond the boundary line was noticed by diplomats and brought instant condemnation. A Western diplomat on the ground indicated that the Russians "seem to have gone beyond the logical stopping point" (the Georgian–South Ossetian border).[26] Russia's invasion of Georgia proper was more manifestly obvious in Abkhazia. On August 10 about 1,000 Abkhaz troops, supported by Russian fighter planes, pushed into the Georgian-controlled Upper Kodori Gorge. That same day Russian forces, mostly airborne assault troops, crossed the Georgian-Abkhazian ceasefire line on the Inguri River and moved deep into uncontested Georgian territory, occupying a series of locations. Russian forces took over the western city of Senaki and seized the military base built there to NATO specifications. It was plundered thereafter, as were other facilities, as Russian forces gathered for themselves over the subsequent weeks "spoils of war."[27]

The fourth phase was the ceasefire agreement and final establishment of a cessation of hostilities. President Nicolas Sarkozy of France, in his capacity as president of the Council of the European Union, flew to Moscow on August 12 and helped negotiate a six-point ceasefire agreement with Russian president Medvedev in the Kremlin. The agreement specified that the armed forces of Georgia should withdraw to "their permanent positions" and that the "armed forces of the Russian Federation must withdraw to the line where they were stationed prior to the beginning of hostilities." In the interim, it allowed that "Russian peacekeeping forces will take additional security measures." This provision became the basis for the establishment of "buffer zones" by the Russian military, acting in a self-designated fashion as "peacekeepers," outside the two enclaves and inside Georgia proper. The geographical extent of the buffer zones was never clear, and it was the subject of considerable international suspicion and tension. The South Ossetian buffer zone overlapped with the already-established security corridor agreed in 1992 but went beyond it. Russian forces established at least eight military posts across Georgian territory, with only the addition of an armband signifying their status as "peacekeepers." The Abkhaz buffer zone extended to the outskirts of Senaki; Russian soldiers also occasionally patrolled in Poti. A letter clarifying that the provision would not apply to populated areas or the main east-west highway was negotiated by international diplomats. The agreement—the six-point plan and accompanying clarifying letter—was eventually signed by Saakashvili on August 15 and by Medvedev the following day. Russian forces slowly began pulling back from their maximum positions within Georgia proper. Many returned to the 58th Army base in Vladikavkaz in North Ossetia. The Tagliavini Report estimated the total number of Russian troops moved into Georgia in August 2008 at 25,000 to 30,000—at least 12,000 in South Ossetia and 15,000 in Abkhazia—and these were supplemented by an estimated 7,000 armed Ossetian and other volunteer militias from the North Caucasus.[28]

The Russian occupation of South Ossetia and parts of Georgia facilitated looting and pillaging of ethnic Georgian villages in South Ossetia by irregular forces from South Ossetia and from across the North Caucasus, who flocked to the conflict as so-called volunteers. Human Rights Watch researchers witnessed the destruction and looting of the eight Georgian villages north of Tskhinval(i) on August 12 (see Figure 4.1).[29] Satellite images of the region showed active fires in ethnic Georgian villages on August 10, 12, 13, 17, 19, and 22, days after the end of fighting in the area.[30] Revenge attacks and freelance ethnicized violence later spread to Georgian property in the villages of Eredvi, Berula, and Argvitsi to the west of Tskhinvali[31] and southwest into uncontested Georgia.[32] Villages north of Gori experienced marauders looting homes and stealing property. Local residents were also kidnapped, beaten up, and murdered.[33] A few Georgian villages and local residents were protected by Russian soldiers but not many.[34]

The final phase was the slow withdrawal of Russian forces from positions within Georgia proper. False starts and confusion over the buffer zones prompted renewed diplomacy by Sarkozy, who negotiated a follow-on agreement with Russia for withdrawal of its forces from areas adjacent to the borders of the enclaves by October 10, and the deployment of at least two hundred European Union monitors. The Russian withdrawal was completed on October 8, 2008, when they removed the last of the checkpoints they had established in the extended buffer zones. European Union monitors began to operate in the area as did Georgian police forces. The Russian forces and their local allies withdrew to the administrative borders of the two enclaves but not beyond that to status quo ante positions. The old Soviet administrative boundaries of the South Ossetian autonomous oblast, not the de facto boundaries of the South Ossetian Republic, were where the victors chose to interpret the clause referring to "pre-conflict lines."

A war that began to restore the territorial integrity of Georgia ended by enabling the de facto Republic of South Ossetia to achieve its claimed territorial integrity. Akhalgori, the main town in the eastern valley that became part of Georgia after 1990, was now incorporated into the Republic of South Ossetia and named once again, or rather renamed in the Ossetian form of its Soviet-era name—as Leningor.[35]

Georgia's Storyline: An Attack on the West

The Georgian government was practiced in couching its actions within the discourse of international norms and Western values. The government's fight against its breakaway regions, thus, was merely a legitimate effort on its part to restore the state's territorial sovereignty. The all-out assault on South Ossetia saw the government use some of the same language invoked by the Bush administration to justify the invasion of Iraq. South Ossetia was a threat that could no longer be tolerated. The people in the region, Ossetians and ethnic Georgians, needed to be "liberated" from criminal separatists in the pay of Russia who were attacking Georgian villages. This geopolitical black hole of chaos could no longer be tolerated by a law and order state. As it quickly became evident that the assault had drawn Russia into the fight and that Georgia would lose, the government turned to representations of the crisis as a premeditated invasion of Georgia by Russia, rather than as a Georgian invasion of South Ossetia. Establishing a favorable spatial lens for the crisis became crucial to the government's public relations strategy, and this required forgetting and eliding the local territorial dimensions of the crisis. The war should not be seen as a war over South Ossetia but rather as a Russo-Georgian war or simply a Russian invasion of Georgia. South Ossetia itself as a place disappeared from the picture; the symbolic visuals of Russian tanks on the move was what was important. The war was scaled up into a more abstract clash of ideographic binaries: freedom versus empire, democracy versus criminality, the West versus the East. This internationalization and axiological upscaling was accompanied by the familiar framing of primary metaphors.[36] Georgia was the little nation seeking its freedom. Russia was the neighborhood bully. In short, the Georgian government sought to project universal affective categories and leave the messy details of the crisis behind.

Aided by effective advocacy networks Saakashvili was able to appear almost daily in the American and European media and place multiple op-eds in top national newspapers. His fluent, accentless English was a major asset in allowing him to make his case (by contrast, his voice was dubbed in a shrill Russian on Russian television).[37] Saakashvili sought to push emotional buttons, to trigger and amplify a range of negative affects (outrage and anger at Russia) and positive affects (identification and support for freedom-loving countries like Georgia). In making his case for the war as an attack on the West, however, he tended to transition into blunt criticism of the West for its appeasement and failure to act resolutely against Russia. In doing so, often in an unmodulated manner, he undermined his own cause. Let us consider a few examples out of the many available.

In an appearance on CNN on August 9, Saakashvili described the reality of the crisis as the small nation of Georgia being brutally attacked by its big neighbor Russia. Casting Georgia's actions as a response to provocations by Russian-backed rebels (their identity beyond that is not mentioned), and to an already-initiated Russian tank invasion of "our sovereign territory," he analogized the conflict to events in Cold War history. "This is exactly the kind of invasion they did into Afghanistan in '79. This is exactly the kind of invasion they did in Czechoslovakia in '68 and then to Hungary in '56." History was repeating itself, with an aggressive Russia using any pretext for a preconceived invasion. Conceding that what was happening was beyond his expectations, Saakashvili pitched Georgia as one of the friendliest toward America in the world.[38] The following day Saakashvili cast the crisis in universal terms: "this is not about Georgia anymore. This is about basic values of humanity, of American values that we always, ourselves, believed in. This is all about human rights. This is all about the future of the world order. And I think there are much bigger things that are at stake here than just Georgia."[39]

This is a theme he amplified in an opinion editorial in the *Wall Street Journal* on August 11: Georgia is fighting for the West; its conflict is about the future of freedom in Europe.[40] On the one hand, these arguments voiced a certain messiah-complex and associated self-aggrandizement; on the other hand, Saakashvili's country was in serious peril and his own life threatened. The visceral shocks of war inflated the stakes. Timuri Yakobashvili told Israeli Radio: "Every bomb that falls over our heads is an attack on democracy, on the European Union and on America."[41] Alexander Lomaia, head of Georgia's National Security Council, cast Georgia as the first domino in a potential Russian takeover of Europe: "If the world is not able to stop Russia here, then Russian tanks and Russian paratroopers can appear in every European capital."[42] Georgia lost its freedom to Russia—to seventy years of communism and slavery. Prime Minister Putin and other Russian leaders "are a product of that system," and they have destroyed Russian democracy. "These people with their KGB backgrounds and with brutal backgrounds will do . . . their best to manipulate the truth to be cynical."[43]

Picking up on a strong statement of affinity for Georgia from Republican presidential candidate McCain, as well as analogies made by the leading U.S. neoconservatives Robert Kagan and William Kristol to Hitler and Czechoslovakia in 1938, Saakashvili warned against the geopolitical imagination that comes with appeasement—the view that Czechoslovakia or Georgia are mere "faraway countries" that are little known. "I heard Senator McCain saying we are all Georgians now. I hope people understand that these are their values at stake. This is freedom in general at stake. This is not some far away remote country in which we know little [sic]. I mean Georgia is very very modern, normal country."[44] In Saakashvili's own mind, Finland was a country that could teach Georgia lessons. "I've read all the books about how Finland fought this kind of war in 1939," he stated, occasionally referring to the Karelia region's experience as comparable to the South Ossetia case (the end state, where most of Karelia was annexed by the USSR, was left unsaid).[45]

Saakashvili offered a catalogue of motives for the Russian actions in Georgia that resonated with classic geopolitical explanations for imperial behavior: control of resources, infrastructure, intimidation, and regime change. Russia's oil riches and desire to assert economic leverage over Europe had emboldened the Kremlin to attack Georgia: "They need control of energy routes. They need seaports. They need transportation infrastructure. And primarily, they want to get rid of us."[46] After speaking with President Bush on August 10, he told Germany's *Rhein-Zeitung* newspaper something similar: "[Bush] understands that it's not really about Georgia but in a certain sense it's also an aggression against

America. The Russians want the whole of Georgia. The Russians need control over energy routes from central Asia and the Caspian Sea. In addition, they want to get rid of us, they want regime change. Every democratic movement in this neighbouring region must be got rid of."[47] The Georgian government charged that the Russians' planes had attacked the Baku-Tbilisi-Ceyhan pipeline at least eight times (already shut down at the time because of a terrorist attack against it by the Kurdistan Workers Party [PKK] within Turkey). The Georgian prime minister Lado Gurgenidze described this as "a direct attack on the energy security of Europe. Militarily it makes no sense."[48] The Russian government denied targeting the pipeline, and in fact it remained undamaged during the war.

"Regime change" was an earlier American contribution to the lexicon of geopolitics, and it was the Bush administration that amplified the Georgian government's charge that this was the ultimate motive of the Russian actions.[49] The Russian government weakly denied the charge, though there was ample public evidence for their contempt for Saakashvili and the desire of some to remove him. This led to considerable speculation about whether Russian troops would march on Tbilisi. Indeed, Saakashvili himself spread panic by stating as much on Georgian television on August 12 as Russian tanks pushed beyond Gori onto the main road to Tbilisi. Earlier that day he spoke at a rally of thousands in Tbilisi, appearing with leaders from four former Soviet republics and Poland, who had flown to Georgia in an act of solidarity. In a CNN interview on August 13, Saakashvili suggested: "[T]heir plan was always to take over the whole of Georgia. Their plan was to establish their own government in Tbilisi. And their plan was to kill our democracy."[50] Georgians, he offered, were feeling let down, "feel exactly like Czechs felt, like Czechoslovakia felt in 1938 after Munich, exactly the same [as] Poland felt after . . . the Soviet and the Germany [*sic*] invasion . . . the murder of the country is reported live." The Bush administration's initial reactions to the invasion "were too soft. You know, Russians don't understand that kind of soft language."[51] Speaking the same day to the right-wing American talk show host Glenn Beck, Saakashvili described how Russians were burning Georgian cities, destroying villages, killing people, and rampaging for food. Georgia was dealing with "twenty-first-century barbarians." Pressed by Beck if this portended the return of the "evil empire," Saakashvili pronounced it an evil with truly global ambitions: "I never thought that this evil would come back again. I never thought the KGB people would again try to run the world. And that's exactly what's happening now."[52] Saakashvili stuck with the theme of evil, and a few days later he told an interviewer that Georgia was "looking into the very eyes of evil."[53] Asked to respond to a Medvedev statement that it is unlikely Ossetians and Abkhazians would ever live together with Georgia in one state again, Saakashvili analogized Abkhazia to the Sudetenland, where a minority group is in charge because the majority has been expelled. South Ossetia is similar, he said, its separatists being "financed, abetted and organized by the Russians."[54]

Russia's Storyline: Genocide and the Responsibility to Protect

President Dmitry Medvedev was on a sailboat on the Volga when Saakashvili ordered the Georgian offensive against South Ossetia. Awoken by his defense minister, Anatoly Serdyukov (who was also on holiday), and told of the offensive, Medvedev was surprised and skeptical about the news. He ordered that the information should be checked, wondering what mental state led Saakashvili to his action and what precisely he was communicating through his actions. When Serdyukov called back verifying the information, Medvedev hesitated to respond and sought to speak with Putin in Beijing. A third call from Serdyukov informing him that an artillery barrage had killed a unit of Russian peacekeepers finally provoked him into ordering an armed response to Georgia's offensive.[55] In the official telling at least, the direct deaths of Russian soldiers convinced him to act. He returned directly to Moscow.

Medvedev's initial cautious reaction to the outbreak of the August War would later become a matter of controversy in Russia. Putin reportedly saw a "lack of resolve" and was angry about it.[56] On the fourth anniversary of the war, a mysterious yet professionally produced video titled *Lost Day* appeared on the Internet, featuring select Russian and Ossetian military officials accusing Medvedev of being "afraid to give the command" for Russia to retaliate.[57] Elaborate plans were in place—in sealed envelopes—should Georgia attack, but an order was never given. As a result, the video alleged, Georgian aircraft and troops were unopposed in the crucial first hours of the conflict, and Ossetians and Russians needlessly died as a result. Residents of the Ossetian village of Khetagurovo (see Figure 4.1), which changed hands a number of times during the war, declared that had the Russians arrived earlier no one would have died there. Marat Kulakhmetov, commander of the Joint Peacekeeping Forces in South Ossetia, pronounced: "The faster had we reacted, the less blood would have been shed by our army."[58] Ossetians, the documentary suggests, were largely left on their own to fight the Georgians on August 8 while the Russian army sat on the Zar road and TransKam. A crucial day in response time was lost, it was implied, because it took Medvedev that long to speak to Putin. The video was a transparent paean to Putin as a leader of heroic action, but it also tapped into the enduring anger of many South Ossetians at Russia's failure to intervene faster.[59] It replayed what became the essential Russian storyline of the August War: an exposed people attacked by a duplicitous fascist regime were rescued by the decisive actions of a powerful strongman.

It was Putin who set the tone of the Russian response to Georgia's attack. Speaking with reporters in Beijing after the news broke, he vowed that Russia would retaliate. In a meeting with French president Nicolas Sarkozy, he gave expression to his fury: "We are going to make them pay. We are going to make justice."[60] After his conversation with Bush at the Opening Ceremony of the Olympics, Putin left immediately for Russia and flew not to Moscow but directly to Vladikavkaz, the capital of North Ossetia, and the base of the 58th Army that was coordinating Russia's counteroffensive. Medvedev led the official response to Georgia's actions. Meeting with the National Security Council, his office released its first statement on the attack after 10 a.m. in Moscow. It was legalistic—the military response was described as "operations to oblige Georgia to restore peace to South Ossetia"—but ended with desire for retribution: "The perpetrators will receive the punishment they deserve."[61] The statement contained three significant geopolitical speech acts. First, Medvedev reproclaimed the right of Russia to maintain a presence on Georgian territory through its lawfully sanctioned peacekeeping mission. To this he added an assertion that went beyond legalism: "Russia has historically been a guarantor for the security of the peoples of the Caucasus, and this remains true today." Second, Medvedev constituted the Georgian action as an act of aggression against Russian peacekeepers and the civilian population of South Ossetia. Georgia's actions were a "gross violation of international law," and its victims were not only Russian soldiers working for peace but also ordinary civilians, the majority of whom were Russian citizens (i.e., Russian passport-holders in South Ossetia). Third, the statement used the first two proclamations to construct a warrant for action. As president of the Russian Federation, he said, "[I]t is my duty to protect the lives and dignity of Russian citizens wherever they may be." These three speech acts—asserting a special role in the region, naming South Ossetians as Russian citizens, and proclaiming a "responsibility to protect"—framed the initial Russian response.[62]

Meanwhile Russian television reports from the region described the desperate situation of Ossetians (and some Georgians) cowering in basements in Tskhinval(i) and the panic of refugees fleeing the region for hospitals and shelter in North Ossetia. Russian-language media circulated estimates of 1,500 to 2,000 dead in South Ossetia and over 30,000 refugees fleeing the conflict zone, figures cited frequently thereafter by Russian policymakers. Medvedev dispatched Emergency Situations Minister Sergei Shoigu to fly to South Ossetia to coordinate assistance to "refugees" there.[63] In a news conference, Russian foreign minister Sergey Lavrov said Georgian attacks on "Russian citizens" in South Ossetia "amounted to ethnic cleansing."[64] What was a "gross violation of international law" in Medvedev's statement on August 8, requiring a Russian response under Article 51 of the United Nations (UN) Charter in self-defense (since Russian peacekeepers and Russian citizens were attacked), soon became a more straightforward claim that Georgian forces were perpetrating "genocide" against Ossetians.

The stimulus for this important shift in framing was Putin's arrival in Vladikavkaz on August 9. There he was taken on a tour of a field hospital treating some of the wounded airlifted from South Ossetia. Russia's state-controlled Channel One covered his visit and presented witnesses recounting a series of atrocity stories. One woman tells Putin that the Georgians "burnt girls alive." Putin asks her: "So it was simply in a house? They were herded into a house?" "Yes, they were herded into a house and set alight." Another woman states: "They stabbed an

18-month baby to death. They stabbed him to death in a basement." Yet another woman declares: "A lad was saying that he himself saw how an old woman was running with two little children, and saw how they were run over by a tank." The broadcast shows Putin replying: "They have gone totally mad. It is genocide, really." A little later, he adds: "This is altogether outside any boundaries of civilized behavior." Another report immediately after showed Putin meeting with refugees, who tell him their hopes are "pinned on you." He tells them in response: "I have no doubt that you will certainly return home, of course." A woman asked: "Will we?" He answered: "Of course, by all means, you certainly will return. We will then help rebuild the city as well as help rebuild homes."[65]

Putin then gave an important speech that marked a significant change in Russian policy toward Georgia's breakaway regions, and toward secessionist movements more generally. In a white jacket and open shirt at the head of a table, seated with Russian and North Ossetian officials (all male), Putin declared that because of its actions Georgia had imperiled its legal claim to South Ossetia:

> The actions of the Georgian leadership in South Ossetia are certainly criminal. And, before everything else, this is a crime against their own people. [This is] a deadly blow against the territorial integrity of Georgia itself and that means massive damage to its national identity. It is hard to imagine how, after all that happened and all that is still happening, they will be able to convince South Ossetia to belong to Georgia. The aggression that has been unleashed has resulted in many victims, including among the civilian population, and has caused, in effect, a real humanitarian catastrophe. And that is, of course, a crime against the Ossetian people.[66]

Putin's speech foreshadowed the decision Russia would take a few weeks later in recognizing Ossetia and Abkhazia as independent states, thus breaking with its sixteen-year policy of officially recognizing both as part of the sovereign territory of the Georgian state. Russia's earlier decision to upgrade its diplomatic relations with the de facto states indicated revision of Russia's policy was already in the works, with the Duma on record as favoring recognition.[67] Putin's speech also provides insight into his initial theory of the August War, which featured Saakashvili as a rogue criminal actor. Russia, he declared, considered Georgia a brotherly nation: "This attitude will be maintained in Russia towards Georgia in the future, in spite of the criminal policy of the current rulers of that country." Georgia's desire to join NATO arose not from a desire to contribute to strengthening international peace. Rather it is "an effort to drag other countries and other peoples into their bloody adventures." Russia's actions, he declared, are legitimate and in keeping with its legal peacekeeping mandate in the area. "For centuries, Russia has played in this part of the world, in the Caucasus as a whole, a very positive, stabilizing role. [Russia] has been the guarantor of security, cooperation and progress in this region. This is how it was in the past and this will be the case in the future. Let no one doubt that."[68]

The next day Putin was back in Moscow, where he met with Medvedev. In a showpiece report for the cameras, Putin declared that the Georgian actions went "far beyond the normal limits of military operations. It seems to me that we are seeing elements of a kind of genocide against the Ossetian people." Putin recommended that Medvedev instruct the Prosecutor's Office to document these atrocities because most South Ossetians were citizens of the Russian Federation.[69] Medvedev did as Putin recommended and also began speaking of genocide. That evening he appeared on television, instructing the head of the Russian Federation Prosecutor General's Office Committee of Inquiry to gather evidence of war crimes for the subsequent prosecution of Georgia. He stated that "[t]here is no other name but that of genocide to describe the forms the Georgian forces' action has taken because these actions have become mass-scale in nature and have been directed against specific people."[70] Then Medvedev spoke with Sarkozy. The readout on the conversation was indicative of the new line emerging within the Kremlin: "Through its acts, the Georgian leadership has essentially caused irreparable damage to the integrity of its own state."[71]

News features in the Russian media provided lurid descriptions of atrocities over the subsequent days and weeks. Stories of Georgian soldiers throwing hand grenades into Tskhinval(i) basements full of civilians and running over people with tanks were staples.[72] Never short of a visceral vignette, Dmitry Rogozin, Russian ambassador to

NATO, said on August 11 that Georgian troops "shot their brother Russian peacekeepers, then they finished them off with bayonets, so we are not going to see them there any more."[73] The Russian media seized on what they said was the title of the Georgian offensive against South Ossetia: "Operation Clear Field."[74] Georgia's intention was to wipe Ossetians "from the face of the earth." The metaphor was repeatedly mentioned by top officials. For example, on August 11, Medvedev declared that the Georgians "used heavy artillery, tanks, aviation and the regular army to literally wipe Tskhinvali, its homes, hospitals and schools, from the face of the Earth."[75] The same day Putin used the metaphor in a set of remarks that underscored his disgust at how U.S. officials were interpreting the August War and responding to it (using U.S. military transport planes to transfer Georgian soldiers from Iraq directly into the conflict zone). Putin added:

> What is surprising is not even the cynicism of such actions, because politics, as they say, is a cynical business in general. What is surprising is the level of cynicism. What surprises is the ability to swap good and bad, black and white, the slick ability to pose an aggressor as a victim of the aggression, and to make the victims responsible for its consequences. But, of course, Saddam Hussein had to be hanged for destroying several Shiite villages. And the present Georgian leadership, who have simply wiped out ten Ossetian villages from the face of this planet, whose tanks were running over children and old men, who have burned civilians alive in sheds—these people, certainly, had to be taken under protection. If I am not mistaken, Ronald Reagan once said about a Latin American dictator: "Somoza is a bastard, but he is our bastard. And we will help him, we will protect him."[76]

As Sarkozy sought to negotiate a ceasefire agreement on August 12, Putin vented his ire, suggesting Saakashvili be hanged like Saddam Hussein. A Sarkozy aide later claimed the French president talked Putin out of pursuing regime change like Bush in Iraq.[77] In contrast to Putin's crude rhetoric of retribution, Medvedev's press conference on the ceasefire with Sarkozy was full of the rhetoric of humanitarian intervention and international law. No sovereign state had the right to do whatever it pleases, he said. "Faced with the killing of several thousand citizens," the Russian state had to take the appropriate action. Georgian forces were perpetrators of "ethnic cleansing." "Under international law these acts are deemed a crime, just as the murder of thousands of citizens is called 'genocide.' There can be no other name for these acts."[78] Asked whether Russia still recognized Georgia's territorial integrity, Medvedev declared that territorial integrity is "decided by people's desire to live in one country." International law "has given us numerous very complicated cases of peoples exercising their right to self-determination and the emergence of new states on the map. Just look at the example of Kosovo." The question left hanging was how much the August War was the Kosovo War redux, a rescue mission in the Caucasus.

The Russian government responded to growing criticism of its occupation of large parts of Georgia by framing its actions within responsibility to protect norms, even though it opposed NATO's evoking these norms in Kosovo.[79] Russian officials went further, arguing that Russia's military response was more virtuous than NATO's 1999 Kosovo intervention, which, Foreign Minister Lavrov explained in a *Wall Street Journal* opinion piece, "degenerated into attacks on bridges, TV towers, passenger trains and other civilian sites, even hitting an embassy." Russia used force "in full conformity with international law, its right to self-defense, and its obligations under the agreements with regard to this particular conflict." Lavrov cited the most infamous failure of UN peacekeeping, a paradigmatic example motivating "responsibility to protect" thinking: "Russia could not allow its peacekeepers to watch acts of genocide committed in front of their eyes, as happened in the Bosnian city of Srebrenica in 1995."[80] Rogozin made a similar case in the *International Herald Tribune* about the need to respond to Saakashvili's order to "wipe Tskhinvali... from the face of the earth." His credibility in making such an argument was poor given that he was a cheerleader for Serbian nationalism in the mid-1990s and was even photographed in Sarajevo a few months after Srebrenica with Ratko Mladić, the Bosnian Serb commander indicted for directing the Srebrenica massacre (Mladić gifted him his command cap). Rogozin also justified Russia's actions by citing nonterritorial conceptions of the Russian state. Because Russian peacekeepers operating legally on Georgian territory were attacked, the Georgian aggression in South Ossetia "should be classified as an armed attack on the Russian Federation giving grounds to fulfill the right to self-defense—the right of every state according to Article 51 of the UN Charter." Also, the "use of force to defend one's compatriots is traditionally regarded as a form of self-defense." The United States, Israel, France, and others have taken military action to protect their citizens overseas.[81] Thus, because Russia had peacekeepers and passport-holders in South Ossetia, Georgia's attack on South Ossetia was really an attack on Russia.

Constituting the meaning of the August War in Russian geopolitical culture were older precedents and analogies from European history (including those also cited by Saakashvili and neoconservatives in the United States). Russia, President Medvedev explained, was enforcing peace in accordance with the United Nations (UN) Charter because one of the lessons of the 1938 Munich Agreement was that one should not appease aggressors.[82] Saakashvili was portrayed as Hitler in some speeches and as Saddam Hussein in others, but most commonly he was framed in the base archetype shared by both, a madman and blood-thirsty lunatic.[83] Announcing the ceasefire agreement on August 12, Medvedev stated that "there are some people who, unlike normal people, once they've smelt blood it is very hard to stop them."[84] The August War was quickly folded into the pantheon of Russian war memory. Medvedev spoke about it at the sixty-fifth anniversary of the Battle of Kursk on August 18.[85] The same day he gave out medals to Russian peacekeepers that were the contemporary Russian equivalent of Soviet and tsarist military honors.[86]

The memory of the Great Patriotic War was consciously evoked by a remarkable spectacle organized in Tskhinval(i) on the evening of August 21, 2008. The internationally renowned conductor Valery Gergiev, an ethnic Ossetian born in Moscow and raised in Vladikavkaz, led a classical musical performance by the Mariinsky Orchestra of Saint Petersburg on the steps of the bombed-out parliament building (see Figure 5.1). Condemning the Georgian aggression and conveying his thanks as an Ossetian for the Russian army's response, he described Tskhinval(i) as a "hero city" that reminded him of pictures of Stalingrad. The last piece played by the orchestra was Shostakovich's Seventh Symphony, a mournful hymn to the suffering of Leningrad at the hands of the besieging Nazis. Broadcast live on Russian television, the concert was a highbrow cultivation of patriotic affect and a clear incorporation of

FIGURE 5.1 Valery Gergiev conducts his Mariinsky Orchestra in front of the destroyed local assembly building in Tskhinval(i), August 21, 2008 [Maxim Shipenkov, EPA].

Tskhinval(i)'s recent experience into a catalogue of Russian suffering and triumphs through adversity.[87]

In the immediate aftermath of Saakashvili's move on South Ossetia, debate resumed within the Russian power structure about Russia's relationship to the de facto states in post-Soviet space. Imperial nationalists had long argued that these should be recognized as independent states and even annexed by Russia. Putin's speech in Vladikavkaz marked the public articulation of a new tilt toward this attitude within the Kremlin. Though he considered the West's use of the "responsibility to protect" creed to justify NATO's intervention in Kosovo as wrongheaded and cynical, Putin decided Russia should nevertheless use the very same creed to advance its own national interests in the Caucasus. The doctrine, however, challenged the absolute sovereignty of the state, which Putin saw as the foundation of international order. It also provided encouragement to secessionist movements, another position that was anathema to Putin. Georgia's actions, however, placed the contradictory policy impulses before Russia's leadership like no other event before. Medvedev debated the issue with his advisers, and in the end, the Russian leadership decided to change official policy and recognize both Abkhazia and South Ossetia as independent states. Transnistria, much to the chagrin of its leadership and supporters in the Duma, was not recognized.

Medvedev announced the recognition on August 26, 2008. In describing Georgian state actions as innately fascist and genocidal (Georgia's attack on Tskhinval(i), for example, was described as a "blitzkrieg"), Russia adopted the central storyline of Ossetian nationalism.[88] "Tbilisi made its choice during the night of August 8, 2008. Saakashvili opted for genocide to accomplish his political objectives. By doing so he himself dashed all the hopes for the peaceful coexistence of Ossetians, Abkhazians and Georgians in a single state." Medvedev explained his decision was based on recognition of the situation on the ground where Ossetians and Abkhazians had expressed their desire for statehood independent of Georgia.

The Kremlin launched a public relations effort to justify the move to the international media. Citing humanitarian concerns as central to his decision, Medvedev told the U.S. television network CNN: "the choice was not easy to make, but it represents the only possibility to save human lives." He cited the legal justification used by the United States and its allies to recognize the independence of Kosovo, namely that it was a unique case. Though Russia rejected this claim, he appealed to the same legal principle, consciously reusing the legal language of the United States on Kosovo: "Our colleagues said more than once that Kosovo was a *casus sui generis*, a special case. But in that case, we can also say that South Ossetia and Abkhazia are also *sui generis*." Asked whether there was not a "double standard" at work in not recognizing Kosovo but recognizing the two breakaway regions, he pronounced the two cases different. The Georgian situation, he explained, "existed for 17 years, during which ethnic cleansing was conducted and cases of genocide took place, both in the early 90s and now it has happened again."[89] Yet they were similar because Russia's main mission, he explained to the BBC, "was to prevent a humanitarian disaster and save the lives of people for whom we are responsible.... We had no choice but to take the decision to recognize these two subjects of international law as independent states. We have taken the same course of action as other countries took with regard to Kosovo and a number of similar problems."⁹⁰ In an interview with the Kremlin-sponsored English-language channel *Russia Today*, he explained the recognition as "designed to prevent genocide, the extermination of peoples, and to help them get back on their feet again."⁹¹

Despite considerable diplomatic efforts, Russia's unilateral recognition of Abkhazia and South Ossetia failed to garner any international support. Furthermore, de facto South Ossetian president Kokoity went off script when he indicated that independence was only a stepping-stone to uniting with North Ossetia and joining the Russian Federation.⁹² Yet, irrespective of these legitimacy failings, Russia's recognition of the two Georgian de facto states was a significant moment in the evolution of Russian revanchism (Figure 5.2). Russia was publicly breaking with the prevailing norms expressed in the Helsinki Final Act and, thereafter, in the 1990 Charter of Paris. At the end of August, President Medvedev gave an interview to Russian television channels in which he outlined five principles guiding Russian foreign

FIGURE 5.2 "Thank You Russia!" Billboard, Tskhinval(i), March 2010. Author photo.

policy. Central to these was the contention that world politics should be multipolar: "We cannot accept a world order in which one country makes all the decisions, even as serious and influential a country as the United States of America. Such a world is unstable and threatened by conflict." Russia would respect the fundamental principles of international law and seek friendly relations with other states, but it would protect "the lives and dignity of our citizens, wherever they may be." Furthermore, Medvedev added: "[A]s is the case of other countries, there are regions in which Russia has privileged interests. These regions are home to countries with which we share special historical relations and are bound together as friends and good neighbours. We will pay particular attention to our work in these regions and build friendly ties with these countries, our close neighbours." Left unsaid was what happened when assertions of "privileged interests" were unwelcome by neighboring states.[93] The question was far from theoretical.

Three years later, Medvedev interpreted the August War as a case of Russia successfully intervening to stop NATO expansionism. Addressing soldiers in Vladikavkaz he declared that their actions had saved lives and curbed a threat coming from Georgia. In unscripted remarks he added: "If we had faltered in 2008, the geopolitical arrangement would be different now and a number of countries attempting to artificially drag themselves into the North Atlantic Alliance would probably be there [in NATO] now."[94] Russia prevented genocide and in so doing it also prevented NATO enlargement.

The U.S. Storyline: Russian Aggression and Georgian Victimhood

Many senior figures within the U.S. government were on holiday or overseas when Saakashvili launched his all-out assault of Tskhinval(i) and the Russians mobilized to respond. Both Dan Fried and Matt Bryza were in communication with the Georgians before their offensive and thereafter. Bryza indicated he did not sleep for four days. He reported to Dan Fried who, in turn, reported to William Burns, John Negroponte, and Secretary of State Condoleezza Rice. These experienced senior officials had many reservations about Saakashvili but were not willing to condemn him publicly. From a holiday cabin in West Virginia, Rice spoke repeatedly with

Saakashvili and other international principals including Lavrov. At the Defense Department, a young staffer who had spent a year in Georgia was at the center of the scramble to gather real-time information from the field. Some came in the form of text messages.[95] He and Deputy Assistant Secretary of Defense for European and NATO Policy Daniel Fata, were Georgia enthusiasts but their boss, Secretary of Defense Gates, who was in Munich, was much cooler and wary of a hasty response.

President Bush first spoke about the Georgia crisis on the evening of August 9 at his hotel in Beijing. He expressed his "deep concern" about the situation and noted that attacks were occurring in regions of Georgia far from the conflict zone. The statement urged "an immediate halt to the violence and a stand down by all troops," as well as for the parties to return to status quo positions. Earlier that day he met with the U.S. women's beach volleyball team. Images of him joking with the female athletes in bikinis while Georgia was at war infuriated some U.S. conservatives.[96] Bush's balanced language in his call for a ceasefire contrasted with Republican presidential nominee McCain's condemnation of "Russia's aggression in Georgia" and call for NATO to review how it can "contribute to stabilizing this very dangerous situation." After the president spoke, Secretary Rice issued a statement, echoing his call for an immediate ceasefire and for Russia to respect Georgia's territorial integrity.[97] At the United Nations, U.S. ambassador John Negroponte condemned the dangerous and disproportionate nature of Russia's actions. Citing disproportionality allowed a storyline of Russian aggression and Georgian victimhood to rebound from the fact that Georgia started the war. (The Russians claimed that their attacks beyond South Ossetia were aimed at crippling the "Georgian war machine," just as NATO attacked Serbia proper during the 1999 Kosovo war.)[98] Admiral Mike Mullen, the chairman of the Joint Chiefs, was one of the few officials who reached his Russian counterpart General Nikolai Makarov. Both men had regular phone calls over the subsequent days that helped reduce tensions. Remarkably, Mullen was able to arrange for U.S. military aircraft to enter the combat zone to return Georgian troops from Iraq before a ceasefire was arranged. He also sought to talk the Russians out of marching to Tbilisi to seize Saakashvili and punish him for war crimes.[99]

Briefing reporters in Beijing the following day, Deputy National Security Advisor James Jeffrey indicated that the White House had made it clear to the Russians that disproportionate Russian actions would have a significant, long-term impact on U.S.-Russian relations. Asked if the United States had plans to send military aid to its ally, Jeffrey said multilateral diplomacy to achieve a ceasefire was where the United States was putting its emphasis.[100] The vice president's office called Jeffrey to confirm if this indeed was the U.S. policy emphasis, as it had received a call from a desperate Saakashvili requesting the United States to send Stinger missiles to Georgia at once. Jeffrey indicated that Bush had not yet ruled out the use of military force or aid.[101] Vice President Dick Cheney decided the White House needed to send a more forceful signal. His press secretary subsequently briefed the media that Cheney told the Georgian president that "Russian aggression must not go unanswered."[102] That evening at the United Nations, U.S. ambassador Negroponte broke diplomatic confidentiality by revealing that Russian foreign minister Lavrov had told Secretary Rice in a phone call earlier that day that Saakashvili "must go." [103] This, he indicated, "is completely unacceptable and crosses the line."[104] The following day, just before his departure from Beijing, a U.S. sports journalist interviewed President Bush. Asked about Georgia, he repeated his policy talking points and then added that he "was very firm" with Putin and Medvedev over the crisis.[105]

Bush returned to a rising chorus of criticism from those who considered him anything but firm. Cheney was critical and wanted an answer to Saakashvili's plea for weapons. Bush's former UN ambassador John Bolton was also critical, later charging that the United States "fiddled while Georgia burned."[106] Its delay in issuing a strong rhetorical response "demonstrated to the Russians that they didn't have any reason to fear any sustained political opposition."[107] There was a return of longstanding tensions within the administration between active interventionism and cautious realism. Bush met with his National Security team at the White House upon his return. Secretary Rice felt there was "a fair amount of chest beating" among officials and "all kinds of loose talk" about a muscular response.[108] To Bush's surprise, his national security advisor stephen Hadley told him he needed to poll his national security advisors as to whether they recommended that the United States put troops on the ground in Georgia.[109] Bush posed the question to his advisers, and they proceeded to discuss a series of military options that ranged from arming the Georgians to bombing the entrance to the Roki tunnel to introducing U.S. ground troops. In the end, no one recommended that the United States use military force (arming the Georgians was a separate matter about which public information has not been released so far). Even Cheney conceded that the use of U.S. force "would be a mistake."[110]

Bush continued to be the target of blistering criticism from former allies. On the evening of August 12, an editorial for the following day's *Wall Street Journal*, an archconservative bastion that had strongly supported Bush during his presidency, was posted online. It charged that the Bush administration "has been missing in action, to put it mildly" over Georgia. Georgia burned and U.S. credibility was on the line while Bush engaged in Olympic tourism. Though the United States was not going to war with Russia over a non-NATO ally, it nevertheless had "forceful diplomatic and economic responses at its disposal." The editorial compared Georgia to Berlin during the Cold War and called for a "Tbilisi airlift" to ferry military and humanitarian supplies to the Georgian capital, with a high-profile administration figure on board one of the planes. Paraphrasing General Lucius Clay (the U.S. military commander in Berlin), it concluded: "Whether for good or bad, how the U.S. responds to Russia's aggression in Georgia has become a symbol of American credibility."[111]

The editorial touched a raw nerve in the administration. For the first time the White House released a "Setting the Record Straight" statement the next day that directly responded to the editorial.[112] Bush's national security team by this point was concerned that the Russians were moving toward Tbilisi and contemplating regime change. Bush met with his advisers on August 13 and thereafter strongly condemned the Russians in a public statement, a move that mollified some conservatives.[113] He ordered Secretary Rice to Tbilisi to further the international diplomacy led by the French to end the crisis.[114] Echoing the Cold War register of critics, Bush said that Secretary Rice would "continue our efforts to rally the free world in defense of a free Georgia."[115] He also ordered U.S. military cargo planes and a warship to deliver "humanitarian aid" to Georgia, a move that seemed like the "Tbilisi airlift" the *Wall Street Journal* editorial had demanded. Mullen worked with the Russians to "de-conflict" the airspace, though they were skeptical the aid was purely humanitarian, and later charged the United States with supplying arms. But the Russians did not advance their positions further. Some administration supporters credited Bush's action with saving Saakashvili, though it is likely Putin made his calculations for other reasons.

The August War became an issue in the U.S. presidential race for only a brief period before being overtaken by a spiraling financial crisis in September. As news of fighting in South Ossetia broke, Randy Scheunemann at McCain campaign headquarters worked the phones to get confirmation that Russian troops had really invaded. He reached the Georgian national security advisor Alexander Lomaia who briefed him on the situation.[116] McCain appeared before the press and read a statement demanding that Russia "immediately and unconditionally cease its military operations and withdraw all forces from sovereign Georgian territory." The United States, UN, and OSCE should immediately "put diplomatic pressure on Russia to reverse this perilous course it has chosen."[117] The Obama campaign's initial statement, written by campaign aide Ben Rhodes, called on both sides to "show restraint"[118] and avoid escalation to full-scale war. This gave the McCain campaign an opening to attack Obama for not immediately taking the side of the victims of Russian invasion.[119] The attack went further. Obama's response to the crisis was "at odds with our democratic allies and yet so bizarrely in sync with Moscow."[120] The McCain campaign sought to leverage the crisis to emphasize their candidate's foreign policy experience relative to Obama.[121] McCain had visited South Ossetia and knew the players in Georgia well. Indeed, McCain was unusually close to the crisis. He had daily phone conversations with Saakashvili, sometimes more than one a day.[122] He famously sympathized with the Georgian president by telling him "today we are all Georgians." McCain also stressed that Moscow's move had larger goals: intimidating Ukraine and imperiling the BTC pipeline.[123] Hitting back, Obama's surrogates made an issue of Scheunemann's past advocacy work for Georgia and charged that McCain's campaign was run by Washington lobbyists.[124]

Senator Barack Obama moved to neutralize the war as a potential campaign issue. At a stop in California, he declared that Russia had "invaded" Georgia. Thereafter he met for the first time with Michael McFaul who made the case that Russia was indeed the guilty party.[125] Obama interrupted his Hawaiian vacation to personally read a long considered statement condemning Russian "aggression," escalation, and violation of the space of another country.[126] After the fighting ended, the war retreated from the headlines but foreign policy experience was an issue anew. Obama sought to neutralize it by choosing Senator Joe Biden, chair of the Senate Foreign Relations Committee, as his running mate on August 22 (Biden had just returned from Georgia and was a strong advocate for immediate reconstruction aid to the country). The unsteady rollout of McCain's vice presidential pick, Sarah Palin, upended the experience critique. When the August War came up in the first presidential debate, Obama allowed no daylight between himself and McCain on the issue. McCain, for his part, warned: "Watch Ukraine. This whole thing has got a lot to do with Ukraine, Crimea, the base of the Russian fleet in Sevastopol... watch Ukraine."[127]

The White House and U.S. Congress unveiled a $1 billion aid package for Georgia in early September and, as the United States grappled with a financial crisis, nevertheless passed it by the end of the month with bipartisan support.[128] The money helped keep Saakashvili and his party in power for a few years until a former supporter, the billionaire Bidzina Ivanishvili, provided the necessary resources for the opposition to challenge his hold on power (Saakashvili's party eventually lost office after a prison torture and rape scandal, with echoes of Abu Ghraib, broke before the October 2012 elections). The United States did not provide new weapons, but it did provide $50 million in security assistance.[129] Bush's administration nevertheless remained committed to Georgia and sought to lock in a long-term commitment by signing a U.S.-Georgian Strategic Framework days before it left office. The cause may have been damaged, but it would go on.

The Georgian authorities record that 413 persons perished on the Georgian side during the August 2008 war: 166 military, 16 policemen, and 220 civilians. Over two thousand were wounded.[130] Approximately twenty thousand Georgians were displaced from their homes in the region. Most now live in the internally displaced settlements built for them with World Bank and other international funding immediately after the war.[131] Russian actions destroyed significant parts of Georgia's military infrastructure and assets. South Ossetia's authorities list the deaths of 364 people during the war, with more than two hundred wounded.[132] Russia listed its casualties as 67 soldiers and 283 wounded.[133] Ossetian villages were looted during the first days of the war. Thereafter Georgian villages were looted and subsequently destroyed. When I visited the Didi Liakhvi valley in 2010, the Georgian enclave north of Tskhinval(i) was a wasteland of gutted houses (see Figure 5.3).[134] These have now been flattened even further. Some ethnic Georgians still live in South Ossetia in the Akhalgori (Leningor) region but they tend to be older residents. In the last number of years, there has been a concerted effort to physically demarcate the boundary line between South Ossetia and Georgia with posts, signs, and razor-wire fencing. Critics label the strategy "borderization" and charge Russia with "creeping annexation" of Georgian territory. With locals living in the borderlands frequently harassed and arrested, South Ossetia is now further isolated from contemporary Georgia than at any time in its modern history.[135]

FIGURE 5.3 Gutted houses, Kurta, former "capital" of Georgian Provisional Government in South Ossetia, March 2010. Author photo.

In January 2016, judges at the International Criminal Court (ICC) at The Hague authorized prosecutors to open an investigation into alleged war crimes and crimes against humanity committed "in and around South Ossetia" from July 1 to October 10, 2008. The judge found that there was a reasonable basis to believe that both categories of crimes had been committed. The list of war crimes included attacks against the civilian population, willful killing, intentionally directing attacks against peacekeepers, and the destruction and pillaging of property. The crimes against humanity named by the ICC included murder, forcible transfer of population, and persecution.[136] In requesting the investigation, the Office of the Prosecutor indicated that "between 51 and 113 ethnic Georgian civilians were killed as part of a forcible displacement campaign conducted by South Ossetia's *de facto* authorities, with the possible participation of members of the Russian armed forces. Between 13,400 and 18,500 ethnic Georgians were forcibly displaced and more than 5,000 dwellings belonging to ethnic Georgians were reportedly destroyed as part of this campaign." While describing war outcomes, the remit of the investigation will encompass the prelude to the launch of the war itself.[137]

The August 2008 war triggered great power storylines premised on rescue missions if not rescue fantasies. The Russians were rescuing the Ossetians from purported genocide; and the Americans were rescuing the Georgians, through a Berlin-like humanitarian aid airlift and subsequent reconstruction aid, from Russian aggression and regime change. In both cases, a plucky little nation battled a neighborhood Goliath. Concealed by the affective terms of these geopolitical storylines was a nasty little war of competing revanchisms.

6

Places Close to Our Hearts

WHEN U.S. PRESIDENT GEORGE W. BUSH first met Russian president Vladimir Putin, he praised him as "an honest, straightforward man who loves his country." Bush indicated that, more than a decade after the Cold War ended, it was "time to move beyond suspicion and towards straight talk."[1] Thereafter, both presidents established a good working relationship based, in part, on candor and frankness. Putin's speech at the Munich security conference did not please his hosts, but it had the virtue of clarifying important differences.[2] Similarly, his speech to the North Atlantic Treaty Organization (NATO)–Russia Council meeting in Bucharest was forthright and blunt. The compromise language of the Bucharest Declaration—Georgia and Ukraine "will become members of NATO"—was a personal rebuke to the Russian leader, for he had made it clear that NATO expansion to these countries was a "red line" for Russia.[3] Two years earlier Russian foreign minister Sergey Lavrov warned publicly that Georgia and Ukraine joining NATO could lead to "a collossal shift in global geopolitics."[4] But those promoting NATO membership for both believed the Russian position amounted to anachronistic sphere-of-influence thinking, and they were determined to prevent what they described as a "Russian veto" on NATO expansion. Putin's remarks on Georgia in Bucharest—discussed in chapter 4—attracted few headlines. His alleged comments on Ukraine, however, were viewed with alarm at the time by some and considered ominously prophetic by many after 2008, and especially so in the spring of 2014. According to an unnamed NATO country official, an irate

Putin turned to Bush and said: "George, you do realize that Ukraine is not even a state. What is Ukraine? Part of its territory is Eastern Europe but the greater part is a gift from us!"[5] Putin reportedly then indicated that should Ukraine join NATO, the state may cease to exist. Russia would then tear off Crimea and eastern Ukraine from the rest of the country. Six years later it appeared Russia was doing precisely this.

As with other notorious Putin quotes, what he said before the NATO-Russia Council in Bucharest in 2008 is more complicated than its media presentation. No official transcript of his speech was ever released. According to the unofficial transcript at least (which he may not have followed in actual delivery), Putin never said "Ukraine is not even a state" or threatened that Russia would take Crimea and eastern Ukraine should the country join NATO. Nonetheless, the unofficial transcript reveals his conviction that Ukraine would disintegrate if it joined NATO. Putin may have believed he was merely describing a geopolitical reality but the audience understandably heard a warning that was easily interpreted as a threat.[6] His remarks are a necessary place to begin comprehending how the 2014 governance crisis in Ukraine ended up precipitating a Russian invasion of Crimea and eastern Ukraine. Both this chapter and the next examine the geopolitics of the 2014 Ukraine crisis. This chapter begins first with some background on Ukraine and the struggles over its geopolitical orientation in the decades preceding 2014. It then provides a brief analysis of the proximate causes of Russia's invasion of Crimea, which emerged from both structural changes in Russia's approach to Ukraine in Putin's third term and fluid contingent circumstances in 2014. Thereafter it turns to how the annexation was presented and staged to the Russian public, the primary audience, and unavoidably to the rest of the world as well. The chapter ends by reviewing how the drama and storylines of the Crimean annexation resonated with ordinary people in Ukraine and southeast Ukraine (with the important exclusion of the Donbas) by reviewing the results of original public opinion survey research from these regions conducted in December 2014.

Ukraine Geopoliticized

Ukraine's fate has always been tied to its geopolitical location (see Figure 6.1). Ukraine might not be interested in geopolitics but geopolitics, to

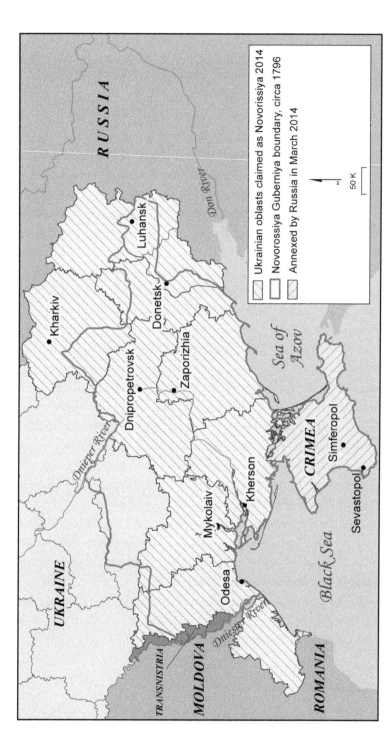

FIGURE 6.1 Map of Southeast Ukraine.

paraphrase one of its famous sons, is interested in Ukraine.[7] Caught between the two most murderous empires in twentieth-century Europe, Ukraine was a space made and remade by the oscillations of imperial power on the European continent. In the wake of the Nazi-Soviet pact of August 1939, the Soviet Union occupied and annexed lands in eastern Poland, later transferring the territory to the banks of the river Bug into the Ukrainian Soviet Socialist Republic. Territories were also transferred from Romania: the Budjak region lying southwest of Odesa along the Black Sea, between the Dnieper and Danube river outlets and Chernivtsi, the northern part of historic Bukovina in 1940; as well as some Black Sea islands in 1945. At the conclusion of the war, the region known variously as SubCarpathia, Hungarian Ruthenia, Carpathian Ruthenia, or Carpatho-Ukraine, part of Czechoslovakia after World War I and then awarded by the Nazis in 1938 to Hungary, was also annexed and became Zakarpattia oblast. As already noted, Soviet Ukraine also acquired Crimea as a "gift" by administrative fiat in 1954. All of these agglomerations of territory were made by a small ruling elite within an imperial power structure headquartered in Moscow, with little consideration of local sentiment or the coherence and legitimacy of the resultant polities. Soviet Ukraine was designed in Moscow from territories conquered and controlled by the Red Army: calculations at the top of the Kremlin power vertical trumped all others.

Wartime genocide and fratricide, followed by postwar population expulsions and transfers, altered Ukraine's ethnic composition significantly. In 1939 there were about four million Russians in Ukraine, constituting about 12 percent of the population; by 1959 that figure had increased to seven million or 16 percent of the population.[8] Decades of socialization within the overwhelmingly Russophonic lifeworld of the Soviet Union induced certain homogeneities and commonalities across Ukraine. However, regional differences remained significant and enduring, with historic Galicia in the west and the Donbas basin in the east representing somewhat cliched geographic opposites on a spectrum of divisions over history, identity, and orientation toward neighboring Russia.[9] Independence in late 1991 enabled a revival of Ukrainian national consciousness and a Ukrainophonic cultural sphere—Ukrainian, historically a lesser-status language, was declared the official state language in 1991— but Russian remained the predominant language of administration in the country.¹⁰ Postwar politicians struggled to bridge Ukraine's regional differences; some presidential election results tended to affirm the idea that there were "two Ukraines," a Ukrainophone west and center that was pro-European and a Russophone southeast that was pro-Russian. The situation was much more complex.¹¹ Not only were there many distinct regional expressions of Ukraine, but within regions there were important contrasts and urban/rural divides.¹² Further Russophones were assumed to be pro-Russian, an unjustified assumption.¹³ Much more consequential was the geographic distribution of the 8.3 million self-identifying ethnic Russians in Ukraine who made up 17.3 percent of the population in the 2001 census (down from 22.1 percent in the last Soviet census in 1989), and the distinct regional identities this shaped.¹⁴ In the nationalizing state that was independent Ukraine, they were numerically the largest ethnic minority in Europe, though consciousness of "minority status" was very much a function of location and the discourse prevailing in Kyiv. Their spatial distribution reveals four distinct Russian ethnogeographies in Ukraine: absolute majority, substantial minority, limited minority, and small minority spaces. Demography was not destiny, however; what also mattered was location and regional particularity.

First, Crimea was the only region of Ukraine in 2014 where ethnic Russians were still a majority of the population. This fact was a consequence of Stalin's expulsion of the peninsula's formerly titular people, the Crimean Tatars; it was also due to postwar migrations to the Black Sea and Soviet naval complex at Sevastopol (incorporating the city of Balaklava from 1957), which remained under direct rule by Moscow even after Khrushchev's 1954 transfer of the peninsula to Ukraine. The region narrowly voted to endorse Ukrainian independence in the referendum of December 1, 1991, which provided a crucial fillip to the breakup of the Soviet Union. Thereafter, separatists' efforts by local ethnic Russians came to naught, despite support from the Russian Duma, powerful mayors like Yury Luzkhov in Moscow and Anatoly Sobchak in Saint Petersburg, and Yeltsin's administration. After the election of Leonid Kuchma as Ukrainian president in July 1994, Yeltsin's government backed off from supporting Crimea's Russian nationalists and prioritized its relationship with Kyiv and the West. It was also preoccupied at this time with its own secessionist challenge in Chechnya.¹⁵ Under the Budapest Memorandum of December 1994, Ukraine facilitated transfer of nuclear weapons on former Soviet bases on its territory to Russia in return for international security assurances about its political independence and territorial integrity. The following year a treaty of friendship was drawn up between Ukraine and Russia, under which Ukraine gave Russia's Black Sea Fleet in Sevastopol a long-term lease in return for Russia's recognition of Ukraine's territorial integrity, but it took two years before it was finally signed in May 1997 during Yeltsin's first visit as Russian president to Kyiv.[16] Despite both these agreements, the issue of Crimea's status rankled many Russian nationalists, both in Russia and Crimea. Pro-Russian sentiment in Crimea remained a latent lever of influence in the Russian-Ukrainian relationship.

Second, after Crimea, the largest minority concentration of ethnic Russians was in the Donbas basin comprising the oblasts of Donetsk and Luhansk. Coal deposits were first developed in this area in the late nineteenth century. Under the Soviet Union, the area became a major center of heavy industry—coal mining, iron ore production, steel production, and machine building—with dense networks of supply chains and markets surrounding it. Captured by the Wehrmacht during World War II, the area saw fierce battles to reclaim it for the Soviet Union. Yet strikes by Donbas miners in the spring of 1991 hastened the institutional crisis of the Soviet Union and its sudden collapse. In the last Soviet census in 1989, 43 percent of Donetsk and 44.8 percent of Luhansk oblasts were ethnic Russian, while ethnic Ukrainians had become a slight majority in both; 66 percent of the Donbas population stated that Russian was their "native tongue" in 1989.[17] The Soviet collapse left the region disoriented, and its riches were soon captured by local mafia clans and political bosses. In the 2001 census, 38.2 percent (over 1.8 million people) of Donetsk and 39 percent (just less than a million people) of Luhansk oblasts declared themselves ethnic Russians. These ethnic Russian populations tended to have higher concentrations in the cities of Donetsk and Luhansk. Due to its modern historical formation as a space forged by Soviet industrialization, the Donbas developed a distinct regional identity that endured after the Soviet collapse. Embedded especially within the Russian territorial economy, it functioned almost as a third space between independent Ukraine and Russia. Partisans of both nations, however, promoted historical myths that claimed it as part of their respective national patrimonies. As violence engulfed eastern Ukraine in 2014 the Donbas' distinctive sense of place was a resource pro-Russian separatists sought to capture and channel.[18]

Third, beyond Crimea and the Donbas, ethnic Russians made up a limited minority—from 25.6 percent to 14.1 percent of the population—in a series of six other oblasts in southeast Ukraine, what can be termed "southeast Ukraine 6" (abbreviated here as SE6). Here the ethnic Russian population in 2001 was just over one in four in Kharkiv and one in five in Odesa, with higher concentrations in the two cities at the center of these oblasts. Southeast Ukraine is often envisioned as a distinct region because it is a majority Russophone region. However, a more disaggregated analysis reveals a strong urban/rural split, with most urban centers predominantly Russian-speaking and the surrounding rural villages largely Ukrainian-speaking. Yet even this divide is misleading, for most Ukrainians are functionally bilingual and many speak a mixed language (derogatorily called *surzhyk*) even if they do not recognize themselves speaking it as such.[19] Ethnicity and language usage, thus, are more fluid, contingent, and negotiated categories than census statistics, with flawed counting categories that are a function of politics, would lead one to believe.[20] Beyond Crimea, the Donbas and SE6 are spaces where ethnic Russians are usually a small minority—a range from 1.8 percent to 9.4 percent—of the population. Geographically these regions are in central and western Ukraine. One important exception is the capital city of Kyiv, where 13.1 percent of the population in 2001 declared themselves as ethnic Russian Ukrainians.

From the outset, geopolitical theorists in the capitals of the former Cold War divide have conceptualized Ukraine as a "prize" in the emergent security architecture of the new Europe. Multiple conceptions of Ukraine existed in Russian geopolitical culture—all imperialistic in the eyes of Ukrainian nationalists. At one end of a spectrum were conceits found in Orthodox and tsarist discourses. These saw Ukraine and Russia as one space and one people, a singular Orthodox Christian people sharing the common ancestral home of ancient Rus. Historically, the geographical territory of the Cossack Hetmanate was known, after its annexation by the Kremlin, as Little Russia. In tsarist discourse its residents were identified as Little Russians next to White Russians (Belarus) and Great Russians (Russia). Soviet discourse had the virtue of at least recognizing Ukrainians as a separate nationality, but tsarist paternalism implicitly informed their common designation as a "little brother" nation to Russia. Coexistent with these imaginations were more troubled visions of Ukraine as an artificial agglomeration of disparate territories. The idea of Novorossiya, examined in greater depth in the next chapter, accented the Russianness of former imperial lands within Ukraine. The notion of a *Russki Mir* (Russian World) foregrounded Russian-language speakers (ethnic Russians, compatriots and "cultural Russians") as a distinct community within Ukraine.[21] Finally, as noted in chapter 2, many figures within Russian geopolitical culture held that Crimea and other territories of Ukraine were eternally Russian. A "red-brown" network of political entrepreneurs on the far right and left in post-Soviet Russia saw the USSR's collapse as a geopolitical injustice against this eternal Russia. Many considered post-Soviet Ukraine an artificial buffer state. Writing in 1997, the most notorious of Russia's revisionist geopolitical thinkers, Aleksander Dugin, declared: "Ukraine as a state makes no geopolitical sense."[22]

The Ukrainian power elite was interested in geopolitics, but mainly to the extent that it gave them leverage with Moscow and further resources to advance their own power, and capital accumulation opportunities, relative to their domestic competitors. Even before the Soviet collapse, Ukraine joined the North Atlantic Cooperation Council in 1991. Thereafter it was the first state in the Commonwealth of Independent States (CIS) to join Partnership for Peace in 1994; and it began to participate in annual Sea Breeze exercises in the Black Sea, despite the geopolitical tensions and paranoia this stoked in Odesa and Crimea (not to mention within Russian military circles, which feared NATO control over Sevastopol, home of the Black Sea Fleet since 1783). Few were as prominent in advocating NATO expansionism into former Warsaw Pact lands (beginning with the incorporation of his own homeland of Poland) as Zbigniew Brzezinski. In a 1994 article in *Foreign Affairs* he underscored the importance of Ukraine's independent sovereignty to the West: "It cannot be stressed strongly enough that without Ukraine, Russia ceases to be an empire, but with Ukraine suborned and then subordinated, Russia automatically becomes an empire."[23] The observation was glib in the extreme (as Chechens and others could attest), but it acquired aphoristic power within Euro-Atlantic geopolitical culture as a statement on Ukraine's geostrategic significance. Secure the "prize" of Ukraine for Euro-Atlantic structures, and Russia becomes a neutered imperial power, supposedly thereafter more likely to turn into a "normal nation-state." To the Russian national security community, who were also reading Brzezinski, the lesson was the opposite: allow Ukraine to join Euro-Atlantic space and Russia's power will be fatally weakened.

Ukrainian soldiers served as part of the Implementation Force (IFOR) in Bosnia-Herzegovina in 1996. A NATO Information and Documentation Centre opened in Kyiv in May 1997. A few months later a Charter on a Distinctive Partnership, establishing a NATO-Ukraine Commission (NUC), was signed at the NATO summit in Madrid. That same year Brzezinski published *The Grand Chessboard: American Primacy and Its Geostrategic Imperatives,* which argued for a strong linkage between the widening of the European Union (EU) and the expansion of NATO, and envisioned Ukraine joining both as part of a "democratic bridgehead" against Eurasia. If Ukraine is to survive as an independent state, Brzezinski averred, "it will have to become part of Central Europe rather than Eurasia, and if it is to be part of Central Europe, then it will have to partake fully of Central Europe's links to NATO and the European Union."[24] Brzezinski wrote elsewhere the same year that his attitude toward Euro-Atlantic expansion is that "we should strive to have our cake and eat it too."[25]

This attitude hints at the contradictory double game wrapped in wishful thinking that surrounded NATO expansion in these years. Russia was to be listened to, yet NATO expansionism should proceed regardless of what Russia said. At the same time as NATO was affirming fears newly independent states had about Russia, it also held that there was no zero-sum game with Russia in the offing. There was, according to this view, no reason for Russia's state defense planners to feel nervous as the Cold War alliance they had trained to fight expanded to incorporate former allies. Russia should accept how NATO presented itself, as a stabilizing presence in Russia's neighborhood. Double games and wishful thinking were met for a while with ambiguity and double games in response. For example, under domestic pressure over corruption and alleged involvement in the murder of an investigative journalist, President Leonid Kuchma announced in May 2002, that Ukraine would work toward the goal of eventual NATO membership at a NUC meeting in Reykjavik, Iceland. Putin understood Kuchma was playing multivector geopolitics to shore up his position. Before going to Reykjavik, he promised Putin Ukraine would become an associate member of the Eurasian Economic Community, an economic trading zone dominated by Russia.[26] At their joint press conference in Sochi on May 17, 2002, just after Kuchma's return from Iceland, Putin even stated: "Ukraine has its own relations with NATO; there is the Ukraine-NATO Council. At the end of the day the decision is to be taken by NATO and Ukraine. It is a matter for those two partners."[27]

The game changed as Russia's relations with the United States deteriorated and the Orange Revolution threatened Russia's power over Ukraine's leadership. A genuine zero-sum contest emerged in 2005 when the victor, Ukrainian president Viktor Yushchenko, joined the schemes and declarations of Georgian president Mikheil Saakashvili to advance Euro-Atlantic integration within their states and beyond. In January 2005 both signed a Carpathian Declaration, and in August 2005 they signed a declaration creating a Community of Democratic Choice to advance Euro-Atlantic integration from the Baltic to the Black Seas. Throughout 2005, NUC foreign ministers worked on an intensified dialogue process to accelerate Ukraine's aspiration to NATO membership. While Yushchenko's Orange coalition was crippled by intense infighting, the cabinet of Yulia Timoshenko nevertheless managed to advance a bid for a Membership Action Plan before the NATO summit in Bucharest in April 2008 despite significant domestic opposition that was led by the Party of Regions.

This brings us to Putin's remarks at Bucharest, which were ostensibly descriptive but revealingly normative (and, thus, understandably received as threatening). According to the unofficial transcript of his remarks, Putin described Ukraine as a "difficult state" because of the agglomeration of territories it acquired by dint of historical circumstances and chance. Adding NATO to the mix would, he suggested, bring the state to the brink of extinction. All powers should act very carefully in Ukraine because of the complexities of the country. While he conceded that Russia had no right of veto, it had significant interests there, citing "seventeen million Russian lives in Ukraine" and Crimea as "ninety percent Russian." Here Putin mistakes the percentage figure of ethnic Russians in Ukraine according to the 2001 census for the actual number (which was 8.3 million out of a population of 48.45 million), and the number of Russian speakers in Crimea for the number of ethnic Russians (which was 1.18 million or 58.3 percent of the total). The slippages revealed a tendency to overestimate the Russianness of southeast Ukraine and Crimea, something that would become consequential in 2014. Since the mid-1990s Russian security planning held that if Ukraine ever seriously sought to join NATO, Russia should exercise its levers of influence in Ukraine and (re)activate separatism in Crimea and Novorossiya.[28] Both spaces were entangled in historic myths, but the first was a clearly bounded peninsula whereas the second was an imagined territory without clear borders.

The interpretation of Putin's remarks as a disturbing expression of a Russian revanchist agenda in Ukraine underscored the clashing geopolitical cultures evident in Bucharest. Whenever Putin talked of Russian security interests in the near abroad, certain NATO officials heard rhetoric that reaffirmed their preconvictions that Russia under Putin was an unreformed imperialist power that had atavistic designs on the territory of Georgia and Ukraine. Other Putin remarks about Ukraine as a land (*krai*) not a country, Russians and Ukrainians as one nation, and Ukraine as a failed state revealed to them a reluctance on his part to accept Ukraine as a fully sovereign independent state.[29] Most considered the subsequent August 2008 war as proof of their convictions. Some, most prominently President Saakashvili himself, prophesied future Russian revanchism in Ukraine. "Crimea is next," he warned.[30] Putin, for his part, was explicit in rejecting this sentiment, which he viewed as an anti-Russian provocation: "Crimea is not a disputed territory," he told German television in August 2008, though he then went on to note the complicated processes going on within Crimea.[31]

In Saakashvili's appearance as an ex-president in Washington, DC, on February 14, 2014, as Ukraine's Euromaidan protests raged, Saakashvili predicted major changes would occur because structural trends were against Putin: looming recession, fatigue with his rule, restive ethnic republics, and declining oil prices. Putin, he told the audience, "knows that he can no longer keep Ukraine, for sure, but he thinks he can keep Crimea. He thinks he can no longer keep Moldova but he thinks he can have Transnistria," which required creating a corridor through southeast Ukraine to connect it to Crimea and Russia. "It's easy to predict, and easy, or doable, to preempt."[32] Later, as the Russian invasion of Crimea unfolded, Saakashvili wrote a series of prominent opinion editorials noting the similarities between August 2008 and March 2014. Putin's behavior, as well as the West's "appeasement" of Russia, he wrote, "evoked a sense of déjà vu."[33]

Putin must have had his own sense of déjà vu looking at the Euromaidan protests against Viktor Yanukovych, the candidate whose fraudulent election in 2004 had sparked the Orange Revolution.[34] Putin had invested considerable personal capital in getting Yanukovych elected in 2004, sending Kremlin political technologists to run his campaign and visiting Ukraine seven times to campaign for him. Yanukovych's defeat was humiliating, and he reacted angrily. However, infighting split the Orange coalition, allowing Yanukovych, whose political image was rebuilt by American political technologist Paul Manafort, to win the presidency in 2010. Now, as his government was preparing for the Sochi Winter Olympics, a project he had worked hard to realize, he was facing organized protesters on the Maidan once again. To add insult to injury, front and center stage at protests were Mikheil Saakashvili and John McCain, two bête noires of Putin and outspoken crusaders for EU and NATO expansion.[35]

The road from the Maidan to the annexation of Crimea, the pathway to crisis and Russian interventionism, was not predestined. The Euromaidan protests did not initially call for Yanukovych's ouster, just a change in policy. The blundering violent response by his government, however, only deepened the crisis and accelerated political polarization across the country. Fearing the wrath of radicalized protest mobs who had demonstrated a capacity for violence, Yanukovych and his security apparatus fled Kyiv, creating a power vacuum and a profound legitimacy crisis. Responding to what he framed as a U.S.-orchestrated coup d'etat, Putin authorized a stealthy Russian invasion of Crimea and disguised Russian support for anti-Ukrainian groups across the southeast. That decision transformed a political regime crisis into a crisis of Ukraine's territorial integrity and of the post–Cold War security order in Europe.

A Contingent Road to Crimea

There is now a large and growing literature examining the motivations of Russian interventionism in Ukraine in 2014. Since supreme decision-making power in Russia is concentrated in the hands of Vladimir Putin, this literature rightly gives considerable attention to proximate causal factors, such as Putin's cognitive psychology, as well as to domestic protests that greeted his announced return to the presidency in September 2011, flawed Duma elections that December, and his inauguration in May 2012.[36] But the broader context of the Ukrainian crisis was years in the making and reflected a deepening alienation of Putin's inner circle from Western political and security institutions. The Russian analyst Dmitri Trenin argues that Putin viewed Medvedev's presidency as an experiment in what could be achieved with the EU and the United States if Russia had a young, modernizing leader who was not from a security service background (i.e., not a *silovik*).[37] While the formally launched "reset" with the United States did yield some positive results—the New Start Treaty of 2010, Russian membership in the World Trade Organization in 2012—relations deteriorated after Russian acquiescence to a UN Security Council resolution establishing a no-fly zone over Libya, and the subsequent use of that resolution to remove Muammar Qaddafi from power (the circumstances of his death, like those of Saddam Hussein, were particularly ignominious). The Obama administration's support for the overthrow of longstanding authoritarian leaders like Qaddafi and Hosni Mubarak in Egypt, and its attempt to oust Bashar al-Assad in Syria, crystalized a fundamental divide, as Putin saw it, between the United States and Russia over the sources of political instability in the world. The United States, whose foreign policy was driven by what Putin once privately referred to as "Obama's women" (Secretary of State Hillary Clinton, National Security Advisor Susan Rice, and UN Ambassador Samantha Power) viewed instability as emerging from repression of people's democratic rights, whereas Putin viewed it as emerging from the United States' reckless promotion of democracy across the globe.[38] Qaddafi's violent demise strengthened the position of hardliners who argued Russia was in a zero-sum struggle for power and influence with the West and that it needed to build geopolitical institutions of its own to counter

"Atlanticism." The ultimate aim of democracy promoters, as some outspoken anti-Putin activists in the West freely acknowledged, was "regime change" in Russia. That few of these figures were close to central positions of power does not seem to have eased Kremlin anxieties about the influence of their message. Deep-seated fears of chaos, collapse, and violent death were in play, experiences familiar to those who lived through the Soviet collapse and its aftermath.

In pursuit of the goal of checking the power of the West, Putin's Kremlin launched a series of initiatives. First, it tilted toward conservative identity politics and passed a wave of legislation from late 2012 to mid-2013 preventing adoptions of Russian children by U.S. or gay parents, criminalizing propaganda for "nontraditional sexual relations," and enabling prosecution of those perceived as insulting religious sentiments or criticizing the Red Army's conduct during World War II.[39] Second, it targeted perceived "foreign agents" of Atlanticism promoting democratization and human rights, what the Kremlin chose to define as "regime change." Third, it sought to consolidate an alternative "Russian world" (*Russki Mir*) to the perceived hegemonic position of Western media and culture in post-Soviet space.[40] Finally, it sought to reanimate the Eurasian Economic Union (EEU) as an explicit competitor to the European Union.[41] Conceived as a counter in a zero-sum struggle, the EEU effectively forced Russia's neighboring countries to choose between alignment with Russia or with the West. For some, the choice was easy. Deeply dependent on integration with the Russian economy, Belarus and Kazakhstan signed up. For others, the choice was unwelcome. After some hesitation, Armenia announced it was ending its quest for a EU Association Agreement and committed to joining the EEU. Moldova and Georgia, both with governments aspiring for stronger relations with the European Union, declined to join.

The country the EEU project created the greatest difficulty for was Ukraine. Though generally viewed as pro-Russian, Viktor Yanukovych sought to extract the best deal possible from the competitive game by playing multivectoral geopolitics, even soliciting funds from China. Russia raised the pressure on Ukraine in August 2013 by imposing new regulations on its imports, a move of manifest geo-economic intimidation. In the end, Yanukovych awkwardly backed out of a proposed Association Agreement as well as a Deep and Comprehensive Free Trade

Agreement (DCFTA), due to be signed at an EU Eastern Partnership Summit in Vilnius, Lithuania, at the end of November 2013. (This was followed by his acceptance of Russian loan financing and gas discounts in mid-December, which many interpreted in ominous geopolitical terms as Ukraine rejecting Europe and joining a Russian-dominated Eurasia.) What unfolded thereafter was a series of eventful contingencies that generated a major geopolitical crisis from an unlikely beginning: protests over a trade deal. This occurred because Yanukovych's decision was inflated into a "civilizational choice" between West and East, rule of law and kleptocracy, freedom and empire. With such aggrandized geopolitical rhetoric inflaming the moment, protests by hundreds of thousands of EU-flag-waving citizens in central Kyiv was inevitably going to raise legitimacy questions in a Ukraine that was too big and too divided to agree.

The Euromaidan protests were a challenge to the institutions of a Ukrainian state hollowed out by more than two decades of rule by predatory oligarchic networks. While Ukraine was not a failed state, it was one where the political and economic system had little legitimacy. Significant majorities of Ukrainians declared their country as "on the wrong track" in public opinion surveys. After the Orange Revolution in 2004, expectations both in Ukraine and abroad were high that Ukraine would break free of oligarchic domination and renew its institutions. This, however, failed to happen and Yanukovych was able to triumph in 2010. With his power base in Donetsk, he proceeded to appoint cronies to key positions within the ensemble of institutions of the Ukrainian state, deepening the hold of corrupt practices over the country's economy and over everyday life. The Euromaidan protests were always about more than a trade deal. Yanukovych's rejection of the EU Association agreement was symbolic of the drift of Ukraine toward greater and greater levels of authoritarian kleptocracy.[42]

A more skillful politician than Yanukovych, with more professionalized police and security forces at his disposal, may well have been able to ride out the Euromaidan protests. But this was not the case with Yanukovych and the coercive apparatus he had at his disposal. These comprised an ensemble of regular police, special police units (the Berkut), hired thugs (the *Titushki*), paid provocateurs, criminal gangs, and, in effect, death squads. Initial rounds of physical violence against protesters on or near the Maidan by Berkut police units disproportionately from Crimea (on November 30, December 1, and December 11) were followed by intimidation and violence against leading activists away from the Maidan, some of whom were kidnapped and murdered. Many protesters were from far-right groups and they quickly met violence with violence and then prepared for more.[43] Yanukovych secretly met with Putin on January 8, 2014, where it was allegedly decided that Ukraine would introduce new repressive laws to crush the Maidan protests.[44] The legislation was rammed through the Rada on January 16, 2014. The draconian laws modeled on those in Russia only brought more people out into the streets while intensifying the international spotlight. Far-right groups brought stones, incendiary grenades, molotov cocktails, and weapons to the protests. While the majority of protesters were nonviolent, a minority fought back in kind against the coercive state apparatus. Events radically deteriorated on February 20, when snipers opened fire on protesters on the Maidan: almost a hundred were killed. An EU-negotiated deal with Russia thereafter, keeping Yanukovych in power, fell apart almost immediately as key figures in his government defected and ran.

That Euromaidan culminated in the violent overthrow of the democratically elected president of Ukraine proved deeply consequential. Most governments and observers in the West sympathized with the protesters and viewed the violence as a regrettable but inevitable part of the "Ukrainian revolution," given the thuggish nature of the regime. The new Euromaidan-inspired government, for example, was quickly recognized. The Ukrainian-born political scientist Serhiy Kudelia, however, argues that the "flagrant use of force by protesters with the tacit support of opposition parties removed the major constraint that had previously kept the political struggle in Ukraine peaceful."[45] Pandora's box was open. Television footage of the spectacle of violence was broadcast to all regions, entrancing, polarizing, frightening, and radicalizing the population. The prior norms governing politics and civic strife were no longer operable.

On the night and early morning of February 22–23, Vladimir Putin met with a small group of his national security principals. At this meeting, they formulated Russia's response to the government collapse in Ukraine. Russian special forces had already been placed on alert days earlier.[46] Yanukovych had agreed to an EU-brokered peace deal on

February 21 but justifiably radicalized crowds on the Maidan, shouting "death to the criminal," rejected the notion that he could stay in office even temporarily. Further, hundreds of riot police guarding the presidential compound disappeared.[47] Yanukovych hastily packed and fled his estate at Mezhyhirya. Crowds of protesters, journalists in tow, overran the estate on the next day looking for him. Despite Putin's well-known contempt for Yanukoyvch, he nevertheless authorized a secret Russian operation to rescue and transport him to Russia. Putin's actions were in keeping with his entrenched antipathy toward revolutions he considered sponsored by Western powers and agents of influence in post-Soviet space. But they were also shaped by his reaction to the Obama administration's decision to distance itself from the Hosni Mubarak regime in Egypt in February 2011. To Putin, that reckless move had resulted in predictable chaos and instability. Unlike Obama, he was demonstrating that he did not abandon allies, even if they were distasteful. Yanukovych was his Somoza. By his own account, Putin and his team then turned to the issue of what to do next. What Putin authorized was a stealthy Russian invasion of Crimea. The operation became visible three days later when teams of unmarked soldiers, soon dubbed "little green men" (or "polite people" in Russia), surrounded and seized a myriad of strategic facilities across Crimea.

An Improvised Rescue Mission

Putin's decision to set in motion Russia's seizure of Crimea was the riskiest move of his career as Russian president. The context within which this fateful decision was made—an all-night meeting to save Yanukovych's life—suggests it was partially an opportunistic reaction to escalating violence and the dramatic collapse of Russia's influence in Kyiv. The prospect that Ukraine's new government would revoke Russia's lease on the naval base was one fear within Putin's team.[48] Putin reportedly made the decision with just four of his advisers present: his chief of staff Sergei Ivanov; the head of the National Security Council Nikolai Patrushev; Aleksandr Bortnikov, the director of the FSB; and Defense Minister Sergey Shoygu.[49] There was no deliberative, consultative policy process nor does there appear to have been much consideration given to the broader strategic consequences of Russia's actions.[50]

Foreign Minister Sergey Lavrov was not consulted. Only Sergei Shoygu, the person who would have to supervise the operation, expressed extreme caution. He was ignored. Like Saakashvili on August 7, 2008, a supreme leader, sleep-deprived and pressured, made what turned out to be a momentous decision in the company of a small circle of mostly likeminded advisers.

Concern over the future of Russia's presence in Sevastopol was reasonable. Only in 1997 did Ukraine and Russia come to terms over the future of the base and historic city, a formerly closed city that was directly administered by Moscow during Soviet times. The 1997 agreements with Russia, in which it acknowledged Ukraine's sovereignty over the area and agreed to a twenty-year lease of the naval base, were controversial and became a bone of contention in domestic politics.[51] Reflecting Ukrainian nationalist sentiment, President Yushchenko sought to raise the cost of the lease while declaring it would not be renewed. During the August 2008 war, he issued two decrees seeking to restrict Russia's deployment of its fleet, suggesting vessels might be prevented from returning to Sevastopol.[52] In the face of fierce criticism, President Yanukovych signed a twenty-five-year extension of Russia's lease a few months after his election in 2010.

It can be assumed that the Russian military had a contingency plan to protect its Black Sea Fleet facilities—which consisted not only of naval facilities at Sevastopol but also two airbases, one of which was north of Simferopol—should they come under threat. Securing these facilities, thus, meant taking control of both Sevastopol (incorporating Balaclava) and Simferopol. Russia likely updated its contingency plan after shortcomings in August 2008 ushered in dramatic reforms in its armed forces. Improved readiness and better coordination with local militias were lessons learned from August 2008. In Crimea, the equivalent local forces were both traditional—Cossack groups from Crimea and Kuban—and new, like the Night Wolves motorcycle gang led by Alexander Zaldostanov, known by the nickname "The Surgeon." Putin and Zaldostanov had a well-publicized first meeting in July 2009, and their meetings thereafter became hypermasculine-image events with Crimea, more often than not, as the backdrop.[53] Russia also had considerable ties to local Russian nationalist groups and to the Berkut forces that fled to Crimea as Yanukovych fled. The Russia Ministry of Defense would later airlift in 170 veterans from Afghanistan and Chechnya, as well as members of motorcycle groups and "patriotic clubs," to play the part of ordinary Crimeans agitating that Russia take control of the peninsula.[54]

According to the Russian investigative journalist Mikhail Zygar, the Kremlin first discussed a specific plan of action with respect to Crimea in December 2013 when the head of the Supreme Council of Crimea, the local Party of Regions leader Vladimir Konstantinov, visited Moscow. He told Patrushev, the National Security Council head, that in the event of Yanukovych's overthrow, Crimea would be ready "to join Russia."[55] Patrushev, reportedly, was pleasantly surprised by the sentiment. He authorized a series of private polls to measure public opinion in Crimea as Ukraine's governance crisis evolved.

It appears that Putin's February 23 decision was about establishing military control to create a menu of future status options—allowing Crimea to claim greater autonomy, supporting a new de facto state on the peninsula, or annexation—rather than about annexation as the endgame from the outset. Zygar reports that liberals and hardliners debated scenarios within the Kremlin.[56] In the absence of crucial information on the closed-door decision-making process, we can only study Putin's public statements. These suggest his decision was not inherently about securing the naval base but about rescuing Crimea, and that affective storylines, more than narrow geostrategic interest, drove the annexation.

A necessary place to start is how Putin described his decision in a made-for-television docudrama titled *Crimea: The Way Home*, which was broadcast on Rossiya 1 to celebrate the first anniversary of the annexation.[57] Like events in August 2008, the Crimean annexation is presented as a reactive rescue mission, one that begins with an anticonstitutional coup in Ukraine that throws the country into chaos and endangers the life of its legitimate president. Putin then approves a dramatic secret operation to save Yanukovych's life. Only at the end of the all-night presidential rescue mission does Putin then authorize a larger territorial rescue mission: "The situation in Ukraine has turned out in such a way that we are forced to begin work on returning Crimea to Russia. Because we cannot leave this territory and people there adrift, under [the] steamroller of nationalists." Vulnerable Crimea required rescue.

The Dramaturgy and Affective Geopolitics of Annexation

Putin's interview for *Crimea: The Way Home* was recorded relatively soon after the events of Crimea's annexation.[58] It goes without saying that it should be viewed skeptically. The extensive reenactment scenes portrayed in the documentary present events as a melodramatic geopolitical thriller, featuring a heroic leader, dramatic missions, daring military maneuvers, and courageous actions by many local patriotic heroes across Crimea. Fact and fiction intermingled in the production of the invasion and annexation as a drama of the paternal rescue. While the theatrics were willful, they were also improvised on the fly, for Putin's move was a shock to many, including some of his own ministers.[59] What is presented as the critical moment of decision, 7 a.m. on February 23, was most likely one in a series of decisions. Many different factors were at work, not least the political passions unleashed by the introduction of unmarked military personnel into a vulnerable location at a sensitive moment. This in itself created an affective wave of expectations and fears that shaped subsequent decisions. The theatrics provided a welcome surge of popularity for Putin, but this was purchased at a high strategic cost. Below I consider four aspects of the Kremlin directed production of the annexation as a drama of affective geopolitics. In exploring Putin's rhetorical framing of the crisis, I also discuss facts that underscore the argument that the annexation can be seen as an improvised rescue fantasy, and that the affective wave generated by this storyline was not simply manufactured in order to disguise something else. The crisis enveloped Putin as much as it was used to envelop ordinary Russians to generate a glorious historic moment. The affective wave acquired a life of its own.

Producing a Mythic Threat

The plot of the Russian storyline on Crimea emerged from the Kremlin's counter-storyline on the Euromaidan protests. It was generated from the same mythic narrative of threat/protection from genocidal fascist forces evoked in August 2008. As already noted, the term "fascist" was a pliable designator of enemies of international communism and the Soviet Union in the 1930s until the Molotov-Ribbentrop pact saw its temporary deemphasis. During World War II, the Great Patriotic War in Russia, "fascism" was a commonplace term applied to the Wehrmacht and its allies. Thereafter, the fascist threat was applied to NATO and the Soviet Union's Cold War enemies. This affective narrative was adapted to the circumstances of post-Soviet space by the de facto states. A Ukrainian version, which focused on the role of the wartime Ukrainian nationalist factions, the Organization of Ukrainian Nationalists (OUN, established in 1929), and the Ukrainian Insurgent Army (UPA, established in October 1942), was long available and used by pro-Russia groups in Ukraine to designate Ukrainian nationalism as innately fascist. In the 2014 crisis, the narrative was activated again to impose a situational frame on events and designate the various parties, irrespective of the empirical complexity on the ground. That there would be little escape from its power was signaled by a meeting with officials overseeing the revision of school textbooks on Russian history in January 2014, during which Putin expressed anger at those who sought to question the received Soviet World War II narrative.[60] Less than a fortnight later Putin, speaking at the end of a Russia-EU summit, suggested radical anti-Semitic and racist nationalism from western Ukraine was infecting the crowds in Kyiv: "this is radical nationalism of a kind that is totally unacceptable in the civilized world."[61]

In a press conference where he sat on a chair, occasionally slouched and annoyed, before journalists at Novo-Ogaryovo on March 4, Putin made clear his situational description of events in Kyiv and Ukraine (the collapse of the EU-brokered deal, Yanukovych's flight and subsequent impeachment and replacement): "There can only be one assessment: this was an anti-constitutional takeover, an armed seizure of power."[62] Putin suggests that the real question is: Why was this done? "What was the purpose of all those illegal, unconstitutional actions, why did they have to create this chaos in the country?" Yet despite his apparent puzzlement about the motive, Putin nevertheless detected the hand of external powers in the "anti-constitutional coup" in Ukraine. Responding in a more revealing way than he may have wished to a question about "local self-defense units" in Crimea, Putin saw their mirror image in Kyiv: "[L]ook how well trained the people who operated in Kiev were. As we all know they were trained at special bases in neighboring states: in Lithuania, Poland and in Ukraine itself too. They were trained by instructors for extended periods.... Did you see them in action? They looked very professional, like Special Forces. Why do you think those in Crimea should be any worse?" Asked who he believed was behind the "coup" in Ukraine, he described it as "well-prepared action. Of course there were combat detachments. They are still there, and we all saw how efficiently they worked. Their Western instructors tried hard of course." Echoing the contempt for liberal "experiments" expressed in his Millennium Message more than a decade earlier, Putin declared that he "sometimes get the feeling that somewhere across that huge puddle, in America, people sit in a lab and conduct experiments, as if with rats, without actually understanding the consequences of what they are doing. Why did they need to do this? Who can explain this? There is no explanation at all for it."

As always, the diagnosis reveals more about Putin's proclivity for psychological projection than it does about the events, as well as his strong attachment to conspiratorial theories of "revolutionary moments" in post-Soviet space. The Soviet Union ostensibly celebrated spontaneous revolution, yet Putin's career as a KGB intelligence officer left him with proclivity toward conspiratology—even if it is not always clear what ends were being served. In theory, the obscurity of ends should call the conspiracy theory into question. It does not, however, because conspiracy is a convenient cognitive shortcut for the analysis of foreign policy for Putin. Putin's public statements indicate he believed the United States was aiding and abetting Ukrainian nationalism as an instrument to challenge Russia's security interests in Ukraine. There is both a far and a near enemy at work in Putin's world. The far enemy is the United States, which is running geopolitical operations against Russia. The near enemy is Ukrainian nationalism, which he allows little identity beyond fascist nationalism. Russia's immediate concern, he stated, was "the rampage of reactionary forces, nationalist and anti-Semitic forces going on in certain parts of Ukraine, including Kiev." Events in Ukraine recalled a familiar past where anti-constitutional coups enabled fascism: "In this kind of situation you never know what kind of people events will bring to the fore. Just recall, for example, the role that [Ernst] Roehm's storm troopers played during Hitler's rise to power. Later, these storm troopers were liquidated, but they played their part in bringing Hitler to power. Events can take all kinds of unexpected turns." Two weeks later, after events had indeed taken turns unexpected by most—though not by Putin, for he had launched his own counter-conspiracy conspiracy—Russia's president justified the Russian annexation of Crimea before an audience in the Kremlin by citing a "they" that blurred external agitators and local fascists. Those who "stood behind the latest events in Ukraine," he declared, "were preparing yet another government takeover; they wanted to seize power and would stop short of nothing. They resorted to terror, murder and riots. Nationalists, neo-Nazis, Russophobes and anti-Semites executed this coup. They continue to set the tone in Ukraine to this day."[63]

Producing Surprise and Confusion

Central to the Russian military invasion plans for Crimea was the element of surprise and, thereafter, of confusion and ambiguity about events on the ground.[64] Disguising an overt "Russian hand," therefore, was a military necessity at the outset. This could be partially achieved on the ground by having the key strategic sites in Crimea seized by specially trained military forces without insignia, and having them mingle with tasked local "self-defense" forces to obfuscate their controlling presence. On February 26, Putin ordered a snap drill of the combat readiness of Russian military forces that provided cover for the movement of ten aircraft of Russian paratroopers from Pskov to Sevastopol airport.[65] At 4:25 a.m. on Thursday, February 27, approximately fifty armed men seized the Crimean parliament and hoisted a Russian flag over the building. They identified themselves as the "Russian-speaking Crimean population's self-defense force."[66] Early Friday morning a convoy of military vehicles and soldiers without insignia arrived at the Belbek airfield and blocked the runway, taking the Ukrainian Air Defense Service in Crimea, and its aircraft, out of action. Simultaneously, a company of armed men seized Crimea's main civilian airport and air traffic control station in Simferopol. The prior Sunday, in Sevastopol, a pro-Russian crowd on the city's main square elected a "people's mayor," a local Russian citizen named Aleksei Chaly, by a show of hands.[67]

At the strategic level, the plan required not only secrecy but also active dissembling by the limited number of figures. Mendacity for military surprise was not new to Putin. In September 1999, at the same press conference where he made his famous "outhouse" remark, he insisted he did not plan a new war in Chechnya. A week later he launched a full-scale invasion.[68] Engaged in conversations with Chancellor Angela Merkel as well as Presidents François Hollande and Barack Obama as the Ukraine crisis unfolded, Putin made the important decision to lie to them about Russia's operations and intentions. Thus, after authorizing the Crimea invasion on the morning of Sunday, February 23, Putin spoke with Merkel where they both reportedly agreed that Ukraine's territorial integrity must be safeguarded. As the actions of Russia's unmarked soldiers became evident midweek, Putin told Merkel on Friday, February 28, that there were no Russian special operations forces in Crimea. It was a blatant lie, and by Sunday, March 2, evidently so. The previous day the Federation Council had approved his request to authorize the use of Russian force in Ukraine. Putin spoke publicly for the first time about the Crimea operation on Tuesday, March 4, after attending the conclusion of the snap military exercises he ordered. He denied there was any Russian military intervention at all in Crimea: "so far there is no need for it, but the possibility remains."[69] Tensions in Crimea had supposedly subsided and there was no need to use Russian armed forces. "[W]e do not intend to interfere," he said. Asked directly if soldiers in unmarked uniforms that strongly resembled Russian army uniforms, blocking Ukrainian military units in Crimea, were Russian soldiers, Putin responded that one can "go to a store and buy any kind of uniform." They were "local self-defense units." Asked if he considered it possible that Crimea might join Russia, Putin demurred and said no: "I believe only the people living in a given territory have the right to determine their own future."

The key unasked question, of course, was what was the relevant "given territory" in the circumstances. The Russian paratroopers that captured the Crimean parliament on February 27 oversaw a change in government, and a motion organizing a referendum on Crimea's status in Ukraine, all at gunpoint. But preparations for this were improvised and haphazard. Oleg Belaventsev, deputized by Shoigu as the commander of Russia's military operation in Crimea, reportedly arrived in Crimea five days earlier to persuade the unpopular incumbent prime minister to step down.[70] He ran into difficulties picking a successor before Vladimir Konstantinov, chair of the Crimean parliament, persuaded him to appoint his friend Sergei Aksyonov as Crimea's new prime minister. The initial proposed date for the referendum was May 25 (the same day as planned Ukrainian elections), and the text did not propose secession from Ukraine but rather that Crimea be seen as a "self-sufficient state." However, the parliament now flew a Russian flag, and the organized pro-Russian demonstrators outside—Russia Bloc and the Russian Movement of Ukraine—chanted "Russia, Russia."[71] On March 1 the date was changed to March 30 and then changed again to March 16. Secession from Ukraine was now explicitly on the ballot. That this was not initially proposed suggests Putin did not immediately decide annexation was Russia's endgame on the morning of February 23, as he later suggested.[72] Rather, he waited for the domestic reaction in Russia, Crimea, and Ukraine. He stated in *Crimea: The Way Home* that he took his final decision to annex Crimea after an opinion poll showed 80 percent of Crimeans favored joining Russia. If true, this places the actual decision to annex just prior to the referendum. A survey by a polling firm run by Natalya Kisileva, a political sociologist at Simferopol State, was most likely the one Putin had in mind. Previously they had conducted a series of telephone polls tracking sentiment in Crimea (possibly those Patruschev ordered). Their last pre-referendum poll was a large face-to-face poll of 2,700 respondents across Crimea conducted from March 8 to 11. It showed 79.7 percent support for annexation (14.5 percent no, 5.5 percent don't know).[73]

Crimea's new leadership presented the peninsula as a "given territory" with its own right to a self-determination referendum, something illegal under Article 73 of the Ukrainian constitution.[74] A 2001 law on the procedure of acceptance into the Russian Federation of new subjects governed Russia's response. This required that Crimea be an independent state that had expressed its will in a referendum. On March 16, the same day as its secessionist referendum, deputies within the Crimean parliament were rounded up and coerced by armed Russians like Igor Girkin to pass a law unilaterally breaking with Ukraine.[75] The following day the parliament declared Crimea an independent state and then requested to join Russia. That evening the Kremlin issued a decree that recognized Crimea as an independent state. Putin's Kremlin address the next day recommended acceptance of Crimea into the Russian Federation, a recommendation approved by the Russian Constitutional Court and the State Duma by a vote of 445 to 1. Crimea was now part of Russia, though only legally according to Russian law.

In deciding to deliberately deceive world leaders he habitually called "our partners," Putin was acting more like a KGB intelligence officer than a state president. Putin allowed his deep contempt for the hypocrisies of the West, most especially the United States, to justify his actions. "[T]hey have lied to us many times, made decisions behind our backs, placed us before an accomplished fact," he declared in justifying Crimea's annexation on March 18.[76] While he protested that Russia acted lawfully, he was well aware that Russia was breaking the rules. But breaking the rules seemed a privilege of great powers, an exception enjoyed by those powerful enough to be above the law in Russia. The rules of international diplomatic society, however, were not those prevailing within Russia. Putin's deliberate deception of his longtime interlocutor Merkel, and other world leaders, severely damaged Russia's standing in the world community. It would lead to him becoming an outcast member of the G8 and a shunned figure at the G20 meeting later that year.

Producing a Scenography of Legitimacy

If the plot summary of Putin's counter-storyline on Euromaidan was "fascist coup," the plot of his initial storyline on Crimea was "humanitarian protection." In this it followed the "responsibility to protect" narrative used during the August 2008 war.[77] Two incidents featured in this storyline, however, occurred after Putin had already decided to invade Crimea. The first was the repeal by the Rada of the law allowing Russian to be an official language at the regional level in Ukraine later in the day on February 23. The move needlessly alienated areas where ethnic Russians were concentrated, and though it was subsequently vetoed by interim president Turchynov on March 3, the damage was done.[78] Russian media had their "proof" of the new government's "fascism." The second was a threat on February 24 by Igor Mosiychuk, a leading figure in the Right Sector, a radical Ukrainian nationalist group that had a prominent role in the fighting on the Maidan, to march on Crimea to prevent any move to dismember Ukraine's territory.[79] The scenario of a "fascist invasion" of Crimea appeared far-fetched, but it played into the preexisting script of Putin and could be hyped using television footage of fascist marches, neo-Nazi symbols, and Euromaidan violence.[80] Indeed, the dominant Russian media storyline the last week of February on Crimea was of reactive moves by local ethnic Russian self-defense forces in Crimea against the "fascist coup" government in Kyiv. Pictures of such forces, some with Berkut uniforms and black facemasks, manning a checkpoint at Armyansk on the Isthmus of Perekop, the narrow strip of land connecting the Crimean peninsula to mainland Ukraine, gave the storyline a scenography that also underscored a broader territorial sovereignty crisis.[81] The seizure of the Crimean parliament and subsequent referendum announcement were represented as legitimate moves in response to the specter of fascism. The Russian-installed Crimean prime minister Sergei Aksyonov was the leader of a small pro-Russia political party (critics alleged he and Konstantinov had ties to organized crime and were involved in illegal real estate deals). Now, as the voice of the "Crimean people," he asked Russia for humanitarian help to ensure "peace and tranquility" in the region, claiming that the new government in Kyiv was unable to keep order.

At the end of the week Putin sought and quickly received authority from the Federation Council to use Russian forces in Ukraine. He also sought another source of legitimacy in the form of a letter he had ousted President Yanukovych sign and backdate to the previous day before the Federation Council vote approving use of Russian force in Ukraine. The letter stated: "Under the influence of Western countries there are open acts of terror and violence. People are being persecuted for language and political reasons. So in this regard I would call on the president of Russia, Mr. Putin, asking him to use the armed forces of the Russian Federation to establish legitimacy, peace, law and order, stability and defend the people of Ukraine."[82] The letter served as another prop in the scenography of legitimacy being assembled by Russia to justify its actions in Crimea. The next day, Russian ambassador to the UN Vitaly Churkin read and waved it before Security Council delegates as proof of the legitimacy of Russian actions there.

As the role of Russian soldiers in orchestrating a takeover of Crimea became indisputable, Russia's informational emphasis shifted to the referendum as an exercise of self-determination by the "Crimean people." The conspiracy projected onto events in Kyiv was the warrant for Russia's actions, which were, to outside critics, themselves the real conspiracy. In Kyiv, according to the pro-Russian media, a U.S.-instigated illegal government had taken power and was now steering Ukraine toward the West and NATO.[83] In Simferopol, by contrast, local defense forces had appointed a new government that was now lawfully steering Crimea toward its destiny with Russia using the legitimate international mechanism of a referendum. That, at least, was the claim. Crimeans were invited to believe and participate.

The new authorities left little to chance. In the Soviet Union, self-determination was a political performance whose controlled use could be an effective mechanism furthering imperial control.[84] Proven techniques returned, beginning with the repression of independent media. Pro-Ukraine television stations were shuttered. The referendum question presented was so framed that a negative vote, or a vote for the status quo, were not options. Instead, voters could indicate if they were "in favor of the reunification of Crimea with Russia as a part of the Russian Federation" or "in favor of restoring the 1992 Constitution and the status of Crimea as a part of Ukraine." The nine-day campaign featured a one-sided advertisement campaign with the slogan "Together with Russia." One poster graphically presented the choice as barbed wire and a swastika on a blood-red cartographic profile of Crimea, or the same profile covered by the colors of the Russian flag (Figure 6.2). Another showed the letters "NATO" crossed-out. Pro-Ukraine rallies were small and sporadic. The Russian military presence by this time was very public and visible. Potential protesters, especially Euromaidan and Crimean Tatar activists, were rounded up, a signal that genuine dissent would not be tolerated. One was the Tatar Reşat Amet (transliterated from Russian as "Reshat Ametov"), who disappeared on March 3. His tortured body was found the day of the referendum.[85] Footage of lines of voters, enthusiastic pro-Russian sentiment, and transparent ballot boxes colored in the scenography of legitimacy. According to figures released by the Crimean authorities after the referendum, 83.1 percent of eligible voters had participated, with 96.77 percent voting in favor

FIGURE 6.2 Crimea's Choice as Nazism or Russia. Referendum Billboard, Sevastopol [Zurab Kurtsikidze, EPA].

of joining Russia.[86] Given the fact that Crimean Tatar and pro-Ukraine organizations had called for a vote boycott, the figures seemed too high to be credible. But the scenography of self-determination was all that counted to those running the show.

Producing a Glorious Historic Moment

As two decorative guards pulled back the tall gold-leaf doors of Saint George's Hall in the Grand Kremlin Palace, President Vladimir Putin confidently strode forward to present an address to members of the Russian State Duma, the Federation Council, and other dignitaries, many wearing black-and-orange Saint George ribbons, emblematic not only of past glories in Russian military history but of pro-Russia forces in Ukraine.[87] Refurbished at great expense (and no small amount of corruption) during Yeltsin's presidency, the Grand Kremlin Palace was a former imperial residence built on the orders of Nicholas I (1796–1855), the tsar who led Russia on a religious crusade that ended in a catastrophic defeat in Crimea in 1853–1856. The refurbishment of its multiple rooms turned it into an opulent space of colonnades, chandeliers, and alcoves, all elaborately decorated in colors recalling tsarist splendor. Putin held his first inauguration in Saint Andrew's Hall, and both it and the adjoining Saint George's Hall served thereafter as favorite locations for Kremlin image-makers to project presidential scenarios of power, with elevated and swooping camera angles visually projecting Putin as tsar and the Duma as attendant and awed boyars. Memories of Nicholas I and Crimea were in the room as Putin began his speech.

Putin's subsequent forty-five-minute speech elicited standing ovations, thunderous applause, flag waving, and tears in the eyes of some. "In people's hearts and minds," Putin declared, "Crimea has always been an inseparable part of Russia." He reviewed its history, tying it to defining moments in Russian history:

> This is the location of ancient Khersones, where Prince Vladimir was baptised. His spiritual feat of adopting Orthodoxy predetermined the overall basis of the culture, civilization and human values that unite the peoples of Russia. The graves of Russian soldiers whose bravery brought Crimea into the Russian empire are also in Crimea. This is also Sevastopol—a legendary city with an outstanding history, a fortress that serves as the birthplace of Russia's Black Sea Fleet [applause]. Crimea is Balaklava and Kerch, Malakhov Kurgan and Sapun Ridge. Each one of these places is dear to our hearts, symbolising Russian military glory and outstanding valour.[88]

It was only through accidents of history that it was separated from Russia: first, illegally gifted to Soviet Ukraine by the personal initiative of Nikita Khrushchev; and, second, ending up in a Ukraine suddenly independent of Russia. The Soviet collapse was reprised as a sudden and intimate disaster of ethnic fragmentation: "Millions of people went to bed in one country and awoke in different ones, overnight becoming ethnic minorities in former Union republics, while the Russian nation became one of the biggest, if not the biggest ethnic group in the world to be divided by borders." Russia nevertheless accepted this unjust state of affairs, expecting Ukraine to remain its friendly neighbor. The violent coup instigated by fascists from the Euromaidan, however, placed Russian-speaking Crimea in jeopardy. "Naturally, we could not leave this plea unheeded; we could not abandon Crimea and its residents in distress. This would have been betrayal on our part [applause]."

Putin's speech was a long recounting of Russian resentment at U.S. actions since the end of the Cold War in Yugoslavia, Afghanistan, Iraq, Libya, and the Middle East. In Ukraine, the United States had undertaken actions aimed against Ukraine, Russia, and Eurasian integration. The sinister other ("they") were perpetually plotting against Russia: "They are constantly trying to sweep us into a corner because we have an independent position, because we maintain it and because we call things like they are, and do not engage in hypocrisy." Here we get a sense of the subjectivity Putin imagines for Russia, a subjectivity conjured from a Soviet ideal of masculine moral rectitude. And, like the plot of many dramas, that masculine ideal has its limits and intolerances. As Putin explains, "[T]here is a limit to everything. And with Ukraine, our western partners have crossed the line. . . . If you compress the spring all the way to its limit, it will snap back hard. You must remember this."

Putin's speech ended with a crescendo of emotive force, and sustained audience applause, as he declared his request to the Federal Assembly to ratify the treaty on admitting to the Russian Federation Crimea and Sevastopol. Thereafter, the theatrics of power changed as Putin's podium gave way to a columned desk. Following three flags carried by goose-stepping guards, Putin, Chair of the Crimean Assembly Vladimir Konstatinov, Crimean prime minister Sergey Aksyonov, and Sevastopol's sweater-wearing "people's mayor" Aleksei Chaly took seats behind the large desk, along with Putin and signed documents formalizing the admission request for Crimea and the city of Sevastopol to join the Russian Federation. A master of ceremonies voice cued the crowd, who provided the men a standing ovation. As some in the assembly shouted, "Rossiya, Rossiya," and the national anthem swelled, Putin shook hands all round while Konstantinov and Chaly pumped their fists in the air (Figure 6.3).

FIGURE 6.3 Russian president Vladimir Putin (2-R) with Crimean leaders (L-R), Prime Minister Sergei Aksyonov, parliamentary chair Vladimir Konstantinov, and Sevastopol mayor Aleksei Chaly sign the formal request for Crimea to join Russia in the Grand Kremlin Palace, March 18, 2014 [Alexey Druzhinyn, RIA Novosti, EPA].

Two hours after the Kremlin ceremonies, the stars of the show were on a different stage, a state-organized concert on Red Square celebrating Crimea's accession to Russia. The event was broadcast live on Russian television while tens of thousands of Moscovites filled the square. In a high-collar black winter jacket, Putin was exuberant, his left arm and clenched fist pumping in emphasis as he reached a rhetorical climax: "[W]e will overcome all the problems, and we will do it because we are together."[89] He ended his speech with a throaty roar: "*Slava Rossiyi!*" ("Glory to Russia").

The invasion and annexation of Crimea proved to be a public relations triumph for Vladimir Putin. His public opinion approval ratings reached over 80 percent and stayed there as Russia became more entangled in war in eastern Ukraine. *Krym Nash* ("Crimea Is Ours") became a mantra of the patriotic wave that engulfed Russia, one quickly satirized by domestic critics but a powerful distillation of patriotic affect nevertheless for most Russians.⁹⁰ Victory Day on May 9, a celebration already central to Putin's rule, saw an elaborately choreographed display of troops and military hardware.⁹¹ The first vehicle to enter the square behind row after row of tight-formation marching soldiers was an armored personnel carrier from a Black Sea Marines brigade: it flew a large Crimean flag. After speaking at Red Square, Putin jetted to Crimea for another choreographed image event, a speech from the naval quay with Sevastopol harbor as his backdrop.⁹² Russian television lovingly covered these events with wide-angle vistas and swooping cameras, creating a kaleidoscope of patriotic imagery and music.

The dramaturgy of the Crimean annexation was completed by its memorialization. Concerts, statues and movies instantly mythologized and "bronzed" the event. On the first anniversary, the Kremlin organized a celebratory music festival on Red Square, with flag, banners, balloons, and top musical acts. Those not able to get into Red Square could watch the concert on a jumbo screen nearby. Putin and his Crimean "band" closed proceedings with a heartfelt rendition of the national anthem (see Figure 6.4). A similar concert was held on the second anniversary.

FIGURE 6.4 Putin addresses the Red Square music concert marking the one-year anniversary of Crimea's annexation, March 18, 2015 [Mikhail Klimentyev, RIA Novesti].

FIGURE 6.5 Monument to Annexation Troops as "Polite People" in Simferopol, Crimea. Photo courtesy of anonymous.

The first bronze statue to the invading Russian forces (the so-called "polite people") was unveiled in May 2015 in the city of Belogorsk (in Russia's far east). Inspired by a widely circulated TASS (Russian News Agency) photograph, it featured a friendly Russian solider (unmasked, unlike the original) returning a lost cat. On June 11 2016 a similar memorial was unveiled in the center of Simferopol (Figure 6.5). A little girl presents flowers to an invading Russian paratrooper as her cat curls at his feet. A rescue myth in bronze, it positioned Crimea as a child grateful for fatherly protection.[93]

Television, inevitably, shaped how most Russians viewed events in retrospect. Prime time broadcast of the docudrama *Crimea: The Way Home* presented reunification as a geopolitical triumph. Asked what he thought of the critics of annexation, its producer, the Kremlin-friendly journalist Andrey Kondrashov, declared: "[M]y feeling is that these people have constructed some kind of illusory world around themselves and do not want to leave it."[94] The real world was the Kremlin's world.

Geopolitical Attitudes in Contested Ukraine

The events of 2014 profoundly destabilized Ukraine and its population. As the country was broken apart and its population became polarized, so also was its media. Pro-Ukrainian media was shut down in Crimea, while in Ukraine Russian state television programming was removed from cable, though those with satellite links could still get it. Russian-language programing by Ukraine-based stations remained. In December 2014, my research colleagues and I organized a public opinion survey in Crimea and southeast Ukraine (SE6) to determine the level of support for the storylines and geopolitical events that had defined the previous twelve months. Over two-thirds of the questionnaire was identical, and those that administered it worked for the same reputable company, the Kyiv International Institute of Sociology (KIIS). Because of the annexation, however, their employees in Crimea could no longer work for KIIS, so the Levada Center, Russia's last remaining independent survey company, took them on as employees. The survey was the first that tried to conduct social science research in the remaining peaceful areas of what had become "contested Ukraine." Because of the fighting in the Donbas, both Donetsk and Luhansk oblasts were excluded from the survey.

The survey in SE6 was of 2003 respondents, and some overall findings are below. Given the salience of ethnolinguistic rhetoric in the 2014 crisis, respondents were divided into four different ethnolinguistic groups based on declared ethnicity and, in the case of those who declared themselves ethnically Ukrainian, the language they indicated they spoke at home. The sample thus had Ukrainians speaking Ukrainian (22.6 percent), Ukrainians speaking Russian (40.7 percent), Ukrainians who declared they spoke both Russian and Ukrainian at home (17.4 percent), and ethnic Russian Ukrainians (11.6 percent).[95] Those who declared themselves mixed or from smaller nationalities were left out of the analysis.[96] The survey in Crimea was of 752 respondents and was conducted in Russian. A few aggregate results are presented here. To determine the salience of identity to attitudes respondents were divided

PLACES CLOSE TO OUR HEARTS 233

simply by declared ethnicity: Russians (63.6 percent), Ukrainians (20.9 percent), Tatars (8.5 percent), and Others (7 percent). This latter category is also left aside. In aggregate the Crimea sample had more ethnic Russian respondents and less Tatar respondents than are estimated to live in Crimea. In reporting the results by declared ethnicity, it must be stressed that the number of Tatars in the survey—only 64 persons—is too small for it to be representative of Tatar community opinion in total. The results presented here for this community are thus informational but *not* ethnonationally representative.

Here I wish to highlight four results that shed light on attitudes toward the events of 2014. Since a desire for Ukraine to become more European was part of the initial motivation for the Euromaidan protests, our survey asked respondents to evaluate whether they considered themselves European or not (Figure 6.6). The aggregate results for SE6 reveal a divided population: 24 percent indicated they were definitely European, 20.9 percent probably so; whereas 23 percent indicated they were probably not and 22.4 percent definitely not. In Crimea, almost 63 percent of respondents said they were definitely not European. Only 3.19 percent said they were definitely European. Asked if they trusted the European Union, over 26 percent of respondents in SE6 said definitely or mostly yes, whereas only 5 percent said definitely or mostly yes in Crimea. Over 27 percent in SE6 and 72 percent in Crimea said definitely not. The overall results reveal how Europeanness was not a

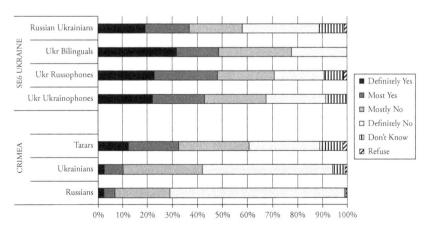

FIGURE 6.6 Self-Perceptions of Europeanness, SE6 and Crimea.

banal geographical identity but a polarized geopolitical one in contested Ukraine at the end of 2014.

Second, the Euromaidan protests were viewed overwhelmingly in a negative light by December 2014 in both SE6 and Crimea. In SE6, only 11 percent indicated that it improved the situation in Ukraine; 66 percent held that it worsened the situation somewhat or a lot. In Crimea the figures are even more emphatic; only 3 percent declared it improved the situation, whereas over 86 percent believed it worsened the situation somewhat or a lot. These results have their ambivalences in that respondents may have supported the protests but viewed the outcome negatively. To the extent the protests hastened Crimea's annexation, Crimeans could have viewed their influence positively though they did not. The anti-Maidan movement was particularly strong in Crimea. The violent scenes that preceded Yanukovych's ouster, and subsequent hasty dissolution of the Berkut, radicalized opinion against Kyiv. Opinion diverges clearly over Putin's counter-storyline about the Maidan protests (Figure 6.7). Here, without naming Putin, we asked respondents about his characterization of Yanukovych's flight from office: "Nationalists, neo-Nazis, Russophobes and anti-Semites executed this coup. They continue to set the tone in Ukraine to this day."

Third, there was a significant divergence of opinion about Putin and the annexation of Crimea. Asked if they trusted Putin, only 9 percent in SE6 said yes, whereas over 79 percent indicated that they did not. By contrast, in Crimea nearly 87 percent of the Crimean sample indicated that they did, more than 59 percent definitely so. Asked if they had an improved or worsened opinion of Putin in the last year, 67 percent in SE6 indicated they had a more negative opinion, whereas only 2.8 percent held a more positive view. In Crimea only 6.5 percent had a more negative opinion, while over 61 percent had an improved opinion (almost 30 percent said their opinion did not change). Both surveys asked respondents how they now considered the joining of Crimea to Russia in March 2014 (Figure 6.8). Over 65 percent in SE6 considered it the wrong step. By contrast, in Crimea sentiment toward the annexation was overwhelmingly positive. Over 83 percent considered it the right move, whereas the rest said it was too early to tell (10.9 percent) or the wrong step (4.79 percent). Not surprisingly, the most conflicted community is the Tatar population, though greater in-depth research

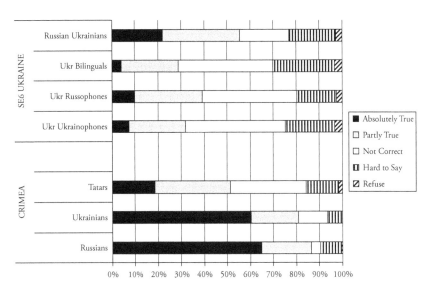

FIGURE 6.7 Attitudes in SE6 and Crimea toward description of Euromaidan as a fascist coup.

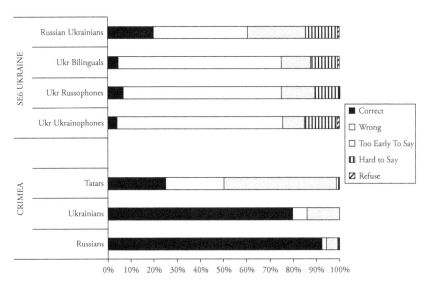

FIGURE 6.8 Attitudes in SE6 and Crimea toward Crimea Joining Russia.

is needed to get a representative indication of their sentiment, now unfortunately an exceedingly difficult task.

Fourth, the Crimean survey asked respondents to evaluate ten statements as to how large a problem they felt the stated problem presented to Crimea. Well-publicized problems like the banks not functioning properly or the transition to Russian laws were ranked as not particularly pressing problems (over 37 percent saw no problem with the banks). The three problems ranked as the largest (very big as opposed to fairly large, somewhat, not a problem, or cannot say) facing Crimea were: the dependence of Crimea on Ukraine for water and electricity (50 percent with another 38 percent ranking it a fairly large problem), the fact that Crimea did not have a contiguous border with Russia (45 percent with 40 percent seeing it fairly large), and the fact that Crimea now had bad relations with Ukraine (30 percent with 32 percent seeing it as a fairly large problem).

Thus the great majority of residents of Crimea in December 2014 may have been satisfied that Crimea was part of Russia, but they had serious concerns about what that meant for their future. Crimea may have had storied places close to Russia's heart, but it was infrastructurally more connected to Ukraine than Russia. This fact was underscored in November 2015 when Crimean Tatar activists blew up the electricity supply lines from mainland Ukraine to the disputed peninsula.[97] Converting affective ties into material infrastructural ties would take time and present significant difficulties. Russia scrambled to lay down new electricity supply lines and to build a nineteen-kilometer rail and road bridge connecting Kerch to Russia, a project that was long-mooted but previously stalled. The current estimated cost of just this one project is between $4.5 and $5.5. billion. Affective geopolitics is costly.

7

The Novorossiya Project

IN BREAKING APART A sovereign territorial state, it is helpful, if not always necessary, to have an alternative geopolitical imaginary at the ready and for this ersatz replacement to have some degree of local credibility and support. When Putin decided to annex Crimea, the move was intuitively presented as a historic Russian territory rejoining the motherland and, further, as the correction of an arbitrary and capricious historical wrong. The demographics of Crimea were such that this storyline resonated with most but not all Crimeans. But when it came to the rest of Ukraine, the Putin administration faced a dilemma. Ukraine's modern history was intimately entangled with that of Russia. Tsarist and Russian Orthodox Christian discourse rendered it the homeland of a common Rus(s)ian people, its capital Kyiv as the mother of all Russian cities, and its land as a Little Russia populated by little Russians. The Bolsheviks recognized Ukrainians as a distinctive nation, constituting it as a fraternal Slavic nation alongside Russians. The Great Patriotic War bound the countries together, first in trial and suffering, and then in redemption and victory. Putin evoked these very discourses—"we are one people, Kiev is the mother of Russian cities. Ancient Rus is our common source and we cannot live without each other"—in his speech recommending annexation of Crimea. Holding that Ukrainians and Russians are one people while, at the same time, seizing territory from Ukraine required a hyperbolic fascist-threat storyline to make sense. According to this scenario, anti-Semitic nationalists from western regions not part of the Russian Empire were Nazi collaborators during the Great Patriotic

War. Now, in the seventieth year of Ukraine's liberation from Nazi rule, these forces were back on the streets and through violent protests on the Maidan managed to oust a legitimate government and seize power in a military coup. Ukraine, as a consequence, was in territorial crisis as ordinary ethnic Russians and Russian-speaking people, concentrated particularly in the southeast, sought protection from the fascist junta now ensconced in Kyiv. In these circumstances, it was understandable that former tsarist and Soviet identities in regions historically close to Russia resurfaced. As an antifascist protector, Russia was obliged to help these people and to allow them to choose their own future. In the Kremlin's storyline, infused with sinister conspiracy and myth, the Ukrainian crisis of 2014 was the return of the existential struggle that characterized the Great Patriotic War. Ukrainians and Russians were one people because they were antifascist, Orthodox, and part of the civilization born from ancient Rus. Outsiders, however, empowered irreconcilable internal fascists. Ukraine, a cobbled-together country, was consequently falling apart.

That was Putin's storyline. But how did the war in eastern Ukraine actually begin? Discerning the origins of secessionism and war in eastern Ukraine poses serious conceptual challenges for scholarly analysts. The conflict is ongoing. It has gone through multiple stages, all of which have features—motives, actions, and events—that are not transparent. An information war pervades coverage of the conflict. Journalistic and political commentary tends to be preoccupied with the question of causality and blame. Many critics of the Kremlin tend to be deeply invested in "hand of Moscow" accounts and explanations, whereas critics of the Ukrainian government tend to frame the conflict as a "civil war." Scholars are generally alert to the complexities, but they too can get caught in framing traps and forced binary contrasts between "domestic sources" and "outside interference."[1] Three starting observations can help analysis go deeper. First, it is important to recognize that there was considerable opposition within the power structures across southeastern Ukraine to the Euromaidan protests and subsequent ouster of President Yanukovych. That power structure rested on networks of patronage and power supervised by regional oligarchs that wove together industrial enterprises, security structures, political parties, and criminal groups. How this power structure operated (and how it was contested) varied from place to place but it was loosely held together by the Party of Regions, with Yanukovych's "Donetsk clan" at its pinnacle. The flight of Yanukovych, the scattering of his coercive apparatus, the seeming triumph of violence as a successful method of political change and the disintegration of the Party of Regions created a legitimacy and governance crisis across the localities where it held power.[2] Place and regional difference came to the fore as local elites scrambled to adjust and recalibrate their interests. As a place with a distinct political economy, demographic makeup, and regional identity, the Donbas posed the greatest governance challenge for the new government in Kyiv.

Second, the activities of various pro-Russian networks and agents of influence on the ground across southeastern Ukraine directly impacted how opposition to Euromaidan and Yanukovych's removal played out within the administrative buildings and across central squares of the region. Translocal actors (those moving back and forth across borders) rode the affective waves of Ukraine's governance crisis, translating it into protest marches, building occupations, secessionist rebellion, and subsequent war.[3] While a governance crisis in Kyiv was always likely to spur secessionist sentiment among certain groups in southeast Ukraine, the escalation of this into armed rebellion in the Donbas and thereafter into irregular and then increasingly regularized war was the result of armed provocateurs from Russia.

Third, this does not mean that the Kremlin was behind all forms of protest against Euromaidan—this is clearly not the case—or that the Kremlin controlled the actions of all secessionist leaders, also clearly not so. Secessionist leaders and later rebel fighters had their own motivations. Having said that, there is considerable evidence to indicate that Russian state security structures worked in partnership with ostensibly private but functionally extended state networks of influence—oligarchic groups, veteran organizations, nationalist movements, biker gangs, and organized criminal networks—to encourage, support, and sustain separatist rebellion in eastern Ukraine from the very outset. The (re)emergence of Novorossiya as a separatist geopolitical imaginary in 2014 was not a local 'bottom up' phenomenon in southeastern Ukraine. It was, instead, a geopolitical project that sought to capture grievances, discontent, and disaffection across Ukraine and translate them into a secessionist revolt that the Kremlin could use to exert leverage over the geopolitical future of Ukraine. Some were true believers but many were not. This chapter examines two central politico-geographic questions. First, what exactly was Novorossiya and how did it return from the past to become a geopolitical project for a few short months in 2014? Second, why did it fail and transmutate into two People's Republics and war in the Donbas? The chapter begins by discussing Novorossiya, first as a historical region and later as a revisionist geopolitical imaginary (mythologized territory). It then tells the story of how over a few short months a war for Novorossiya became a war in the Donbas. The chapter concludes with how ordinary citizens in SE Ukraine 6 (SE6) and Crimea viewed the project at the end of the tumultuous year of 2014.

Novorossiya as a Revisionist Geopolitical Imaginary

In defeating the Ottoman Empire in a series of battles between 1768 and 1774, the Russian Empire acquired control over a series of territories north of the Black Sea in a region formerly controlled by the Crimean Khanate known as the "wild fields." *Novorossiya* was the name given to a province established by Catherine's imperial bureaucracy to rule this new territory. Like equivalent names in the United States, it signified an aspirational imperial project to domesticate "savage" frontier lands into territories ordered by enlightened reason. The province was subdivided into three regions run by Cossack regiments. Administrative structures were formed around military units raised from the local population, and militia members were simultaneously farmers and soldiers. The territory was constantly growing, and in time it expanded to include Russia's newly acquired lands around the Azov Sea. In 1803, Novorossiya was split into three smaller provinces but, until 1874, they were integrated within Novorossiya's general governorship. The port city of Odessa, founded in 1794, was the political center of Novorossiya for most of its history.

There is no fixed geographic definition of Novorossiya. Some historical renderings exclude the Crimean peninsula and the Kuban and Stavropol territories on the east side of the Black Sea.[4] At other times Crimea was seen as part of Novorossiya. During the Russian Civil War, three aspirational Soviet republics were proclaimed across the lands of former Novorossiya and sought inclusion into the Russian Soviet

Federal Socialist Republic (RSFSR).⁵ None of the subsequent political and administrative boundaries created by Communist rule corresponded to any historical delimitation of Novorossiya. In Soviet times, Novorossiya signified the imperial past and only appeared in specialized books or historic novels. Its connotations were with imperial adventure, the glorious victories of Russian troops over the Ottoman Empire, and the deeds of famous generals, such as the tsarist heroes Alexander Suvorov and Mikhail Kutuzov.

Novorossiya's emergence as a revisionist geopolitical imaginary corresponds with the growing governing crisis within the Soviet Union in 1990. In July 1990 the newly elected deputies of the parliament of Soviet Ukraine proclaimed the supremacy, independence, integrity, and indivisibility of Ukraine's state sovereignty within the boundaries of its territory. Like similar "sovereignty proclamations" by constituent republics in Yugoslavia and other parts of the Soviet Union, the move triggered a reaction among those groups who were majority populations at the federal level but minority populations within certain republics. Ethnic Serbs in Croatia and Bosnia-Herzegovina, and ethnic Russians in Ukraine, Kazakhstan, Belarus, and other successor states, shared this structural characteristic. In August 1990, a group of pro-Russian intellectuals in Odessa sought to revive the idea of Novorossiya. They formed the Democratic Union of Novorossiya, whose main spokesperson was Oleksii Surylov, a professor at Odessa State University. He argued that the inhabitants of southern Ukraine were a distinct ethnos—a melting pot of settlers from ten neighboring nations—separate from ethnic Ukrainians. Because of its unique identity and history, Novorossiya should become an autonomous region within a federated Ukrainian state.⁶ The notion, however, was more than a vehicle for an autonomy claim within Ukraine, for its partisans soon became entangled with the crisis in neighboring Soviet Moldova. There, as noted earlier, moves by cultural nationalists to "reunite" with Romania provoked countermobilization by Russophonic elites and workers, especially in Transnistria. In September 1990, a self-constituting assembly in Tiraspol proclaimed the Pridnestrovian Moldovan Republic (PMR). Unlike Ukraine, separatism based on imperial identity soon became a viable cause in Moldova.

Within Russia, the notion of Novorossiya was not central to the search for a new "Russian idea" as described in chapter 2. None of the prominent intellectual figures of this time used it. Nevertheless, the idea of restoring the Soviet Union, or at least creating a larger Russia that incorporated historic imperial lands where ethnic Russians were now supposedly "stranded," had many supporters. A catalyzing event, as we have noted, was the short war in Transnistria from March to July 1992. Ideologically engaged young Russians in search of action, like Alexander Borodai and Igor Girkin, traveled to Tiraspol to join the separatist Transnistrian forces and adopted Novorossiya as a romantic ideal to justify and aggrandize their cause. The notion was appealing because it constituted Transnistria as part of the patrimony of the Russian Empire. Girkin considered Crimea and Novorossiya as the "crown jewels" of the empire. But the idea pushed by Surylov and others to unite Transnistria with southeast Ukraine made little headway.

Novorossiya remained a romantic lost cause in imperial nationalist circles in Russia and Ukraine over the subsequent two decades. Vladimir Zhirinovsky's Liberal Democrats' surprise showing in the December 1993 Duma elections deepened fears within Kravchuk's Ukraine about Russia's intentions. Kravchuk's chief spokesperson indicated, "Ukraine can't help being worried, especially since Zhironovsky's group came to power with demagogic slogans, including the restoration of Russia's imperial borders."[7] Russian state officials came to see Novorossiya as a reserve card in their geopolitical relationship with Ukraine. As noted already, some privately threatened at this time that if Ukraine took a strong anti-Russian stance in its foreign policy—such as seeking to join NATO—the Russian government would "activate" the Russians in Ukraine in such a way as to divide and possibly break up the country.[8] Ukraine's territorial integrity remained uncertain until the Budapest Memorandum a year later and the Russia-Ukraine Friendship Treaty of 1997. Thereafter, relations between the two states improved.

Because Ukraine was characterized by strong regional identities, moments of political polarization tended to produce calls for local autonomy before there was any discussion of outright separatism. During the Orange Revolution, for example, the Donetsk oblast administration scheduled a referendum on autonomy for the region.[9] Separatism, to the extent that it was expressed, tended to be localized and enframed in Soviet nostalgia, referencing the short-lived Soviet Republics of 1917–1918 rather than more expansive tsarist-era territories as justificatory historic precedents. The borders of Novorossiya, after all, were vague and uncertain. A year after the Orange Revolution, a few marginal pro-Russia nationalists created the Donetsk Republic as a secessionist movement claiming heritage—visually by appropriating its black, blue, and red tricolor flag—to the short-lived Donetsk–Krivoy Rog Soviet Republic of February–March 1918. The first public appearance of the movement's banners was during a protest at a NATO informational event organized at Donetsk National University in October 2005.[10] It attracted little support and languished on the margins of political life in Donetsk, which was dominated by the Party of Regions, behind which were Rinat Akhmetov and Yanukovych's entwined power networks, until the spring of 2014.

The idea that separatism was a Russian "lever" in Ukraine endured within the Russian state apparatus. In January 2008 as Ukraine's Orange government made a bid for NATO membership, some within Kremlin circles reportedly considered the idea of creating a Moscow-friendly buffer state from Crimea to Odesa that would secede from Ukraine and join Transnistria to form a new state called Novorossiya.[11] The schema was never encouraged, for it was apparent that the Orange movement was in trouble. While Yushchenko was outspoken in his support for his friend Saakashvili in August 2008, public opinion in Ukraine was more supportive of Russia than Georgia during the war.[12] Yanukovych's triumph in the presidential elections in February 2010 removed the threat of Kremlin encouragement of regional division within Ukraine until the crisis in late 2013 and the spring of 2014.

During this time the Kremlin sought to exercise soft power in Ukraine by promoting the idea of the Russian world (*Russki Mir*). Conceptualized originally by liberal political technologists as a way of rebranding Russia, the expression was at the nexus of different Kremlin concerns during Putin's reign—promoting Russian language and culture overseas, managing relations with "compatriots" in the near abroad, and articulating a vision for Russia as a distinct civilizational space, along with aspirations for a meaningful geopolitical sphere of influence. By Putin's return to the presidency in 2012, the concept also had religious overtones and became part of his articulation of Russia as a conservative civilizational bastion opposed to morally suspect liberal Western values. The French analyst Marlene Laruelle notes that the "inflection of the Russian World does not call into question the independence of Russia's neighbors per se, but rather their geopolitical orientations."[13] In this sense, the concept was an aspiration to translate a perceived lever of soft power—ethnic Russians, compatriots, and Russian-speaking populations in the near abroad—into something that could have hard power consequences. The notion invested Russophone language practice, and majority Russian-speaking communities, with a geopolitical significance that was to prove misplaced, because Russian-speaking did not necessarily translate into a pro-Russia or pro-Putin orientation. Like the Anglophone Irish, many if not most Russophone ethnic Ukrainians had views distinct from the metropolitan power.

The conflict in eastern Ukraine can be broken down into a series of different phases separated by important turning points. Our concern is with the period between mid-February and mid-May 2014, characterized by four different phases, which is when the Novorossiya project was first activated on the ground in southeast Ukraine and subsequently became territorialized in compromised form as an alliance of two separatist entities within Donetsk and Luhansk oblasts fighting against Kyiv. As is well known, the separatist aspirations of Novorossiya activists failed to gain traction across southeast Ukraine with the exception of the Donbas. The death of many pro-Russia activists in Odessa in horrific circumstances on May 2, 2014, was also a symbolic death of the Novorossiya project. The war for Novorossiya had become the war in the Donbas.

Activating Novorossiya

The first and most occluded phase of the conflict is the period before Yanukovych's flight, when it was apparent that his government was struggling to contain and suppress the Euromaidan protests. These protests challenged Putin's strategy of locking Ukraine into the Eurasian Customs Union and thus into a Russia-centered pole of power. As already noted, these protests were interpreted in the Kremlin as a Euro-Atlantic operation to prise Ukraine from Russia. Great-power centrism and threat inflation predominated. How the Putin administration's policy emerged and evolved is a subject for future historians to detail. Putin's decision that Russia's best response to Yanukovych's flight from power was to authorize an invasion of Crimea was momentous. That decision was part of a broader response that saw Russia publicly threaten military intervention in Ukraine while covertly authorizing reserve security assets and irregular agents to aid pro-Russian forces across southeast Ukraine as a means of undermining Kyiv's new government. Russia's annexation of Crimea was the first outright annexation by one state of another's territory since Saddam Hussein's attempted annexation of Kuwait. Putin understood the decision as historic in Russia-centric terms—and not as an act, the first annexation on the continent of Europe since World War II, that revived the memory of the Nazism he claimed to be fighting.

In deciding to annex Crimea, Putin was shifting the intellectual foundations of his foreign policy practice from great-power geopolitics (competitive statecraft conducted within the existing territorial order) to revisionist imperial geopolitics (competitive statecraft that seeks to remake the existing territorial order). The shift was a victory for revisionist geopolitics in Russian geopolitical culture, a tradition we discussed in chapter 2. Many of the champions of Novorossiya in Russia were part of the Izborsky Club, an intellectual pressure group established in 2012, which sought to revitalize the "national-patriotic" network of the early 1990s that had brought together "brown" (ultranationalist) and "red" (Soviet and Communist) intellectuals to support the Supreme Soviet in its struggle with Yeltsin. Three distinct revisionist networks converged around the Novorossiya idea.[14] The first was the "red" tradition of thinkers like Alexander Prokhanov, the longtime editor of *Den* and *Zavtra*. This tradition portrayed Novorossiya as an anti-oligarchic project that would reclaim the great achievements of the Soviet Union—the factories of the Donbas—and renationalize them so they served the interests of workers. Novorossiya was to be a renewed form of the Soviet Union. Both Alexander Borodai and Igor Girkin, veterans of the Transnistrian war, wrote for Prokhanov's publications and were part of his network, as were younger figures like Pavel Gubarev, formerly a member of the neofascist Russian National Unity Party and subsequently of the Progressive Socialist Party of Ukraine (a small pro-Russian party based in southeast Ukraine). Girkin had also fought with Serb ethnonationalist forces during the Bosnian War and with Russian forces in Chechnya. He was reportedly a colonel in the

Federal Security Service (FSB) in 2013.[15] The second was the "brown" tradition of far-right figures like Alexander Dugin, a figure we have already noted. A creative and entrepreneurial propagandist over the decades, Dugin's geopolitical schemes were already well known throughout the Russian security apparatus. Dugin viewed Novorossiya as a welcome gathering of territories that would expand Russian territory and create a new "Large Russia" (*Bol'shaya Rossiya*) as the anchor of a Eurasian great-power pole in opposition to the West. The third was a "white" tradition that held a romantic vision of the Russian Empire under Romanov rule. Associated with the movement known as "political orthodoxy," it combined conservative orthodox religious values with imperialist nostalgia. Putin's conservative turn in 2011 made common cause with this tradition. It viewed Novorossiya as a sacred cause, and much of the iconography of Novorossiya drew upon tsarist sources. The Novorossiya flag, a blue Saint Andrews cross on a red background, was reportedly created by Gubarev after the tsarist naval flag and was said to signify the central role the navy played in establishing Novorossiya. This flag was later replaced by a flipped white-yellow-black tricolor, the flag of the Russian Empire, but the change was not widely noted or adopted. To the "founding fathers" of Novorossiya on the ground in Ukraine, figures like Gubarev and later Borodai and Girkin, the different ideological colors that infused Novorossiya were a complementary matrix of inspiration rather than contradiction.[16] The result was a postmodern geopolitical imaginary, a simulation of authentic territory created from a collage of disparate signs and symbols. The first invented map of Novorossiya used by the rebels, for example, featured a claim to eight oblasts in southeast Ukraine including Kharkiv, which was never part of historic Novorossiya. Enframing the map was the invented flag of the Donetsk People's Republic

Putin's decision to annex Crimea and encourage Novorossiya were, as already noted, longstanding reserve Russian response options should Ukraine move decisively to integrate with Euro-Atlantic institutions. There is no evidence that Prokhanov or Dugin directly influenced Putin's radical decision to pursue these options. What they did provide, however, was networks of activists that could be used to organize and mobilize imperial nationalist vanguards to achieve them.[17] Unlike Prokhanov or Dugin, the "orthodox oligarch" Konstantin Malofeev is alleged to have had direct policy input into the Kremlin's policy-planning process prior to Yanukovych's flight.[18] A lawyer by education, Malofeev made important business connections with figures in the Orthodox Church that allowed him to create Marshall Capital Partners, an investment fund specializing in telecommunications that allegedly grew by "raiding" its competitors. In 2012 he entered politics after publicly campaigning for a "clean Internet" and conservative values, safe and politically useful positions in the firmament of the Kremlin's managed democracy. Malofeev's investment firm hired Borodai for public relations work and also reportedly Girkin. He also has personal ties to Dugin.[19]

Understanding the flow of communication and command between the presidential administration, the FSB and GRU (Main Intelligence Directorate), and private actors like Malofeev, Glazyev and others, at the upper levels of the power vertical, and then their relations with figures like Borodai, Girkin, and Igor Bezler (another military veteran), on the lower operational levels of the power vertical, is very difficult for the exercise of power is hidden and the activities covert (until, and if, exposed). Nevertheless, many analysts have connected the dots in general terms, while leaked documents and recordings hint at how the process likely worked.[20] Rather than a direct controller, Putin's presidential administration appears to have been more the regulator of an emergent field of geopolitical enterprise in which certain venture capitalists (oligarchs and officials interested in further ingratiating themselves with the Kremlin) presented plans, anticipated interests, and funded promising "startups," namely ambitious agitators and violence entrepreneurs on the ground. The presidential administration likely had different factions vying to influence Putin's decision-making. As the field grew, and local entrepreneurs, some with ties to Ukrainian oligarchs, emerged alongside those sponsored by Russian networks, the Kremlin's advisers and regulators (known in Russian as *kurators*) migrating to and from the field would occasionally intervene to direct the general enterprise and its various players.[21] Later the Kremlin's involvement became much more direct and forceful, deploying Russian military formations in Ukraine to defeat the Ukrainian army and special operations forces to eliminate wayward and troublesome warlords within the secessionist enterprise zone. Over the course of the Ukraine crisis, the presidential administration's chief political technologist Vladislav Surkov emerged as the chief Kremlin *kurator* on Ukraine.

In February 2015, the Russian newspaper *Novaya Gazeta* published a memo it claimed was presented to the Kremlin in mid-February 2014 as the Euromaidan protests raged.[22] Malofeev was suggested as one of its authors though he denied it.[23] The document presented a conspiratorial reading of the Euromaidan, claiming that protesters "from all appearances are controlled not so much by the oligarchic groups [in Ukraine] but to a significant extent by Polish and British intelligence services." It suggested the European Union (EU) and the United States were willing to allow Ukraine's disintegration. "Given this, Russia is obliged to interfere in the geopolitical intrigue of the European Union, aimed against the territorial integrity of Ukraine." To save the Russian position in Ukraine, Russia should "play on the centrifugal aspirations of various regions of the country with the purpose, in one form or another, of initiating the annexation of its eastern regions to Russia. The dominant regions for applying efforts must be the Crimea and Kharkiv Region in which exist fairly strong support groups of the idea of maximum integration with the RF [Russian Federation]." While the takeover of Crimea and "several eastern oblasts" would create burdensome budgetary expenditures, "from the geopolitical perspective, the win will be priceless: our country will gain access to new demographic resources, [and] highly-qualified cadres of industry and transportation will be in its possession." The memo stressed that "the process of the 'pro-Russian drift' of the Crimea and eastern Ukrainian territories must be created in advance to lend this process political legitimacy and moral justification." "[A] PR strategy must be built to accentuate the forced, reactive nature of the corresponding actions of Russia and the pro-Russian-minded political elite of the south and east of Ukraine." The memo suggested a choreography of referendum and annexation: "it is necessary to prepare the conditions for conducting in the Crimea and Kharkiv Region (and later in other regions) referendums raising the question of self-determination and further possibilities of annexation to the Russian Federation."

The significance of this memo, relative to others yet unseen, is at this point unclear. What is noteworthy about its claims are (i) its deeply conspiratorial attitude toward the actions of the European Union and the United States; (ii) its conceptualization of Ukraine as a state on the brink of disintegrating into "western regions plus Kiev" versus "eastern regions plus Crimea"; (iii) its articulation of the need for Russia to look toward the annexation of Crimea and several eastern oblasts; (iv) its privileging of geoeconomic rather than spiritual, ethnonationalist or primordial reasons for Russia to do so; and (v) its emphasis on the projection of "political legitimacy" and "moral justification" as Russia moved to facilitate annexation of Crimea and eastern regions. The memo does not mention Novorossiya but views Crimea and the eastern oblasts as a singular policy challenge. Also noteworthy is how wrong the memo was analytically: the specter of sinister others engaged in geopolitical intrigue overdetermines the need to responsibly analyze empirical facts and details. There was no outside control of the Euromaidan protesters, no Euro-Atlantic plan to break up Ukraine, and no serious secessionist movement in western Ukraine though revolts against state institutions had taken place there in early 2014. Its assumptions about pro-Russian integrationist sentiment in eastern Ukraine proved to be largely incorrect. The Donbas, which it viewed as too dominated by Yulia Tymoshenko's *Batkivshchyna* (Fatherland) Party and by the oligarch Rinat Akhmetov, turned out to be the center of pro-Russian secessionist sentiment, whereas Kharkiv/Kharkov, which it saw as ripe for "integration," curbed local separatism quickly. Further, its proposal to use the EU vehicle of "Euro-regions" to facilitate annexation was bizarre. Whatever its actual policy significance, the memo is nevertheless interesting in revealing how some Russian analysts viewed the Ukraine governance crisis (even before Yanukovych's flight) as a geopolitical opportunity to annex a series of territories from Ukraine, not just Crimea.

The Ukrainian-born economist Sergey Glazyev, a longtime supporter of Russians in the near abroad (see chapter 2), is another senior figure frequently cited as presenting plans to the presidential administration to realize Novorossiya in southeastern Ukraine. The Russian investigative journalist Mikhail Zygar claims that "[i]t was Glazyev who more than anyone else promoted the concept of recreating 'Novorossiya'... Glazyev's vision was for Novorossiya to join Russia, as Crimea had." According to him, Putin did not want to take any decisive action and repeatedly told Glazyev that the inhabitants of eastern Ukraine should be the ones to make the first move.[24] Subsequent evidence supports this claim.

In August 2016 the Ukrainian Prosecutors Office released a series of audio recordings made between February 27 and March 6, 2014, featuring Glazyev conspiring with Konstantin Zatulin, a Russian Duma leader born in Batumi, and others to mobilize pro-Russian groups in three southeastern cities: Odessa/Odesa, Zaporozhye/Zaporizhia (Glazyev's birth city), and Kharkov/Kharkiv.[25] Glazyev declares on a March 1 recording, to an unknown activist called Anatoly, that he has "an instruction to raise everybody, to raise the people. The people should gather in the square, and ask Russia for help against the Banderivtsi."[26] He adds soon thereafter: "I have the direct instruction to raise people in Ukraine where we can." The tapes were redacted by the Prosecutors Office so we do not have the full record, and their overall significance may not be apparent for some time. Nevertheless, they do have scholarly value. First, they provide a glimpse into how Crimea and southeast Ukraine (Novorossiya is not mentioned on the released segments) were entwined at the outset in the minds of certain Russian-based power brokers trying to shape events in Ukraine. Both were seen as theaters for secession and possible follow-on annexation. Second, they reveal the improvised and amateurish nature of this one vector of attempted coordination from Moscow. Various pro-Russian activists are called and call up Glazyev and Zatulin to ask for instructions and money. Both men discuss their financing of these groups. The supplicant nature of relations is apparent as are general conditions of chaos and confusion. Third, the recordings hint at what would become a manifest gap between the dream of Novorossiya harbored by revisionist geopoliticians in Moscow and sociopolitical realities on the ground in southeast Ukraine. Glazyev's plans may have been great but his power to effect change on the ground was limited. For example, Glazyev asked Anatoly: "Why is Zaporozhye quiet?" He discusses the storming of the Odesa regional administration building with a local pro-Russian activist (Denis Yatsyuk), but recounts that their subsequent attempt on March 3 was an abject failure.[27] Fourth, the tapes make clear that Glazyev and Zatulin viewed the creation of a credible public scenario for intervention the most important goal. Speaking to Yatsyuk he explains: "It is very important that people appeal to Putin. Mass appeals directly to him with a request to protect, an appeal to Russia, etc." With Putin authorized to use Russian troops in Ukraine by the Federation Council (March 1) and soldiers massing on the border, Glazyev sounds like a theater director trying to arrange a cast of unruly players to perform the first act of a classic rescue fantasy play.

The Crimean annexation and the Novorossiya project shared a similar theatrics of secession at the outset. The script features an ominous fascist threat (referred to as the "Banderivtsi"), which justified local government occupation by "self-defense forces" (advised and supplemented by agents of the Russian government). The new leadership then proclaims a referendum of self-determination and issues an appeal for protection to Russia. Whereas Crimea was successful, the Novorossiya movement in southeast Ukraine faltered from the outset.

Crimea had certain innate advantages. It had powerful affective appeal to ordinary Russians both as a storied sacred place in Russian history and as a beloved vacation spot that held happy memories for many people. As a peninsula, it was geographically separate and distinct, by virtue of its ethnic Russian majority, from other parts of Ukraine. Furthermore, it was home to strategic Russian military bases and had Russian forces deployed there already. Controlling it over all other places in Ukraine was the priority of the Kremlin. The Ukrainian forces there were ordered not to fight and subsequently many voluntarily joined the new order. By contrast, southeast Ukraine was a much larger space that had no uniformly clear physical geographic dividing line from the rest of Ukraine. It had many distinctive places forged by tsarist and Soviet power, but no oblasts where ethnic Russians were a majority. Parts of it bordered Russia but others, like Odesa, were far from Russian territory and potential supply lines of Russian activists (who took trains and buses across the border for demonstrations within Ukraine).[28] As a result, it was a more challenging terrain for the operationalization of an assisted secessionist project.

The military tactics the Russian security apparatus adopted were different in southeast Ukraine. Whereas Crimea featured command control by deployed Russian military forces over local "self-defense" units, the situation in the Donbas featured no overt Russian military presence. Agency, as the Glazyev tapes make clear, was in the hands of local and translocal groups (those with ties to similar organizations inside Russia).

The theatrics of secession followed the same script as in Crimea. Indeed, this script was a reflection of what Russia projected onto Euromaidan: a foreign power directs local agitators; they initiate a geopolitical coup in the name of "the people" by occupying public spaces and provoking violence; this leads the legitimate lawful authorities to defect or flee; the coup leaders subsequently appeal to their sponsors for aid and protection, which they receive. Euromaidan and anti-Maidan were similar geopolitical projects in the minds of those directing Russia's state security apparatus. In an even more disturbing irony, fascist-like putsches were launched using the rhetoric of antifascism. Indeed, many of the leading figures in the pro-Russian movement were genuine neo-Nazis.[29]

Localized Power Struggles

Prior to Yanukovych's flight on February 21, 2014, pro-Maidan protests greatly outnumbered anti-Maidan protests across Ukraine. One study documented 3,721 protest events associated with Maidan between November 21, 2013, and February 20, 2014.[30] Only 402 of these could be attributed as anti-Maidan protests. Prior to Yanukovych's flight, the Party of Regions, and to a much lesser extent the Ukrainian Communist Party, tended to be behind these protests. In late 2013 these featured denunciations of the European Union Association Agreement and support for the Eurasian Customs Union. Subsequently, demonstrations expressed solidarity with the government and law enforcement agencies while denouncing violence attributed to Euromaidan protesters. In general, anti-Maidan rallies were a lot smaller than pro-Maidan rallies and more likely to feature demonstrators whom organizers paid to attend. The geography of protest events is significant. Almost two-thirds of all Maidan protest events occurred in western and central Ukraine. Less than a fifth of all pro-Maidan protests were in the south and east. By contrast, over 40 percent of all anti-Maidan protests were in the Donbas. Nearly half of all anti-Maidan protests occurred in just five cities (Donetsk, Kharkiv/Kharkov, Odesa/Odessa, Luhansk/Lugansk, and Mykolaiv/Nikolayev).[31]

The intensity and size of anti-Maidan protests increased considerably after Yanukovych's flight. The most catalyzing event was the appearance of "little green men" in Crimea on February 27 to "rescue" residents there from what was presented as a looming fascist invasion. This "protection" of Crimea from the new government in Kyiv changed the rules and transformed expectations. Pro-Russian activists in southeast Ukraine and Russia were energized. The combination of the legitimacy crisis of the Ukrainian state, the Euromaidan example of how to seize power, and Crimea's takeover by Russia created a political environment for previously marginal political forces to come to the fore. Instead of the Party of Regions or Communists leading anti-Maidan protests, a plethora of Russian nationalist groups and parties—Russian Bloc, Russian Unity, Rodina, Oplot, Slavic Guard, Odesskaya Druzhyna (Squad), different Cossack groups, and militias—seized the moment. On March 1, Russia's Federation Council swiftly passed Putin's request for authorization to use force to protect Russian lives in Ukraine. A second rescue mission seemed imminent.

Southeast Ukraine did indeed see an influx of Russian activists from various Russian nationalist groups, some no doubt agents of the Russian security services. They swelled the numbers at rallies, particularly on weekends. Ukraine's new interior minister, Arsen Avakov, whose own power base was Kharkiv, dubbed them "professional touring provocateurs;" others simply referred to them as "protest tourists."[32] But no Russian military forces alighted from helicopters to invade southeast Ukraine; the troops massing at the border remained there. Instead, city by city and town by town, there developed a power struggle between previously marginalized Russian nationalist groups, emboldened by the takeover of Crimea and boosted by volunteers from Russia, and the existing political power structures in these towns. The latter comprised political parties, oligarchs, mafia networks, and local police and security structures as well as civic activists and social groups, some of whom, like the young men in Shakhtar Donetsk's fan club (ultras), were pro-Maidan and radicalized in opposition to Russia's invasion of Crimea. Whether the local power structure was coherent and led by a powerful figure that came out early against the separatists or whether it was divided internally—or, just as important, whether power brokers felt they could play both sides for advantage and thus sat on the fence—proved crucial in deciding the fate of the locality.

Russian Special Forces seized the parliament in Simferopol on February 27, with Igor Girkin among their number. That day they installed Sergey Aksyonov as the new prime minister of Crimea. The very next day Pavel Gubarev, as equally marginal as Aksyonov, gave a fiery speech before the deputies in the Donetsk regional government building. The following day (March 1), in emergency session, these deputies voted to make Russian the official language in Donetsk and to hold a referendum on its future. Outside, Gubarev addressed a pro-Russian rally of a few thousand demonstrators in front of the building and, railing against the corrupt deputies inside, proclaimed himself the people's governor of Donetsk.[33] Gubarev, who was in telephone contact with Aksyonov, returned the next day with a group of activists and replaced the Ukrainian flag on the building with the Russian tricolor. On Monday, March 3, 2014, they stormed the building, evicted the deputies inside, and proclaimed themselves a People's Committee who were now the new authorities. "I am for Donetsk being a part of Russia," he told the gathered media. Seated on a dais behind the Donetsk Republic flag, an orange-and-black Saint George's flag, and the Russian tricolor, the faux governor proclaimed: "There is no chance that people here in Donetsk will want to continue being part of Ukraine. The people of Donetsk support us. We shall address the Russian authorities and ask them to bring a peacekeeping force here. It is the only way to keep order."[34]

A more violent storming of the administrative building in Kharkiv the same weekend left two dead and nearly a hundred injured as a pro-Russian mob, reportedly paid by Russian intelligence agents, evicted Euromaidan supporters already within the administration building.[35] A young activist from Russia replaced the Ukrainian flag with a Russian one atop the building, taking a selfie in the process.[36] The same weekend several thousand demonstrators in Mykolayiv/Nikolaev called for that region's secession from Ukraine and unification with neighboring Odesa and Kherson. Organized by a pro-Russian group called the People's Front, demonstrators chanted, "Russia will help us," "We support federalization," and "Odessa, Nikolaev, Kherson."[37] Altogether, similar protests were recorded in eleven different cities across the southeast, including Crimea, that weekend. Superficially there appeared to be little difference between what was happening in Crimea and beyond it. In a remarkable gesture, the Twitter account of the Permanent Mission of Russia to NATO circulated a map of the eleven locations across southeast Ukraine and Crimea with Russian flags superimposed on each.[38]

The new government in Kyiv responded to the lawlessness by appointing oligarchs Igor Kolomoyskyi as governor of Dnipropetrovsk and Serhiy Taruta as governor of Donetsk. Dnipropetrovsk had seen both pro- and anti-Maidan demonstrations. Kolomoyskyi moved quickly to buy off and suppress those in opposition to the new government. Together with local councilors Gennady Korban (later charged as head of an organized crime group) and Boris Filatov (later mayor), he established and financed a militia called the "Dnipro Battalion" that was used to defend the region against separatists, securing and advancing their own business interests in the process. Kolomoyskyi, an Israeli citizen, also funded a series of other pro-Ukrainian militias, including some with far-right beliefs. Taruta was not nearly as effective in Donetsk. The dominant oligarch and power broker in Donetsk, Rinat Akhmetov, had been offered the job of governor but turned it down. Given his extensive business interests across the extended region (including in Russia) and overseas, he was hedging his bets as best he could. By contrast, Yanukovych's network, led locally by his son Oleksandr, was bent on sabotaging Kyiv's power. Another powerful figure was Aleksander Yefremov, a former governor of Luhansk/Lugansk and parliamentary head of the Party of Regions. In the summer of 2016 Kyiv charged him with direct involvement in establishing the Lugansk People's Republic.

Thereafter, as the spectacle of Russia's annexation of Crimea dominated television screens and digital media ecosystems, pro-Russia rallies and pro-Ukraine counter-rallies proliferated across southeast Ukraine, peaking on weekends. Participants sometimes clashed and fought, as happened in a mass brawl in Donetsk on March 5. The day after, the SBU (Ukrainian Security Service) managed to arrest Gubarev and spirit him away to Kyiv. (He was released two months later in a prisoner exchange, by which time he had lost out to other secessionist leaders.) The following Sunday (March 9) was the two hundredth anniversary of the birth of the celebrated Ukrainian poet Taras Shevchenko. The boxer Vitali Klitschko visited Donetsk and called for unity in the face of provocations by separatists. But pro-Russia gangs rampaging through the city center attacked those gathering around Ukrainian flags to mark the anniversary.[39] The commemoration was soon abandoned. Further clashes between rival demonstrators—ones carrying Russian flags and chanting Putin's name and ones brandishing the Ukrainian flag and denouncing Russia's takeover of Crimea—in the center of Donetsk left a pro-Ukrainian activist dead and many injured.[40] On March 22 thousands rallied, crying "Russia" and then "Crimea! Donbas! Russia." The following week saw a diminution in the intensity of protests, though tensions and fear of invasion remained high in Donetsk and other regions.[41] The iconographies of the protests were well established by this stage. Pro-Russian protesters carried Russian, Soviet, Orthodox, Saint George, Donetsk Republic, and Novorossiya flags, ribbons, images, and banners as well as placards denouncing the United States, Obama, NATO, and fascism. Pro-Ukrainian protesters carried the Ukrainian flag as well as that of the European Union and also, occasionally, banners, imagery, and slogans associated with far-right political movements in Ukraine.

The Divergence of the Donbas

During March administrative buildings were variously stormed and occupied by protesters, then retaken by officials before being stormed again and reoccupied temporarily. On April 6 the conflict entered a new phase when crowds of lightly armed pro-Russia protesters, featuring women and teenagers, seized the administrative buildings of Donetsk, Luhansk, and Kharkiv. Instead of pressing demands on the existing authorities they unilaterally declared themselves as new permanent authorities, People's Republics, who were from thereon formally replacing Ukrainian state structures. They then declared they would hold referendums on May 11 on joining Russia, and appealed to Putin to send troops to protect them from Ukrainian reprisals. Outside they erected barricades of tires and razor wire to secure their permanent occupation of the buildings. The next day the group in Donetsk expanded their occupation by seizing the local SBU building in the city. In Luhansk, SBU officers had resisted the storming of the administrative building on April 6 for six-and-a-half hours. No one sent any help and by the evening they capitulated to the crowd, who subsequently seized an arsenal of weapons. An armed police unit was later dispatched to liberate the building, but the operation was then canceled.⁴² Multiple factors were at play: institutional weaknesses and an absence of will, but also fears that protesters wanted violent confrontation in order to create a visually compelling outrage that would trigger Russian intervention in the guise of a rescue mission. From the outset, officials in Ukraine were keen to avoid repeating what they saw as Saakashvili's blunder in supposedly taking Putin's bait in August 2008 and then losing badly.⁴³

There was no storming of administrative buildings in other prominent southeast cities. In Odesa, a stronghold of Party of Regions officials, pro- and anti-Maidan groups were active but strived to limit violence. After Yanukovych's flight, pro-Russian activists were greatly energized and set up a tent camp on a central square to demand federalism, at a minimum, and Novorossiya as an ideal.⁴⁴ Rallies were also held in Izmayil, the major town in Ukrainian Bessarabia, and rumors about Transnistria unifying with the region in a new Novorossiya abounded. Pro-Russian activists proclaimed the creation of a People's Republic of Odessa on April 16, but they lacked sufficient strength and organization to seize any administrative buildings. Pro-Maidan forces were also gearing up to fight should Russia invade.⁴⁵ Significantly, in Kharkiv, the local power structure did respond to the April 6 occupation by organizing the forceful removal of the protesters. Local mayor Gennady Kernes, normally a rival of Interior Minister Avakov and a prominent supporter of Yanukovych and Russia in the past, played a crucial role in outflanking the Russian nationalist "fight club" group Oplot in the city.⁴⁶ Three weeks later he was shot in the back by a sniper but survived. While Odesa and Kharkiv were divided cities, they remained in Ukraine.

From that moment on the fate of the two major urban centers in the Donbas diverged from that of other cities in southeast Ukraine. The Donetsk Peoples Republic activists were led by Denis Pushilin, another of the marginal local politicians, suddenly thrust in the spotlight as head of the pro-Russia movement in the city. In the nineties Pushilin had worked for a Russian ponzi scheme company. Around him gathered a motley crew of locals, many of them minor criminals and unemployed young people hoping to better their condition by joining the separatist enterprise early, and some pensioners nostalgic for the Soviet Union.⁴⁷ Pushilin, who was rumored to have ties to Akhmetov, sat behind a desk in the Donetsk administrative building while the oligarch bargained with both sides behind the scenes. On the wall of Pushilin's office was a map of Ukraine upon which someone had drawn a line halfway through the country. The eastern side of the map was overwritten with just one word: Russia.[48]

Filibusters from Russia

Another significant turning point came within a week of the occupations in the Donbas. This involved an invasion of a group of armed Russian militants who were previously active in Crimea. Beyond its more common meaning as a legislative maneuver, a filibuster is also the name of a figure who undertakes nominally unauthorized military expeditions into a foreign territory to foment or support revolution. In U.S. history, filibustering is associated with the U.S. frontier pioneers who incited revolution in Texas that led to it becoming an independent state before its annexation by the United States. U.S. filibusters, most famously William Walker, were also active in Central America. On April 12, some sixty kilometers from Donetsk, a group of modern-day Russian filibusters seized the police station and SBU building in Sloviansk as well as the police headquarters in nearby Kramatorsk. The Interior Ministry building in Donetsk was also occupied. Igor Girkin, the FSB veteran with ties to Malofeev, led the group of fifty-two fighters. Girkin and his men had helped in Russia's takeover of Crimea and were now to apply some of the same military muscle to create a breakaway entity in eastern Ukraine. He later explained:

> When the Ukrainian government was disintegrating before our eyes, delegates from areas of Novorossiya started to arrive in Crimea, wishing to repeat what was happening in the Crimea. It was the clear desire of everyone to continue the process. Delegates planned rebellion in their respective districts and asked for help. Aksyonov, because of his work load, working 20 hours a day, asked me to take charge of the northern territory. And he made me an adviser on this issue. I began to work with all the delegates: from Odessa, Nikolaev, Kharkov, Lugansk, Donetsk. Everyone was confident that if the uprisings developed, then Russia would come to help.[49]

Directly north of Donetsk, Sloviansk was a strategic point on the road north to Kharkiv. Rebels also seized the towns of Kramatorsk, Artimivsk, and Debaltseve south of Sloviansk. Igor Bezler, yet another Russian Security Forces veteran turned rebel, seized the police station in Horlivka, northeast of Donetsk. The gunmen proceeded to build checkpoints outside the towns with sand and tires and a Russian flag to signify their allegiance. Here was the first instance of the creation of barriers and checkpoints with armed men publicly designating pro-Russian space on Ukrainian mainland territory beyond Crimea (Figure 7.1). Girkin expected that Russian forces would arrive to back them up.[50]

Kyiv decided they needed to respond immediately to the armed seizures and open defiance of Ukrainian territorial sovereignty. Up until this point they had feared being lulled into a trap by responding with force. Not having done so in Crimea, however, had enabled Russia to annex the territory "without firing a shot," as Putin later boasted.

FIGURE 7.1 Pro-Russian protesters standing in front of an occupied police station, in Sloviansk, Ukraine, April 13, 2014 [Anastasia Vlasova, EPA].

This time they would fight back. On April 13, Kyiv launched an Anti-Terrorist Operation (ATO) that sought to surround the rebels and drive them out. Implementation of the ATO was in the hands of a poorly equipped Ukrainian army and fractious auxiliary militias, neither trained in counterinsurgency methods.[51] The ATO soon alienated many eastern Ukrainians skeptical of all sides. In meeting lawless separatist violence with indiscriminate state-sanctioned violence, it further fanned the flames of war. In time, what looked to many like a civil war began to grip the region. Residents previously agnostic and indifferent to the conflict were soon swept up in the escalating violence and militarism. People were forced to take sides or flee for their lives. The spark for all of this was Girkin, the filibuster. He later took pride in this: "It was me who pulled the trigger of war. If my unit hadn't crossed the border, everything would have ended as it ended in Kharkov, in Odessa.... The wheel of war that is still going was put into motion by our unit."[52]

Amplifying Novorossiya

A few days after the invasion of the modern-day Russian filibusters, the cause of Novorossiya as a separatist geopolitical imaginary received a huge boost when Russian president Putin addressed the subject on *Direct Line*, an annual call-in show his administration has long used to present him as an accessible, competent, and benevolent tsar-like father of the nation.[53] A month after the Crimean annexation, this television event lasted a remarkable three hours and fifty-four minutes, during which Putin answered a total of eighty-one questions and appeals. It was in this context that Putin first spoke of Novorossiya. The stimulus was a rather obsequious question from the former Russian politician Irina Khakamada, who asked if there was a possibility of a compromise solution in Ukraine with the United States, which she indicated was the real money power, more than the EU, in Kyiv. Putin's response did not challenge the conspiratorial premise of the question but the idea of a great power deal over Ukraine. The essential issue, he argued, is how to ensure the legitimate rights and interests of ethnic Russians and Russian-speakers in the southeast of Ukraine. He then provided his audience with a history lesson:

> I would like to remind you that what was called Novorossiya [New Russia] back in the tsarist days—Kharkov, Lugansk, Donets'k, Kherson, Nikolayev and Odessa—were not part of Ukraine back then. These territories were given to Ukraine in the 1920s by the Soviet government. Why? Who knows. They were won by Potemkin and Catherine the Great in a series of well-known wars. The center of that territory was Novorossiysk, so the region is called Novorossiya. Russia lost these territories for various reasons, but the people remained. Today, they live in Ukraine, and they should be full citizens of their country. That's what this is all about.[54]

Three aspects of the Putin response are worth noting. The first was how Putin localized the geopolitical crisis by, disingenuously, describing it as a crisis of Ukraine as a state. His storyline was organized around the presumption that "ethnic Russians and Russian-speakers in southeast Ukraine" were under threat from a fascist government and in need of protection. This attributed a coherent identity to both a very broad sociolinguistic category and a much smaller self-ascribed ethnic group. The effort in the Ukrainian Rada on February 23 to repeal the Yanukovych law allowing the Russian language official status at the regional level was held as prima facie evidence of a radical threat to all Russophones. Second, in calling out the southeast of contemporary Ukraine as previously part of Russia, Putin further challenged the territorial integrity of Ukraine even as he ostensibly articulated respect for it. Ukraine is implicitly an arbitrarily assembled collection of regions rather than a country with its own long and storied past, like Russia. Russia is implicitly equivalent to the tsarist empire. The contingency of Ukraine is underscored here by Putin's exclamation: "Why? Who Knows?" Ukraine, by this logic, is something of a historic accident. The third, most significant feature of Putin's response was his very use of the word Novorossiya. Emerging in a spontaneous, free-flowing dialogical context, his recognition and amplification of a previously little-known revisionist geopolitical imaginary was a speech act that made news around the world.[55] In naming Novorossiya, Putin conveyed public legitimacy upon a separatist cause at the very moment armed pro-Russian filibuster groups were fighting under its banner in the Donbas. The rhetorical move legitimated provocations while ostensibly disavowing them. Revisionist geopolitics was articulated and then followed by platitudes respecting Ukraine's territorial integrity.

By the time he spoke, it should have been clear to Putin that most of southeast Ukraine was not rallying to the Novorossiya cause but instead mobilizing against perceived Russian invaders and fifth columnists. For Crimea, the Kremlin had Kisileva's public opinion polling to guide its decisions. For Ukraine, the Kremlin had Glazyev, a macroeconomist with a cause, not a sociologist with favorable polling data. Preconvictions not facts about public sentiment in southeast Ukraine drove the Novorossiya project.

Evidence about what people there really believed was soon public. A few days after Putin's *Direct Line* appearance, a public opinion survey of the eight oblasts in southeast Ukraine claimed by Novorossiya rebels was released. Funded by the oligarch Victor Pinchuk (for the news magazine *Mirror Weekly*) and conducted via phone and face-to-face interviews with 3,232 respondents during April 8–16 by the Kyiv International Institute of Sociology (KIIS), the survey revealed how little support pro-Russian separatism had in most regions. This was evident in two crucial questions in the survey. The first asked respondents: Do you support the introduction of Russian troops into Ukraine? In the eight oblasts as a whole, just over 11 percent responded affirmatively (definitely yes, 5.6 percent; mostly yes, 6.1 percent). Another 11 percent were neutral. The majority of the population answered definitely no (58.4 percent), with the rest answering mostly no (15.9 percent) or refusing to answer (2.2 percent). There was regional variation, with Zaporizhia the most definitely against and both Luhansk and Donetsk standing out as the most supportive of Russian intervention (Figure 7.2). However, in these oblasts, that sentiment was still a minority one: 19.3 percent; while 66.3 percent of respondents in Donetsk and 53.4 percent in Luhansk did not support the introduction of Russian troops.

A similar pattern was evident in the second crucial question: Do you support the idea that your region should secede from Ukraine and join

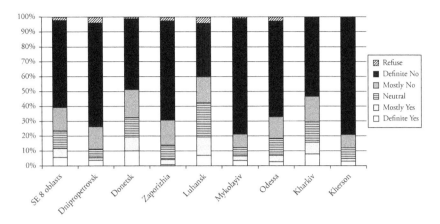

FIGURE 7.2 Support in SE8 for the Introduction of Russian Troops into Ukraine, Mirror/KIIS Survey, April 2014.

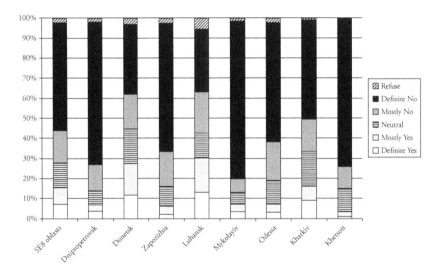

FIGURE 7.3 Support in SE8 for Secession from Ukraine. Mirror/KIIS Survey, April 2014.

Russia? (Figure 7.3). In the eight oblasts as a whole, 15.4 percent supported joining Russia whereas 69.7 percent were definitely against this proposition. The highest levels of support for secession and annexation were in Luhansk (30.3 percent) and Donetsk (27.5 percent), with Kharkiv

(16.1 percent) recording the next-highest level of support.⁵⁶ In short, opposition to separation and annexation was overwhelming in six of the eight oblasts, and even within the Donbas it was majority sentiment.

Abandoning Novorossiya, Saving the Donbas

The Putin administration's relations with the filibuster fighters operating in eastern Ukraine were characterized by a dual-track policy after April 2014. In practical material terms, various agencies within the Russian state were actively aiding the fighters, providing financial assistance to the separatist authorities, and distributing humanitarian aid. Yet in rhetorical terms the presidential administration kept its distance from the emergent rebel leaders and warlords in its public statements and actions. With revolt against Kyiv consolidating itself in the Donbas rather than in other parts of southeast Ukraine, the Kremlin, for the most part, stopped using the term "Novorossiya."

Was it ever a goal of the Kremlin to annex Novorossiya or the Donbas?⁵⁷ What is apparent is that Putin proved reluctant to undertake a large-scale invasion deep into Ukraine to create a land bridge to Crimea or to go farther toward Odesa and potentially also create a corridor to Transnistria. NATO Supreme Commander General Breedlove stated on March 23 that the Russian forces massed on Ukraine's eastern border had the capability to run to Transnistria if the decision was made to do so.⁵⁸ The events of May 2 in Odesa provided Putin with a justifiable scenario to launch such a land invasion, but he did not act. That day a pro-Ukrainian march by football fans was attacked and six people were killed. In response pro-Ukrainian forces set alight the pro-Russia protest camp in front of the Trade Unions House. These protesters sought refuge in the building, but it was also set alight amid shooting and Molotov-cocktail-throwing by both sides. Local police did not intervene. In the end forty-eight civilians died, along with forty-two pro-Russia activists in the Trade Unions House fire. The "Odessa massacre" was headline news in Russia, Ukraine, and elsewhere for days. Even as evidence of Right Sector involvement in the deaths was revealed in Russian media, Putin declined to authorize a rescue mission scenario.⁵⁹ By this time, international pressure was growing on the Russian leadership to lower tensions and pull back its support for rebel fighters in the Donbas.

The following Wednesday, after talks with the president of Switzerland, Putin publicly requested that the rebels postpone the referendums scheduled for that Saturday.[60] When the rebels went ahead anyway on May 11, 2014, the Kremlin indicated it "respected" the results. It ignored the call by Donetsk Peoples Republic officials for Moscow to "absorb" the entity.[61]

Meanwhile armed militias were growing in power as the area descended into lawlessness. Thousands fled for the relative safety of other parts of Ukraine.[62] After the spectacle of self-determination referendums in Donetsk and Luhansk on May 11, the results of which were never in doubt, Alexander Borodai became the new prime minister of the Donetsk People's Republic (DPR). Borodai was a longstanding friend of Girkin. Like him he had come from Crimea to the Donbas. Girkin became defense minister. As both "retired" Russian Security Force veterans in post-Soviet space, Girkin and Borodai sought to professionalize the work of creating a separatist state. In this they received help in the form of a militia calling itself "Vostok" (East). Its connection to a disbanded GRU-controlled Chechen unit of the same name, a unit involved in ethnic cleansing in South Ossetia after the August War, was unclear.[63] Some of its fighters were Russian citizens from Rostov and the North Caucasus but some were local Ukrainians. The figure leading it was Alexander Khodakovsky, previously head of a Special Forces unit under Yanukovych. For a short while he became the "security minister" of the DPR. Vostok appeared at the end of May and cleared out the ragtag group of people squatting in the Donetsk administrative building. Some of its members initiated what was to become a protracted battle over Donetsk airport, suffering losses—over thirty Russia nationals were estimated killed, many from "friendly fire"—after Kyiv counterattacked with helicopter gunships. In an increasingly lawless environment, power was devolving in the hands of militias and warlords. Three dominated the scene in Donetsk: the Russian Orthodox Army, the Vostok Battalion, and Oplot, which its leader Aleksandr Zakharchenko had transformed from sports club to armed militia. Akhmetov retained influence. Members of Vostok dispersed a crowd trying to break into his Donetsk mansion and started guarding the building. Akhmetov finally decided to come out publicly against the rebels, issuing a statement to that effect on May 19.[64] The factories that he controlled, which required access to international markets, remained so important to the future of the separatist republics, however, that it was in their best interest to placate him. This they did. Despite the fighting that followed, Akhmetov's factories on both sides of the front line continued to operate, and their output continues to be sold to buyers across the world.[65]

The summer of 2014 did not go well for the rebels. While they managed to take over Ukrainian border posts and open up supply lines for weapons and volunteers, Novorossiya was not a cause catching fire locally. In Sloviansk, Girkin complained about his lack of manpower and local volunteers. On July 5, in the face of imminent Ukrainian army encirclement, he and his men retreated from Sloviansk and then Kramatorsk. They would also lose Artimivsk and Debaltseve and fall back to Horlivka. Girkin's retreat deepened fissures within Russian imperial nationalist circles.[66] He felt abandoned:

> Initially I thought that the Crimean scenario would be repeated—Russia would enter. It was the best option. And people strived for it. Nobody was going to come out for Lugansk and Donetsk Republics. Initially everyone was for Russia. And the referendum was conducted for Russia, and they went to fight for Russia. People wanted to join Russia. Russian flags were everywhere. I had the Russian flag at the headquarters and everyone else too. And people received us with the Russian flags. We thought, the Russian administration will come, logistics will be organized by Russia and there would be another republic within Russia. And I didn't think about any kind of state building. And then, when I realized that Russia will not take us in; this decision was a shock for us.[67]

The downing of the Malaysia Airlines passenger jet MH 17 by the rebels, and the international revulsion it engendered, followed soon after. By late July it looked as if the Ukrainian forces had the upper hand and would surround and crush the separatist forces. Kyiv announced it was "preparing for the final stage in the liberation of Donetsk."[68] But by this point, in the wake of the territorial losses around Sloviansk, Putin's government had decided it could not afford to abandon the cause it had ignited in eastern Ukraine.[69] It also needed to dampen rising levels of criticism from imperial nationalists within Russia.[70]

Increasingly heavy and sophisticated weaponry was made available to the rebels (a contributing factor to the downing of MH 17). Vladimir Antyufeyev, for decades the head of "state security" in Transnistria, was brought in mid-July to replace Khodakovsky.[71] Borodai was recalled to Moscow. (Antyufeyev became acting prime minister in his absence.) After Borodai's return to Donetsk on August 7, he announced his resignation as de facto prime minister, declaring his work as crisis manager at the DPR "start up" at an end (though he remained as deputy prime minister).[72] Zakharchenko, the local Oplot commander, replaced him as the public face of DPR. Girkin, who had become outspokenly critical of Vladislav Surkov's role in dialing back the Novorossiya project, was forcefully removed from the scene and returned to Russia.[73] At the end of the month, he was pictured in camouflage fatigues meeting with Konstantin Malofeev, Alexander Dugin, and others at the Valaam Monastery in Russian Kerelia.[74]

As it obfuscated the cause of the MH 17 downing, Russian media began to amplify the genuine plight of civilians caught in the spiraling violence of the Donbas.[75] The Kremlin then launched a high-profile rescue mission in the form of "humanitarian convoys," army KamAz trucks repainted bright white and loaded with food, blankets, and medical supplies.[76] Materiel not on prime time television was also flowing across the compromised state border. So also were increasing numbers of regular Russian troops, euphemistically described as "vacationers" since the official Russian government storyline was that they were volunteers on vacation that had gone to fight for Novorossiya.[77] In late August the Russian-backed separatists, supported by artillery fire from within Russian territory and using Russian-supplied tanks, captured the town of Novoazovsk on the Black Sea and advanced to the outskirts of Mariupol, a city both sides had tussled over since April. They also targeted the Ukrainian forces near Horbatenko and Ilovaysk that sought to drive a wedge between the two rebel strongholds of Donetsk and Luhansk. What happened next was a turning point in the war. Surrounded by reinforced rebels, Ilovaysk became a "kettle" for the Ukrainian forces, regular army, and militia groups. An agreement of their retreat was negotiated. Moscow was involved because on the early afternoon of August 29, 2014, the Kremlin released a statement addressed to the "Novorossiya militia" that called on "militia groups" to "open a humanitarian corridor for the Ukrainian service members who have been surrounded" so that they could leave the combat area unimpeded.[78] But there was no immediate retreat. Eventually the Ukrainian forces moved out, along with their heavy equipment and weapons. Between Ilovaysk and the village of Novokaterinivka, the retreating column was ambushed by a large contingent of rebel forces. A "humanitarian corridor" became a "meat grinder," with hundreds of young Ukrainians decimated in horrific circumstances.[79] The following day Ukrainian defense minister Valeriy Heletey declared that the ATO was over for Ukraine was now facing a "Russian full-scale invasion."[80] In early September, Russia and Ukraine, under the auspices of the Normandy Four, agreed on a ceasefire in Minsk, Belarus. It would gradually unravel over the winter, forcing another agreement, Minsk II, in February 2015, which largely stabilized after separatists recaptured the railroad town of Debaltseve a week later. An estimated nine thousand people have died in the fighting in eastern Ukraine, and thousands more have been wounded. More than a million people have been displaced. The fighting has continued since at variable levels of intensity.

Attitudes to Novorossiya in Southeastern Ukraine at the End of 2014

In the same December 2014 survey in southeast Ukraine discussed in the last chapter, my colleagues and I posed a series of questions about what ordinary people in six of the eight oblasts in southeast Ukraine (i.e., SE6) and Crimea thought about the legitimacy of Novorossiya and the Donetsk People's Republic and Lugansk People's Republic referendums on May 11, 2014. Remember, it was not possible to reliably survey in Donetsk and Luhansk/Lugansk. As might be expected, the results revealed a clear divide in attitudes between SE6 and Crimea, and some notable internal variations. Since Novorossiya has both historical and revisionist connotations, we posed three questions to elucidate attitudes, for it is entirely possible that a respondent could accept that Novorossiya was indeed a historical province but reject the implicit proposition that this was thus the basis for a legitimate contemporary separatist claim. Accordingly, we asked first about history—implicitly triggering questions about legitimacy—and then posed two political questions about what Novorossiya meant for contemporary Ukraine.

The first question asked interviewees: "Is Novorossiya a myth or a historical fact?" Overall, 52.3 percent of respondents in SE6 categorized it as a myth, 24.2 percent as a historic fact, with a high ratio of 22.1 percent opting for "don't know"; and 1.3 percent refusing to answer. In Crimea, by contrast, only 15.5 percent of respondents saw it as a myth, whereas 67.4 percent saw it as a historic fact while 17 percent said they didn't know. The high "don't know" response rate can be the result of many factors—genuine confusion about the competing messages from television and other media sources, a result of self-perceived lack of historical knowledge, inability to choose between two blunt options, or avoidance of a sensitive question.

A follow-up question asked those (n = 970 in SE6 and n = 635 in Crimea) who did not choose the "myth" option: "Is it possible that this historical fact can be used as a basis for separation of Novorossiya from Ukraine?" In SE6 14.7 percent answered "yes, it's possible," 44.1 percent answered "no, it means nothing now," with 37.0 percent answering "don't know" and 4.2 percent refusing to answer. In Crimea, by contrast, 60.8 percent of the subsample indicated that the historic fact was a basis for separatism, 27.7 percent indicated that it does not mean anything, while 11.1 percent indicated it was hard to say and only two respondents (0.31 percent) refused to answer.

The third question presented polarized options to elicit how respondents felt about Novorossiya. It presented four options: "Is the use of the term 'Novorossiya' either a) Russian political technology to destroy Ukraine, b) the expression of the struggle of residents of South-Eastern Ukraine for independence, c) don't know, and d) refuse." Overall, in SE6 52 percent chose the Russian political technology option with 18 percent seeing Novorossiya as reflecting an independence wish, and again with a high ratio of "don't knows" at 18 percent and 2 percent refusal. In Crimea, just 10 percent considered it Russian political technology, whereas over three-quarters of respondents viewed it as reflecting a legitimate desire for independence; and 14.2 percent indicated it was hard to say while three respondents refused to answer.

A similar divide is evident in responses to a question about whether respondents agreed or disagreed that a referendum allowing the Donbas to join Russia was an inalienable right of the people of that region to express an opinion about the future. In SE6 those who strongly or mostly agreed with this made up just over one in four of the respondents (9.8 percent strongly agreed, 17.3 percent mostly agreed). The great majority disagreed: 20.8 percent disagree, 33.9 percent strongly disagree, with 17.1 percent indicating they could not say. By contrast, in Crimea 56.8 percent definitely agreed while 26.7 percent largely agreed; barely 10 percent disagreed while 6.1 percent indicated they did not know.

Figures 7.4 through 7.7 reveal the breakdown of these results by ethnicity in Crimea and ethnolinguistic group in SE6. What is evident from these graphs is that a minority—roughly 15–25 percent —in SE6 are, in general, supportive of separatist sentiment and the idea of Novorossiya. Analysis by oblast reveals that this community tends to be concentrated in the oblasts of Kharkiv and Odesa. What is also evident is that Vladimir Putin's assertion of a coherent two-part interest group—ethnic Russians and Russian-speakers—which constitute the basis for a distinct and bio-political Russian interest in Ukraine is a conceit of Russian geopolitical culture, not sociodemographic analysis. There is no correspondence between Russian-language use and pro-Russian sentiment. Statistical analysis reveals that those declaring themselves ethnic Russians in Ukraine (itself a somewhat political act that may be independent of their actual family history) are more likely to be supportive of secessionist sentiment than the general population, but this does not mean that all self-declared ethnic Russians necessarily advocate this. Further analysis reveals that those who are older, less educated, and more nostalgic for the Soviet Union are likely to be sympathetic to Novorossiya and the People's Republics.[81] Place makes a difference, because ethnic Ukrainians in Crimea are more likely than those in SE6 to be sympathetic to secessionism and to annexation by Russia.

Though Novorossiya was an ideal that lived more in the imaginations of marginal geopolitical revisionists than ordinary people before 2014, it became a passionate cause that drove people to protest and take up arms for a few weeks in the spring of that year. The horrific deaths in Odesa on

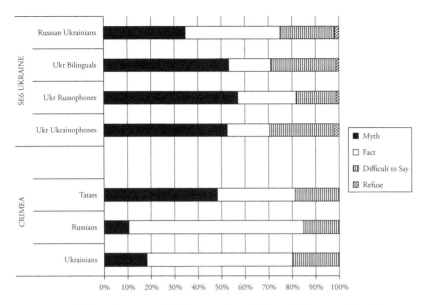

FIGURE 7.4 Attitudes toward Novorossiya: Historic Fact or Myth?

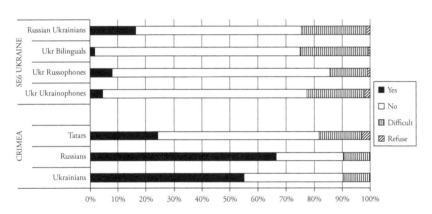

FIGURE 7.5 Attitudes toward Novorossiya: Basis for Separatism?

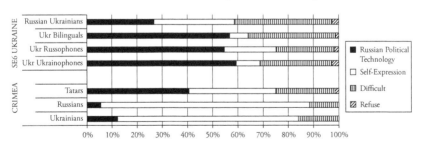

FIGURE 7.6 Attitudes toward Novorossiya: What Is Novorossiya?

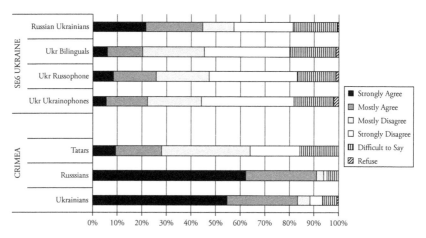

FIGURE 7.7 Attitudes toward Donbas Joining Russia Referendum, SE6 and Crimea.

May 2, 2014, marked the beginning of the end for the dream of a secessionist confederate state encompassing all of southeast Ukraine, one that would eventually join Russia. It took time for those who pinned their careers on it to come to terms with its demise. Girkin saw it as a noble cause betrayed by slippery advisers and corruption around Putin. At the end of 2014, Alexander Borodai conceded that "Novorossiya doesn't exist. We all, of course, use this term, but it is a false start, to be honest. Novorossiya is an idea, a dream. This is an idea that hasn't been realized."[82] For him the dream was still alive in the territorial form of the two People's Republics, which he hoped would be springboards for a future Novorossiya. On the first anniversary of the People's Republics, a series of statements by separatist leaders in eastern Ukraine acknowledged that Novorossiya as an active geopolitical project was dead. Pavel Gubarev formed both a militia and party named Novorossiya but he lost out in the power machinations in Donetsk. He survived an assassination attempt, while his party was banned from the November 2014 elections in the Donbas on a technicality. Oleg Tsarev, the only member of the Ukrainian Rada to defect to the separatists, had also formed a Novorossiya party and even became speaker of a proposed Novorossiya confederation assembly, but he too ended up on the margins. In May 2015 he declared both suspended until further notice. The ostensible reason was that the movement and parliament were incompatible with the peace process led by the Normandy Four that resulted in two separate ceasefire agreements in Minsk.[83] Aleksander Kofman, a businessman from Donetsk acting as a "Minister of Foreign Affairs" for the Donetsk People's Republic, was more forthright, conceding that Kyiv had successfully thwarted its efforts in the southern regions of Ukraine. Novorossiya as a project was now "closed due to the fact that its supporters in Kharkov and Odessa were successfully suppressed by official Kiev."[84] The project failed but the cause, and the war, live on.

8

Geopolitics Thick and Thin

ON NOVEMBER 24, 2015, a Turkish F-16 fighter jet shot down a Russian Sukhoi Su-24M aircraft on the Syria-Turkey border. For seventeen seconds the Russian aircraft crossed the southern tip of a salient of Turkish territory that Syria claimed rightfully belonged to it. Two Russians ejected from the plane over Syria. A local Turkmen militia, commanded by a Turkish citizen, fired at the aviators, killing one. A second Russian serviceman was killed during a rescue mission to save the surviving aviator. The incident, recorded on radar systems by many countries and partially captured on video camera, was the first time since the Korean War that a North Atlantic Treaty Organization (NATO) country's fighter jet destroyed a Soviet/Russian Air Force aircraft. Fortunately the event did not escalate into a full-blown NATO-Russia crisis, although with tensions high over the Ukraine crisis and two authoritarian leaders at loggerheads, it could well have done so. There were background accusations. Turkish president Erdoğan was aggrieved that Russia was bombing co-ethnic kin in its southern near abroad while aiding Kurdish separatists, while Russian president Putin saw Turkey as an accomplice of international terrorists. Entwined territorial and terrorist anxieties, as well as near abroad insecurities, preoccupied both men. Had Russia responded with force against Turkey, this could have triggered Article V of NATO's Washington Treaty, and NATO members would have faced the prospect of war with Russia over a tiny piece of territory in the Middle East most knew nothing about.

Relations between the NATO alliance and Russia are now at their lowest point since the collapse of the Soviet Union. Airspace violations, incidents at sea, military training exercises, and hybrid war hysteria have kept tensions high.[1] After Crimea, NATO moved to strengthen its capacity to respond to perceived Russian encroachment on the Baltic countries. The Obama administration's European Reassurance Initiative was launched in June 2014 with a $1 billion budget for training and temporary rotations. In a speech in Riga in September 2014, President Obama declared: "We'll be here for Estonia. We will be here for Latvia. We will be here for Lithuania. You lost your independence once before. With NATO, you will never lose it again."[2] Talk of Russia as the greatest strategic threat to the United States returned.[3] In a November 2015 speech, the new U.S. secretary of defense Ashton Carter declared: "Russia has been violating sovereignty in Ukraine and Georgia and actively trying to intimidate the Baltic States." Its actions in Syria and "nuclear saber-rattling" raised concerns about the commitment of Russia to strategic stability and norms against the use of nuclear weapons. Declaring that the United States did not want to make Russia an enemy, Carter nevertheless issued a warning: "But make no mistake; the United States will defend our interests, and our allies, the principled international order, and the positive future it affords us all."[4] In January 2016 the U.S. Department of Defense announced that it would raise spending on its European Reassurance Initiative to $3.4 billion in order to substantially increase the deployment of heavy weapons, armored vehicles, and other equipment to NATO countries in Central and Eastern Europe.[5] With the potential for military confrontation between Russia and NATO a real one, it is vital that we ask probing questions about how U.S. geopolitical culture and the West more broadly understand and frame the current geopolitical crisis and its territorial spaces. That is what this concluding chapter seeks to do.

This book began by identifying the irony of how structurally similar affective storylines in U.S. and Russian geopolitical culture produced mutual incomprehension, not understanding. It is a striking fact that, in the heat of the crises generated by Russia's invasions of Georgia and Ukraine, both Russian and U.S. geopolitical culture drew upon the same archetypal narratives to frame the meaning of the crises for their populations. Russian and U.S. leaders presented themselves as fighting "empire" and defending vulnerable populations victimized by imperial schemas of the other power. Both states posed as defenders of the freedom and self-determination of smaller nations (just as in the Cold War). Sections within both cultures presented their adversaries as fascists pursing polices that were reminiscent of Nazism. The need to respond to "fascist" violence justified Russia's invasions of Georgia and Ukraine. Prominent U.S. politicians and neoconservative writers also promoted Nazi analogies. Former U.S. secretary of state Hillary Clinton described Putin's annexation of Crimea as reminiscent of Hitler's actions in the 1930s.[6] Both Obama and Putin were redrawn as Hitler by protesters on each side. Imperial Russian nationalists conflated Nazism and NATO, while "Putler" became a popular portmanteau on Ukrainian social media and beyond.[7]

These shared archetypes are connected to very different affect-imbued collective memories of the onset of World War II, of the tremendous suffered during that war, and of the Cold War division of Europe that came thereafter. The Soviet victory over Nazism is a sacred memory of collective suffering before the glory of victory. (Putin's elder brother Viktor, it should be recalled, died during the siege of Leningrad.) That suffering, visceral, personal, and beyond debate in a totalitarian system, was used to justify the Soviet domination of Eastern Europe. U.S. geopolitical culture saw the Cold War's end as its victory and the Soviet collapse as a triumph of its policies and values. Russian geopolitical culture, born of catastrophic collapse, sought the restoration of Russian power in Europe, the Caucasus, and Eurasia. Mythologized memories of the Great Patriotic War provided a storehouse of "victim," "enemy," and "hero" affect for expressing this desire.[8] Caught between European Union (EU) and NATO enlargement and Russian revanchism was an arc of states north and east of the Black Sea—from Tallinn to Chișinău and from Sukhum(i) to Baku—struggling to establish their place in the world but divided within by language, memory, identity, and culture. For some of these states—Moldova, Georgia, and Azerbaijan—there never was a stable and settled security order after the Cold War. From the untidy ends of these states, and in former Yugoslavia, would develop resentments, competition, and an exponential security dilemma that would eventually unravel the Cold War settlement between Russia and the West.[9]

There are three geopolitical frames used in U.S. and, more broadly, Western geopolitical culture concerning the current standoff with Russia that I want to isolate and challenge in this chapter. As frames previously attached to Cold War systems of meaning, they provide popular answers to three questions about the current geopolitical crisis. The first is the question of how the archipelago of frozen conflicts and separatist territories that now characterize Russia's borderlands are to be understood and described. A popular answer in U.S. geopolitical culture, an amplification of the prevailing frame in the countries concerned, is that these are "occupied territories." The second question concerns what Russia wants in its near abroad. An automatic response, also prevalent in the region, is a "sphere of influence," a desire that is condemned as both archaic and incompatible with the ideals of U.S. and, more broadly, Western geopolitical culture. As already noted, this finds expression in the claim that Russia should never be allowed a "veto" over the security preferences of its neighboring states. The third question concerns the debate in U.S. geopolitical culture over U.S. support for Ukraine as it faces multiple structural crises and Russian-supported territorial separatism. Many within U.S. geopolitical culture amplify the Kyiv-based rhetoric that Ukraine is "fighting for freedom" and for the "free world," with powerful factions advocating that the United States should provide advanced weapons to Ukraine's military so it can take the fight to Russia and its proxies in occupied Ukraine. In sum, I wish to tackle three popular geopolitical frames that describe the current confrontation between Russia and the West: "occupied territories," "sphere of influence," and the "free world."

In considering these three questions, I want to draw out the broader critical geopolitical argument implicit in this book by utilizing an ideal-type distinction between thin and thick geopolitics.[10] The distinction between the two concerns the quality and depth of geographical understanding and worldliness. Thin geopolitics thinks in universal abstractions and operates with only the most superficial regional geographical understanding. Its primary mode of representation is moral dichotomization. Ideographic and geo-graphic contrasts—empire versus freedom, democracy versus authoritarianism, Russia versus the West—are its lexicon. Heterogeneous places, states, and crises on the world political map are forced into homogenizing categories. Local conflicts are "scaled-up" into global contrasts; moments of crisis are "amped-up" into high-stake credibility tests. Geographical particulars are overridden by metaphors, analogies, and affective storylines that appear universal. Cultural and religious fables, tabloid images, movie plots, and patriarchal aphorisms about bullies and aggressors are popular expressive vehicles for thin geopolitical thinking in U.S. geopolitical culture (they have, of course, long been popular in other geopolitical cultures as well).[11]

The rearticulation of the traditional hemispheric terms of U.S. national exceptionalism as a universal messianism, proposed by President Woodrow Wilson to justify U.S. intervention in World War I, placed thin geopolitical reasoning at the center of U.S. geopolitical culture in the twentieth century.[12] While the language about a unique American civilizing mission was tempered and challenged after Wilson's presidency, it reemerged strongly after World War II as U.S. and Soviet tensions escalated into what became a global Cold War.[13] Instability in Greece and Turkey was amplified by President Truman into a moralized placeless commitment to "support free people who are resisting attempted subjugation by armed minorities or by outside pressures."[14] Supporting 'freedom' required geopolitical discourse to be, in the words of Truman's secretary of state Dean Acheson, "clearer than the truth."[15] Containing Communism in Europe spread to encompass the ends of the earth, despite the ostensible intentions of its chief strategist.[16] Foreign policy was driven more by cultural axiom induced by domestic politics than strategic calculation.[17]

The domino theory is perhaps the most infamous instance of thin geopolitics in U.S. geopolitical culture. It was only one of a series of metaphors giving expression to U.S. national security state anxieties about the spread of Communism across the world map. Images of "falling" countries displaced not only geographic knowledge but also the ability to differentiate between vital and nonvital state interests. The Cold War was an everywhere war.[18] Because anti-Communist outbidding in U.S. domestic politics and media culture made these metaphors pervasive thinking tools, the United States ended up intervening in a series of nonstrategic locations across the world. In many of these sites, most tragically Vietnam, it was soon out of its depth. But to politicians and ideologues seeking applause lines and political advantage from tough talk and moral posturing, the discourse of thin geopolitics was irresistible.[19]

By contrast, *thick geopolitics* rests on recognition of the importance of spatial relationships and in-depth knowledge of places and peoples. Grounded in the messy heterogeneity of the world, it strives to describe the geopolitical forces, networks, and interactions that configure places and states. It recognizes that local conditions matter, that agency is rarely singular, that power is exercised geographically, and that location, distance, and place influence its operation. While moral-legalistic talk remains the powerful public transcript in U.S. geopolitical culture, foreign policy failures in Iraq, Afghanistan, Libya, and Syria have brought home the thickness of geopolitics to the U.S. diplomatic and security community. The worldly skepticism and sense of limits in thick geopolitics, however, are in tension with the normative core of U.S. geopolitical culture. Thick geopolitics, thus, is more felt than articulated, more a private transcript than a public one, because most US politicians prefer, and wish, geopolitics to be "clearer than the truth." Let us now turn to the three interpretative challenges.

Occupied Territories: Understanding Russia's Contemporary Geopolitical Archipelago

With its intervention in Ukraine, Russia has significantly added to the archipelago of territories that it helped create and now supports across what was formerly post-Soviet space but now is the frontline of a Russia-NATO standoff. To the de facto states of Abkhazia, South Ossetia, Transnistria, and Nagorny Karabakh has been added Crimea and the two People's Republics in eastern Ukraine. How should one understand this collection of territories and what created them? For Russia's longstanding adversaries in the "frontline states," the answer predictably involves empire. Writing after Crimea in the *Weekly Standard*, the magazine edited by prominent U.S. neoconservative William Kristol, former Georgian minister Timuri Yakobashvili argued that Russia has paid no price for its "continuation of the 20 percent rule—to buy your freedom from their sphere of influence, you should be prepared to sacrifice 20 percent of your territory to the empire—while the West continues to have no response but endless negotiations to maintain the peace inside an occupation."[20] In keeping with Georgia's storyline during the August War, the narrative is not about the places themselves but about

Russia violating the territorial integrity of neighboring states and the West's feckless response to its occupation of the sovereign territory of its neighbors. Some policy officials and commentators in U.S. geopolitical discourse echo this language of "occupied territories." For example, former U.S. deputy assistant secretary of defense for Russia/Ukraine/Eurasia Evelyn Farkas argued that Russia "cannot be allowed to prevail in Georgia, Ukraine, and Moldova, where its military occupations have led to so-called frozen conflicts that serve as a Russian veto on the policies of those countries."[21] A *New York Times* feature reduced Russian military intrusion to one clear pattern: "interventions that inflame conflict and create permanently tense and unstable 'frozen zones,' allowing Russia to exert influence and confound its opponents and, often, its rivals in the West."[22]

The argument in this book is more contingent and contextual, one that underscores the heterogeneity of these spaces while situating them within the contradictions of Putin's project of restoring Russia as a great power. As argued in chapter 2, Russian geopolitical culture under Vladimir Putin became concertedly revanchist in conception and goal. That revanchism did not have a teleological territorial end: it was not about re-creating the territory of the Soviet Union or about expanding the territorial expanse of the Russian Federation. Instead, it was about restoring existing Russian state strength and territorial integrity. Ground zero for this project was Chechnya, Russia's inner abroad, and it was Putin's use of force there that launched his political career and gave impetus to the project of "making Russia great again." Subduing Chechnya, however, was no easy matter. The strategy of Chechenization required Putin to concede near-comprehensive internal sovereignty to a proxy there, Akhmad Kadyrov, and, after his assassination in May 2004, to his son Ramzan.[23] Chechnya is the first space of exception in a Russian geopolitical archipelago of dependencies.

Putin inherited the Russian dependencies of Abkhazia, South Ossetia, Transnistria, and, indirectly, Nagorny Karabakh. These spaces gave Russia influence over political life in these neighboring states, but they also complicated life for the Kremlin as these territories became a symbolic cause for Russian imperial nationalists and revisionists though, as nonethnic Russian spaces, they were less central to their imagined Russian geobody than Crimea, northern Kazakhstan, and

Novorossiya. Because of perceived U.S. encroachment in the Caucasus and Saakashvili's very public campaign to take Georgia into NATO, supporting Georgia's breakaway regions became a firm national security interest for Putin's government by the summer of 2004. The escalation and clashing revanchism described in this book, which was fueled by deep personal animosity, propelled Putin into making these states explicit Russian dependencies—fictional independent states—following what it saw as the United States' and European Union's similar sponsorship of Kosovo as an independent state. In certain ways, this outcome was undesirable for Russia as it hardened attitudes and limited its capacity to pose as an influence broker in Georgia. Many, including President Medvedev, however, represented it as a victory in that Russian intervention appeared to check the expansion of NATO to the Caucasus.

Putin's major geopolitical goal on returning to the presidency was the creation of the Eurasian Economic Union, envisioning this as a possible counterweight to the European Union, China, and the United States in world affairs. It was a fanciful notion that created great difficulties for Russia's preferred leader in Ukraine, Viktor Yanukovych. With his fall, Putin made the most significant decision of his presidency, and authorized the invasion and subsequent annexation of Crimea. The move was played as a triumph in Russia, and many in the West bought into the official Russian narrative of this as a bold, victorious move. But taking Crimea was a significant strategic blunder for many reasons.[24] With this one move, Putin instantly alienated the majority of public opinion in Ukraine, most significantly Russian-speaking Ukrainians, not to mention doing serious damage to Russia's position in the international community. He also made it virtually impossible for a future pro-Russian leader to be popularly elected in Ukraine. Putin doubled down on the error by fanning already existing grievances in southeast Ukraine and giving Novorossiya activists free reign. But here the aspirational goal failed quite publicly except for in the Donbas, where a nasty war took hold that soon required Putin to sacrifice Russian soldiers and materiel. Facing defeat of even this project, Putin sought a third-best outcome, the creation of de facto states along the lines of Transnistria that would give Russia a degree of leverage over the geopolitical orientation of Ukraine. This was security through vandalizing the territorial order in Ukraine, the supposed creation of a sphere of influence through destruction and extortion. Instead, it was the creation of further spaces of dependency and instability at the border.

The result is that Putin has now helped produce a series of disparate Russian dependencies in its near abroad. These spaces are heterogeneous and diverse, but they have three common characteristics. First, they are conflict zones, some still ravaged by the war fighting that swept through the region twenty years ago (Abkhazia) and also by more recent war fighting (South Ossetia). The Donbas remains a simmering war zone. Chechnya has been reconstructed thanks to an estimated flow of at least $14 billion in reconstruction funds since 2001, yet conflict still simmers below the surface.[25] Those regions that have largely avoided conflict—Transnistria and Crimea—have nevertheless suffered from the broader protracted and unresolved nature of the conflict there. While supposedly enhancing the geopolitical power of Russia, this collection of territories brings with them considerable headaches. Abkhazia and South Ossetia feature toxic legacies of mass displacement, property destruction, and, to date, impunity from any accountability for war crimes. Crimea brings with it unresolved and politically sensitive questions concerning the Crimean Tatar population. This community ties Crimea to other Tatar communities in Russia and to Turkey, a historic paternal power to this Turkic people (Abkhazia is also tied to Turkey by historic displacements and diasporas). In the Donbas, displacement, theft, and criminality are major challenges. Identifying and propping up leaders who are skilled at maintaining local legitimacy and power, without manifest abuses of power, is a significant and ongoing challenge for the Russian *kurators* (advisers) working in the region. At present, Putin's presidential administration official Vladislav Surkov, an ally of Ramzan Kadyrov, is the top *kurator* managing the challenging portfolio of dependencies for the Kremlin.

Second, these regions are fully dependent upon subventions from Moscow (Nagorny Karabakh is an indirect and more complex dependency). Many regions within Russia are in the same situation, but these regions are particular drains on the federal budget because they have little prospect of realizing any competitive advantages for economic growth due to their geopolitical isolation. Subvention flows tend to make local corruption easier, which deepens the multiple legitimacy challenges unrecognized regions face.[26] Reading these regions as simply criminal havens, however, is too glib.[27] In the past, Putin's sponsorship of territories that were seen as not fully Russian generated a backlash among narrow Russian ethnic nationalists, such as those that mobilized in October 2011 around the slogan "stop feeding the Caucasus."[28] While Russia and Putin personally enjoy high approval ratings in these regions, should the subventions be cut back significantly this may change.[29] Chechen forces associated with Ramzan Kadyrov were linked not only to militia groups operating in the Donbas but also to the murder of the Russian opposition politician Boris Nemtsov in February 2015.[30] Kadyrov has demonstratively, if violently, positioned himself as Putin's leading defender in Russia. But with the Russian economy ailing and the incorporation of Crimea into Russia a new priority megaproject for the Kremlin, Kadyrov's claim to continued Kremlin largess is under threat. In an era of diminished resources, Russia's dependencies are in competition with each other. In September 2015 Russia began paying pensions, allowances, and state salaries in rebel-controlled Donbas. It is estimated that these territories now cost the Kremlin over $1 billion a year in subsidies and aid.[31] Putin's Russia, however, is unlikely to stop funding its expanded archipelago of dependencies because of loss aversion and sunk costs.

Finally, Russia's current geopolitical archipelago is made up of diverse spaces that have unusual forms of sovereignty (Table 8.1).[32] Chechnya is a powerfully autonomous inner abroad run by a ruler who raises his own taxes and has security forces that are personally loyal to him. Crimea is an annexed space that is becoming more integrated into Russian security and defense systems. New missiles and troops have been introduced to the peninsula. South Ossetia and Abkhazia are nominally independent states where Russian bureaucrats and security officials play central roles. For example, Anatoly Khrulev, the commander of the 58th Army wounded in South Ossetia on August 9, 2008, now serves as chief of the General Staff of Abkhazia. Both de facto states host Russian military bases while Russian border guards patrol their frontiers. Russian military forces are also on the ground in the Donbas in advisory capacities, while they have long been in Transnistria as peacekeepers.

Given this, it is understandable that parent states would present these spaces as "occupied territories" to Western audiences.[33] The binary

TABLE 8.1 Russia's Geopolitical Archipelago

Dependency	Estimated Size, km²	Current Population Estimate	Estimated % Population Change since 1989	Parent State	Sovereign Status according to Russia	Sovereign Status according to West
Abkhazia	8,432	240,100	−55	USSR/Georgia	Independent State	Russian-supported frozen conflict on Georgian territory
Chechnya	17,300	1,268,989	+16	Russia	Russian Republic	Recognized Russian territory
Crimea	27,000	2 284,400	−6	Ukraine	Russian Republic	Russian-occupied Ukrainian territory
Donbas	17,000	2 250,000	−50 (since 2014)	Ukraine	Unresolved legitimate secessionist region within Ukraine	Russian-supported simmering conflict on Ukrainian territory
Kaliningrad	223	941,873	+8	Russia	Russian region	Recognized Russian territory
Nagorny Karabakh (indirect)	4,400	120,000	−38	USSR/Azerbaijan	De Facto State	Unresolved frozen conflict on Azerbaijani territory
South Ossetia	3,900	30,000	−70	USSR/Georgia	Independent State	Russian-supported frozen conflict on Georgian territory
Transnistria (PMR)	4,163	475,000	−36	USSR/Moldova	De Facto State	Russian-supported frozen conflict on Moldovan territory

trap that accompanies this charge, however, inhibits understanding. Rather than occupied/not occupied, these spaces are *more than occupied territories*. All these spaces are essentially contested territories with both longstanding distinctive regional identities and ties to Russia. The ugly ethnicized warfare that accompanied the independence of Georgia and Moldova were costly early failures of state- and nation-building. Russian forces never really left Transnistria, while they soon returned to Georgia's breakaway regions as internationally sanctioned peacekeepers. Russian military personnel never really left Crimea either, but Ukraine managed to avoid the fate of Georgia or Moldova by steering clear of exclusivist nationalism. In contrast to Cold War understandings of "occupied territories" as nation-states military occupied against their will by the Soviet army, the majority of the residual residents of these territories supported the presence of Russian troops on their territory, sometimes overwhelmingly so. Minority communities—ethnic Georgian Mingrelians in Gali and Tatars in Crimea—hold different views. In a March 2010 survey in Abkhazia, my colleagues and I found more than three-quarters of ethnic Abkhaz, Armenians, and Russians strongly supported the Russian base at Gudauta.[34] We also found overwhelming levels of support for Russian troops in South Ossetia and strong but lesser levels in Transnistria.[35] Our survey in Crimea in December 2014 asked respondents whether they believed Crimea should be demilitarized or a home for Russian military forces. Over 88 percent of respondents chose the latter. The rhetoric of occupation tends to define disputed territories exclusively in terms of their Russian military presence. This can lead to the misperception that locals are mere puppets or dupes of the Russians with no agency or agendas of their own. More perniciously, it can lead to the perception of Russia supporting local residents, people who have been born and lived their lives in these locations, as fifth columnists. Discourses about Soviet occupation, sustained by museums, can stigmatize generations of non-core ethnic groups as outsiders, illegal migrants and threats. In implicitly mobilizing a language of insiders and outsiders, purity and aliens, the discourse of occupation expresses rather than analyzes the conflict over disputed territories. One does not need this language in order to acknowledge the injustice and suffering caused by violence and forced displacement in these regions. Indeed, using the suffering of genuine victims to amplify core nation victimhood, and moralized geopolitical campaigns may not serve the human security interests of forcefully displaced persons at all.[36]

People who live in these regions have varying aspirations. Majority sentiment in South Ossetia wants to be united with North Ossetia and incorporated into the Russian Federation. By contrast, the population of Abkhazia is split, with ethnic Abkhaz largely content with independence whereas the region's ethnic Armenians and Russians favor incorporation into Russia. Those ethnic Georgians still living in Abkhazia, predominantly in the Gal(i) district, have difficult lives and complex loyalties.[37] Nagorny Karabakh depends on Armenia for its security, which in turn depends on Russia, and Armenia is part of the Eurasian Economic Union. Majority sentiment in Transnistria favors joining Russia and its Supreme Soviet requested such an outcome in 2014.[38] Public opinion in the Donbas is difficult to determine—many of the region's residents no longer live there—but anecdotal evidence suggests that the rebel leadership, mired in criminal activities, is not popular.[39] Dubbing the territory "occupied" against the will of its residents, and those critical of Kyiv as traitors, may be morally satisfying but it is hardly the best place to begin healing a divided country. For better or worse, Crimea and other Russian-supported dependencies are now symbolic sites of Russian power, affective spaces with outsized figurative meaning. On both sides of the current geopolitical standoff, thin geopolitics obscures a much more complex reality.

Sphere of Influence: Proximity and Paranoia

The second interpretative challenge is explaining why Russia invaded Georgia in 2008 and Ukraine in 2014, and why its forces remain on the internationally recognized territory of both states. As I reviewed at the outset, there are two predominant interpretative traditions—that which understands Russia as an unreformed and reinvigorated imperial power, and that which understands Russia as a great power pursuing its interests like all great powers. The former interpretation is hegemonic in U.S. geopolitical culture and elsewhere. Glibly summarized, it is the view that expansionism is in Russia's DNA. Russia's actions are offensive not defensive, and its creation of a sphere of influence on the

European continent started the Cold War. While not denying the imperial character of the Russian state or its use of military force, political realists like John Mearsheimer view these actions as defensive reactions to encroachment policies pursued by NATO and the European Union. Great powers, he claims, are "always sensitive to potential threats near their home territory."[40]

The empirical history and narrative argument in this book complicates both of these interpretative traditions. Russian geopolitical culture was characterized by fierce debate between competing visions of Russia in the wake of the Soviet collapse. For historic, strategic, political, and humanitarian reasons, the Russian state aspired to the role of protector of vulnerable ethnic minorities and compatriots in its near abroad. A consistent interventionist policy toward the near abroad, however, never developed. Geopolitical entrepreneurs competed with each other to push the state toward more muscular policies and revisionist schemes, but their interests and power waxed and waned. Putin's revanchism evolved through eventful processes, which contingently activated revisionist potentialities that may otherwise have remained latent. As the Bush administration pursued polices that Putin saw as revisionist—unilaterally abandoning treaties, invading a sovereign state and deposing its leader, and seeking further expansion of NATO to include foundational regions of the USSR, namely Georgia and Ukraine—he moved toward more interventionist positions in the near abroad. Russia would not allow Saakashvili to disturb the status quo for its dependencies in Abkhazia and South Ossetia. It informed NATO that its expansion to Georgia and Ukraine was a "red line" issue because it involved proximate states where Russia had vital national security interests. Those interests were commonly described in terms of people—Russians, Russian-speakers, and passportized compatriots. But there were hard security interests at stake too—projecting power in the South Caucasus, protecting the Black Sea Fleet—as well as an affective geopolitics involving identity, status and memory. When these objections were largely ignored at Bucharest, and Saakashvili gambled in attacking South Ossetia, Russia's leadership showed little hesitation in responding with disproportionate military force. It then went further and changed the status quo in post-Soviet space by recognizing Abkhazia and South Ossetia as independent states. "If they are revisionists, we can be revisionists" was the prevailing attitude. It was a geopolitics fueled by cynicism.

The invasion of Ukraine was different both because NATO's expansion there was a dormant issue and because it was a country whose history was even more entwined with that of Russia than Georgia. Here Putin's pursuit of the Eurasian Economic Union as an imagined counterweight to the European Union created a zero-sum dynamic that had baleful effects in Ukraine. Ukrainians legitimately aspiring for a different future elevated Yanukovych's refusal to sign a trade agreement with the European Union into a "civilizational choice" they wanted to reverse. When the Euromaidan protests spiraled out of control and Yanukovych fled, Putin took the radical decision to change the status quo and authorized the Russian invasion of southeast Ukraine and subsequent annexation of Crimea.

To many, Putin's threat storyline about a military coup and fascism menacing ethnic Russians and Russophones in Ukraine was wildly inflated, a scenario without evidence, cynically deployed to justify territorial aggrandizement. Mearsheimer's counterargument, however, needs to be taken seriously. Both can be reconciled if one takes the position that there is nothing predetermined or objective about great powers being sensitive to potential threats near their home territory. The process is thoroughly contingent and constructed by prevailing geopolitical discourses and entrepreneurs. There are plenty of precedents for proximity inducing aggressive and paranoid geopolitics by great-power leaders.[41] In 1960, for example, the United States began organizing, funding, and training Cuban exiles to overthrow Cuba's new revolutionary government. The sponsored paramilitaries invaded the Bay of Pigs in April 1961 and were soundly defeated. In 1976, Ronald Reagan revived his bid to become the Republican presidential candidate by using the Panama Canal as a symbol for an unapologetically revanchist geopolitics: "We bought it, we paid for it, we built it, and we intend to keep it." To a radio audience that year he explained: "The loss of the Panama Canal would contribute to the encirclement of the US by hostile naval forces and thereby [threaten] our ability to survive."[42] Four years later Reagan was elected U.S. president. Soviet- and Cuban-sponsored subversion in Central America and the Caribbean were an obsession of his administration.

Besides aiding the military government in El Salvador, his administration sponsored counterrevolutionary paramilitaries—gathered, in part, from former security officials of the ousted dictator Somoza—to overthrow the Nicaraguan Sandinista government. Nicaragua, Reagan once warned, was only a two-day drive from the Texas border.[43] At the time, the country had a population of only three-and-a-quarter million. In a national television speech in 1984, Reagan explained that San Salvador, El Salvador's capital, "is closer to Houston, Texas, than Houston is to Washington, DC. Central America is America. It's at our doorstep, and it's become the stage for a bold attempt by the Soviet Union, Cuba, and Nicaragua to install communism by force throughout the hemisphere."[44] Reagan later authorized a U.S. military invasion of the Caribbean island of Grenada to end a perceived threat to U.S. students studying there. His successor, George H. W. Bush, invaded Panama in December 1989 to protect threatened American service personnel, overthrowing its leader Manuel Noriega in the process.[45] Given this history of intervention, and subsequent U.S.-led military interventionism, it is difficult for U.S. leaders to frame Russian interventionism in its "backyard" as anachronistic and reprehensible "sphere of influence" behavior without generating countercharges of hypocrisy and double-standards.[46]

Past U.S. interventionism in its near abroad, however, does not justify current Russian interventionism. There is a broader structural dynamic involving larger powers and their neighboring regions at work here that is worth recognizing.[47] Geopolitical entrepreneurs in competitive power struggles within large states can use borderland instability to amplify national anxieties for their own purposes, steering geopolitical cultures in one direction and not another. This is purposeful paranoia, convenient conspiracy construction featuring a sinister other, a foreign "hand," plotting against home and homeland. In both the Reagan and Putin cases, this activation of fear over the loss of historic military bases stirs up nationalist passions, which they intensify and channel for their own domestic political purposes. The United States/Russia becomes "great again" by recovering places of imperial glory and using demonstrative force or proxies against perceived enemies in the near abroad. The longer-term strategic contradictions of the policy, not to mention the lives lost and treasure wasted, are of little concern. The nation is symbolically remasculinized, and the entrepreneurial strongman leader consolidates his popularity and power.

The current Russian case, of course, is quite different in its details from Reagan's fixation with demonstrations of U.S. power in Central America and the Caribbean. Russia today is an authoritarian state where one branch of the state apparatus (the FSB) exercises inordinate power. A small inner-circle executive can exercise its will without the countervailing power of parliament and law (unlike Reagan, who pursued diversionary means to subvert the law to finance the Contras). Regime preservation and survival are pressing matters for Russia's current ruling elite, which has enriched itself enormously during its decade-and-a-half in power.[48] With the economic basis of Russia's relative prosperity during this period at an end, the game of regime preservation through revanchism has had to adapt. The broad strategy of creating spectacles of national greatness and glory jumped from a legitimate to an illegitimate path in 2014.[49] (The 2018 World Cup is under a cloud of uncertainty.) Putin brought the Sochi Olympics to an end the very day he decided to invade Crimea. Because of the international sanctions imposed in the wake of the Crimean invasion, annexation, and subsequent sanctions after MH-17, Russia economic decline was more pronounced than dramatically falling oil prices—down 70 percent in early 2016 since summer 2014—would warrant. In 2015, the Russian government estimated that gross domestic product (GDP) dropped 3.7 percent. Capital flight rates surged. In the two years since the Crimean invasion, the ruble fell by 50 percent against the U.S. dollar. Inflation, as a result, has spiked significantly. Real wages fell by 10 percent between December 2014 and December 2015. In global terms, Russia remains a middle-income country with a per-capita gross national income of $13,220, a figure that hides vast disparities of wealth within the country.[50] The economic foundation of Russia's position as a great power requires it to have cooperative economic relations with the rest of the world. Its major companies need to be able to borrow on international capital markets and access modern technological systems. Putin's actions in Ukraine have undermined that foundation, the affective highs of the Crimean invasion purchased at the expense of the structural prosperity of the state. As many analysts have noted, Russia cannot rebuild its economy without Western investment, technology, and market access.

Putin has backed the country into a corner and undermined its bargaining position with China. Russia needs a new growth model that is not dependent upon global commodity prices, and to have any chance of creating that Russia needs the West.[51]

Whether NATO likes it or not, NATO is perceived by Russia as a U.S.-led sphere of influence encircling its borders. It is delusional to think Russia, or for that matter China, see NATO the way the alliance wants to be seen. All of this presents a serious geopolitical dilemma that could become a classic security dilemma. Should the United States and its allies continue to pressure Russia with systemic sanctions at the risk of further weakening it? Should NATO pursue deterrence through military exercises and personnel and equipment deployments near Russia's borders? Is there a danger that Russia's leadership will feel cornered by the West and, like imperial Japan in 1941, escalate militarily because it sees this as the only path to ensure its survival and avoid a humiliating defeat?[52] This historical analogy, citing a lesson less familiar, competes with more habitual slogans about standing up to bullies, and not permitting spheres of influence. The desire for moral clarity is strong in U.S. geopolitical culture but it is too often acquired by refusing worldly knowledge and the moral complexity that comes with it.

The "Free World": Ukraine as a Cause in U.S. Geopolitical Culture

This brings us to the third question: Is Ukraine's fight against Russian invasion the United States' fight also? The Soviet/Russian threat was, of course, the foundational anchor of U.S. geopolitical culture during the Cold War. Politicians socialized into its terms operationalize its frameworks and catchphrases without much deliberative effort. When Barack Obama defeated John McCain in 2008, however, the United States acquired its first genuinely post–Cold War president, someone not actively socialized as a politician into the conflict's preconceptions, habits, and ways of thinking. It also acquired a president with a more reflective form of masculinity and awareness of tacit identity privileging dynamics.[53] His administration's subsequent pursuit of a "reset" in U.S.-Russian relations was driven by recognition of how the hierarchy of interests and priorities in U.S. foreign policy had become distorted by special relationships and causes. The policy successfully refocused both states on areas of common interest, and it yielded important successes like a new START arms control agreement, Russia's accession to the World Trade Organization, and improved cooperation on Afghanistan, Iran, and North Korea. The two states agreed to disagree on Georgia.[54]

Critics of the Obama administration, former Bush administration officials, and Saakashvili supporters were nevertheless predisposed to see the reset initiative as a betrayal of Georgia. One figure still actively pushing Georgian membership in NATO was Damon Wilson, the Bush administration official who helped drive initial U.S. support for a MAP for it and Ukraine. Wilson joined the Atlantic Council, a longstanding bipartisan establishment think tank in Washington, DC, first as director of its international security program and later as its executive vice president. Together with Frederick Kempé, its president, they significantly expanded the organization's finances during Obama's presidential terms, securing increased support from foreign states, corporations, and wealthy individuals.[55] The Atlantic Council's Eurasia Center, for example, is named after a Romanian billionaire oligarch. Wilson and others authored policy briefs pushing NATO membership for Georgia.[56] The issue had bipartisan support on Capitol Hill and within a series of Washington, DC, think tanks, most prominently the Jamestown Foundation, the Heritage Foundation, and the new McCain Institute, established in 2012. The Obama administration was also officially supportive—it did not renounce the Bucharest Declaration—but the longstanding opposition of France and Germany remained.

The Ukraine crisis of 2014 renewed debate about how the United States should provide aid to help Georgia and Ukraine escape Russian interference. The White House, concerned more about terrorism in the wake of the Boston marathon bombing, was late to focus on Ukraine. It has informally granted policy leadership to the European Union and its Eastern Partnership Association Agreement. When this failed and the country became a crisis, U.S. politicians went to Ukraine to lend their support for the protesters. As already noted Senators Chris Murphy (D-Conn.) and John McCain spoke on the Maidan. McCain told the crowd: "This is your moment... the free world is with you, America is with you, I am with you."[57] A few days earlier U.S. assistant secretary of state Victoria Nuland had visited the Maidan and was pictured distributing cookies to the protesters. Behind the scenes U.S. diplomats worked on brokering the settlement that was supposed to end the crisis. Nuland's efforts thereafter, in conversation with U.S. ambassador Geoffrey Pyatt, were famously recorded and released, most likely by Russian intelligence.[58] The incident greatly strengthened Russia's storyline that the United States was interfering in Ukrainian affairs and picking its leaders (within the Russian media environment at least). But the subsequent government of Petro Poroshenko (president) and Arseniy Yatsenyuk (prime minister) after presidential and parliamentary elections was a democratically elected one, with a level of legitimacy Russian state officials could not dispute (the Russian media was a different matter). The United States, the European Union, and the International Monetary Fund welcomed its reformist agenda and personnel, some of whom came from Georgia (most notably former Georgian president Mikheil Saakashvili, who was appointed governor of Odesa), the Baltic countries (the new minister of the economy, Aivaras Abromavičius, was from Lithuania), and the United States (the finance minister was Natalie Jaresko, an investment banker and former U.S. State Department official). But by the time Ukraine's reformist government was established, the country was at war and in desperate need of a financial bailout.

Because it was invaded by Russia, in violation of the 1994 Budapest Memorandum, Ukraine became a cause célèbre in Washington, DC, and Brussels. The United States was a party to this security assurance deal and some officials, like incoming Defense Secretary Ashton Carter, had worked on it. Before the end of March, the U.S. Congress had approved $1 billion in loan guarantees to Ukraine on a strong bipartisan basis. The money was modest next to the European Union's $15 billion in loans and the International Monetary Fund's $18 billion. These emergency infusions helped the interim government avoid default and financial collapse. It legitimated two veteran Ukrainian politicians—Poroshenko and Yatsenyuk—whose fluent English helped them sell the idea of a "new Ukraine" to international audiences in the years thereafter. Their script followed that pioneered by Saakashvili after the Rose Revolution. Couch Ukraine's struggle within the abstract ideographs of Euro-Atlantic geopolitical culture, most especially the words "freedom" and "democracy." Scale up the crisis: emphasize that the stakes are global not local. Marginalize awkward internal questions about exclusionary nationalism by externalizing blame. Project victimhood, implicitly that experienced by a unitary nation ("the Ukrainian people"). Publicly identify with Israel. Seek U.S. financial aid, military training, advanced weapons systems, and NATO membership. Adding a surreal element to the déjà vu was that Mikheil Saakashvili was leading efforts on Capitol Hill for the new cause, his newly adopted country Ukraine.[59]

The public speeches the Ukrainian leadership delivered in Washington, DC, largely followed this script. In September 2014 Poroshenko was invited to address a joint session of Congress. Despite the already generous levels of support championed by the White House, he used the platform to plead for lethal U.S. military aid: "Blankets and night-vision goggles are important. But one cannot win a war with blankets!" Poroshenko argued that the war in Ukraine "is not only Ukraine's war. It is Europe's, and it is America's war, too. It is a war of the free world—and for a free world! Today, aggression against Ukraine is a threat to global security everywhere." "Just like Israel, Ukraine has the right to defend her territory."[60] The choice before the world today was a choice between civilization and barbarism. Poroshenko's speech, which came before a defiant speech by Benjamin Netanyahu to Congress in March 2015, did not win him plaudits from the White House.[61]

Yatsenyuk's rhetoric was similar. In his first meeting with Obama on March 12, 2014, he explained that the crisis in his country was "all about the freedom."[62] A year later in June 2015 he appeared before the American Jewish Council's Global Forum. Ukraine, he declared, "is defending not only Ukraine. We are defending Europe and we are defending international law and order."[63] "It's not just about us" but about global order. "Ukraine matters for the unity of the free world for, today, the free world is in jeopardy." He added: "This is a war between the past and the future, between the dark and light, between freedom and dictatorship." Visiting Washington, DC, at the same time, Poroshenko reiterated the idea there was a singular Ukrainian nation fighting on the frontlines of Europe: "We have shown the world the true face of our nation, one that fights for European values and defends European security on its frontiers."[64] The reality was a lot more complicated. Some of the units fighting on the frontline were ordinary Ukrainian soldiers; some were part of neo-Nazi militias.[65]

There were, as Poroshenko's speech before Congress complained, limits to U.S. aid. The issue that caused division and contention within the U.S. foreign policy community was whether the United States should provide lethal military aid to Ukraine.[66] In Congress, a bipartisan group championed this type of aid and it became part of a Ukraine aid bill, the Ukraine Freedom Support Act, signed by the president in December 2014. Crucially, this left the final decision about whether to actually send lethal aid in the hands of the president. A bill sponsored by Republican senators only, the Russian Aggression Prevention Act of 2014, provided major non-NATO ally status for Ukraine, Georgia, and Moldova for the purpose of transferring military aid and "defense services" to them. However, it did not make it out of committee.[67] To pressure the president to release military aid, a number of prominent experts on Russia and Ukraine from established think tanks in Washington, DC, released a report under the auspices of the Atlantic Council in February 2015 calling for the United States to provide $1 billion in military aid to Ukraine.[68] A bipartisan group in Congress has continued to agitate for offensive weapons for Ukraine since.[69] In his confirmation hearing in February 2014, Ashton B. Carter, the U.S. secretary of defense, told senators that he was "very much inclined" to support increased military assistance to Ukraine, including the sale of lethal arms.[70]

Despite the bipartisan support in Congress, the consensus of many (though far from all) Russia and Ukraine establishment experts, and support among many senior Obama administration officials, President Obama did not authorize lethal U.S. military aid to Ukraine.[71] The reason why is that from the outset Obama and his inner circle of advisers in the White House concluded that Ukraine was much more important to Russia's national security than it was to the United States. There was a fundamental asymmetry of interests. Furthermore, U.S. foreign policy had overextended itself in Iraq and Afghanistan. In Obama's thinking, the United States needed to avoid the mistakes of the past, mistakes that led it to become entangled militarily in faraway conflicts it could not meaningfully solve. This position was evident from the beginning when Obama chose not to match and validate the inflated rhetoric of the new Ukrainian leadership about the crisis. Obama, instead, emphasized how Russia was pursuing policies that would end up making it weaker, not stronger.[72] In the face of widespread geopolitical aggrandizement of the crisis in Euro-Atlantic geopolitical culture, Obama chose to "right size" the challenge. Russia, he explained in response to a reporter's question at the end of March 2014, "is a regional power that is threatening some of its immediate neighbors—not out of strength but out of weakness."[73] Somewhat controversially, Obama's inner circle chose to deliberately avoid describing Russia's action as an "invasion of Ukraine," a situation description in international law some administration lawyers argued the United States should avoid.[74]

Obama explained how he saw the Ukraine crisis to a journalist in 2016: "Putin acted in Ukraine in response to a client state that was about to slip out of his grasp. And he improvised in a way to hang on to his control there." Obama recognized that location matters. "The fact is that Ukraine, which is a non-NATO country, is going to be vulnerable to military domination by Russia no matter what we do."[75] Russia and the United States had asymmetrical interests in Ukraine and asymmetrical means of realizing those interests. Russia would always retain escalatory dominance in any proxy war. Obama, however, did not view Russia's actions as effective expressions of power. "Real power means you can get what you want without having to exert violence. Russia was much more powerful when Ukraine looked like an independent country but was a kleptocracy that he could pull the strings on." Obama, thus, saw Putin's decision to resort to violence in Ukraine as a sign of weakness and declining hegemony. His position on Ukraine was in keeping with his broader rejection of establishment foreign-policy thinking in Washington, DC. This he derided as the "Washington playbook," the tendency in U.S. geopolitical culture to gravitate toward the use of military options to respond to crises across the world.

Obama's position on lethal aid to Ukraine was an acknowledgment of the power of geographic location in world affairs. Proximity inevitably brought influence by major powers over their weaker neighbors. Conceptualizing power as most effective when structural and soft, Obama's position recognized that Russia had a predominant national security interest in the fate of states in its near abroad. This was not recognition that Russia, therefore, had the right to a sphere of influence there. Rather, Obama observed that because it had to resort to violent means and interventionism in its near abroad, Russia was undermining the necessary soft-power basis for a sphere of influence over its neighbors. "Ukraine has been a country in which Russia had enormous influence for decades, since the breakup of the Soviet Union," Obama explained. "And we have considerable influence on our neighbors. We generally don't need to invade them in order to have a strong, cooperative relationship with them. The fact that Russia felt compelled to go in militarily and lay bare these violations of international law indicates less influence, not more."[76] Obama's argument, expressed not in a public speech but in a magazine interview in his last year in office, met with predictable criticism. His reasoning was at odds with the public sentiments of many within his administration. The impulse to universalize to a placeless moralized plane—the "principled international order"—is still a powerful habit in U.S. geopolitical culture.

Conclusion

The United States has paid a high price over the last decade and a half for the thin geopolitics that drove its foreign policy after 9/11, a geopolitics that organized the world in moralized binaries without regard to the depth and complexities of geography and history. That complexity rebounded upon the United States in Iraq and Afghanistan but also in Georgia in 2008. Since then U.S. geopolitical culture has moved in small but significant ways under President Obama to acknowledge the messy complexities of geopolitics, and the limited ability of the United States to address intractable problems far beyond its borders. Obama has avoided geopolitical frames like "occupied territories," "sphere of influence," and "free world." However, American exceptionalism and the civilizing mission of liberal hegemony run deep.[77] Most U.S. politicians still believe their country is, in Abraham Lincoln's words, the "last best hope of earth."

It is understandable that nationalizing state elites on Russia's borders instrumentalize the language of liberal hegemony to promote their interests and ends. The resultant moralized language of freedom, however, has a capacity to lead the United States astray. Frames like "occupied territories," "sphere of influence," and "free world" may resonate with longstanding Cold War commitments to support captive peoples, but they risk inducing a self-righteous bubble of understanding that is too far removed from ground-level actualities in post-Soviet space. Furthermore, these failures of understanding can lend support for wrongheaded policies that exacerbate rather than defuse the current geopolitical crisis.

In conclusion let me return to the three critical geopolitics concepts introduced at the outset of this work, and how each can help promote deeper forms of geopolitical understanding and analysis. I want to make three normative points about post-Soviet space as a *geopolitical field*. First, post-Soviet space cannot be reduced to a moralized story of empire and freedom. The historical and spatial evolution of Russian revanchism on its western and southern borders needs to be grasped. It is simpleminded to blame this *solely* on innate Russian aggression or overdetermined logics of reimperialization. Russia's policies were developed within a context that was shaped by the actions of terrorists (domestic and international), neighboring nationalizing state-political dynamics (especially "colored revolutions"), NATO expansionism, European Union enlargement, post-Soviet de facto state dynamics, global economic conditions (especially oil and gas prices), and domestic power-structure machinations. Developing a deeper understanding, as noted at the outset of this study, is not justification. In fact, it enables more effective analysis and response. The pursuit of renewed Russian power and prestige in world affairs by the Putin administration has produced significant internal and external contradictions. Chechnya was pacified at the cost of empowering an unpredictable warlord. Relations with neighboring states were poisoned over territorial disputes. Crimea was annexed at the cost of integration into Western capital markets. Short-term glory was achieved at the cost of accelerating long-term economic decline. Russia has failed to establish a sphere of influence over Georgia and Ukraine; instead, it has strengthened their determination to escape Russian influence.

Second, the effort of modernizing groups within Georgia and Ukraine that are struggling to build democratic institutions accountable to citizens, and economic structures that serve the public interest, deserve the support of international institutions.[78] International nongovernmental organizations, funded by Western governments and private foundations, do vital and necessary democratization, governance, and peace-building work with local partners in these countries. But embracing Bonapartism in the Caucasus or shoring up select Ukrainian oligarchs, no matter how good a game they talk, is not "support for freedom." Western institutions need to be fully aware of the moral hazards of their ideals being captured and used by self-serving elites leading intolerant forces in nationalizing states. NATO members need to be fully conscious of the dangers of what the American political scientist Barry Posen terms "reckless driving" by states who have become confident in U.S. commitments to them.[79] Furthermore, NATO itself has a manifest problem with backsliding on democratic institutions, practices, and culture by a number of its members, most especially Hungary, Poland, and Turkey. The danger of NATO empowering member states to air historic grievances rather than seek reconciliation about historical traumas is real and greatly complicates the current geopolitical antagonism.[80] NATO's open expansionist policy, and its mantra of a "Europe, whole and free," needs critical reevaluation. In practice over the last two decades, it has become the rhetoric of a civilizing mission, not a security alliance.[81] Reflexive use of the term "the West" in opposition to Russia is unlikely to disappear as this reified binary is deeply engrained (and yes, the subtitle and language of this book is inevitably complicitous with this commonplace dichotomy). Nevertheless this moralized binary is frequently misleading and open to abuse by promoters of primordial divides and civilizational clashes.

Third, the international community needs to make a concerted effort to address the unresolved territorial conflicts in Ukraine, Moldova, and the Caucasus. Each conflict is different and will require customized conflict management and resolution strategies. Debate about the applicability of the principle of "self-determination" needs to occur. The United States and many other states recognized this principle in the territorial conflict over Kosovo. Through its diplomatic efforts as part of the Minsk Group, the United States has also recognized it as an important element of any peace settlement of the still volatile Nagorny Karabakh conflict.[82] Other conflict resolution practices—partially shared sovereignty, United Nations soft law principles on restitution, and return issues—need to be considered also. World politics has long been characterized by contentious struggles over space and place.

Rhetoric that reflexively reiterates "state sovereignty" and "territorial integrity" as seemingly straightforward and "principled" international norms disguises that complexity. State formation, and state collapse, is often violent and brutal. As George Kennan recognized long ago, the "national state pattern is not, and should not be, and cannot be a fixed and static thing. By nature, it is an unstable phenomenon in a constant state of change and flux."[83]

The second core critical geopolitics concept in this work is *geopolitical culture*. I have stressed the power of affect in the conceptualization and practice of geopolitics in Russia's near abroad. This should not be surprising. Messianic visions and nationalist passions, cartographic fantasies, and hubris characterize the exceptionally tragic twentieth-century history of this region. Affective commitments are often presented and experienced as positive and worthy ideals: the desire to liberate, rescue, and modernize; to expand freedom; to prevent genocide; and to overcome evil. But pursuit of these ideals can often be a will-to-power disguised as virtue. It is a mistake to dub affect as simply "emotion" or "irrationality." The deliberative and affective are entwined and codependent. Heightened awareness is needed on the various ways in which geopolitical thinking is embodied, of how unconscious habits, gender norms, and discursive frames shape how people see the world in culturally blinkered ways.

While recognizing that geopolitical cultures have many elements in common, regime type and state power structure—economic, security, ideological, and political networks—condition their operation. It is much easier for authoritarian regimes to manipulate geopolitical cultures and use administrative resources to promote particular visions and theories in public debate. From the very outset, and following prior Soviet practices, Vladimir Putin and his inner circle have used state-controlled television to create captivating and affirming spectacles of Russian geopolitical success, first in the inner abroad of the North Caucasus, and thereafter in the near abroad.[84] As in the Soviet Union, images of power and glory strive to mask the actualities of stagnation and decline. Words and images are made to serve political needs, to construct the realities required at the moment. President Erdoğan of Turkey is a partner, then an enemy, and then partner again. The presentation and circulation of mendacious theories blaming Ukrainian military actions for the downing of the Malaysian passenger jet MH-17 in July 2014 is one of many disturbing examples of how Russians are deliberately misled about significant events in the world.[85]

U.S. geopolitical culture is also shaped by embedded power structures—state bureaucracies, transnational corporations, political parties, and ideological movements—as well as the power of money in shaping political discourse. Wealthy states and advocacy networks persistently seek to instrumentalize U.S. geopolitical culture to serve their ends. Washington, DC, is full of lobbying firms and think tanks that produce partisan policy findings to serve those that pay. This money funds research institutes, policy reports and experts to promote causes, policies and perspectives, as well as frames and feelings, as expressions of U.S. values and U.S. national interests. These efforts to condition and capture U.S. geopolitical culture have significant influence over how Congressional members and staff view the world.[86]

Despite being a large disputatious democracy, foreign policy debate in Washington, DC, is characterized by bipartisan agreement on the positive value of democracy promotion, free trade and NATO. Careerist calculations, social conformism and groupthink reinforce this worldview. During the 2016 U.S. presidential election, however, this consensus was attacked, most vociferously by Donald Trump. He threatened to repudiate U.S. trade deals and security reassurance for the Baltic states, wondered whether NATO was obsolete, repeatedly expressed admiration for Vladimir Putin, and suggested that the U.S. should consider recognizing the annexation of Crimea. As Trump closed in on the Republican Party nomination, he hired Paul Manafort to run his campaign. Manafort later left under a cloud of suspicion about his financial dealings in Ukraine. Trump's surprise victory in November has thrown the future U.S. role in the liberal international order it helped create into question. The nationalist universalism that underpinned liberal hegemony—"American values are universal values"—may well splinter into a narrow "American first" form that rejects the costs and constraints associated with the dominant "imperial" form, the vision of America as an "indispensable nation." At present, Trump's questioning of liberal hegemony and friendliness toward Putin is at odds with longstanding sentiment in his Republican Party, now the predominant party in the U.S. Congress.

This brings us to the third core concept, the geopolitical condition, and the impact of ongoing transformations in communications and media.[87] There can be little doubt that the rise of social media platforms as expressive infrastructures have changed how people experience and communicate about politics. Twitter, for example, can be a vehicle for protest but it is, more often than not, a vehicle for the diffusion of affective (geo)politics, immersing users in a customized stream of information that invites, indeed rewards, inflamed reactions and responses. Donald Trump turned the platform into an instrument to diffuse incendiary political claims and insults. This generated media spectacles that he was able to use to outflank traditional political candidates. Trump's vow to build a wall on the U.S.–Mexico border, and have Mexico pay for it, expressed not only ethno–cultural revanchism but also desire for humiliation. America would become great again by bullying its neighbor in the near abroad. Similar nostalgia for imagined past greatness was evident in the United Kingdom's BREXIT debate. A specter now haunting the European Union, as near and far abroad wars send waves of desperate people to its borders, is revanchist nationalism. Rising desire across the West for hard borders, traditional hierarchies, and "walled sovereignty" is a profound challenge to the liberal international order.[88] The double shocks of BREXIT and Trump in 2016 were triumphs of "post–factual" politics. Tabloid media offering xenophobic solutions to the frustrations of a precarious world remain powerful engines of affective geopolitics.

Xenophobic tendencies are flaring at a time when the common security threats facing all states—climate change, global pandemics, nuclear proliferation, information system vulnerabilities, and transnational terrorism—are acute and require greater levels of global cooperation and sovereignty pooling. Profound structural changes across the planet, and in the balance of power between major states, pose serious challenges to the embedded liberal order created by the United States and its allies after World War II. Liberal orthodoxy and habits of predominance in U.S. geopolitical culture will have to adapt. Collective effort and patient diplomacy are required to upgrade world order institutions to meet these planetary security challenges. Like it or not, we are all in this together.

NOTES

Introduction

1. President Putin, "Statement on Terrorist Attacks in the United States," September 11, 2001, http://en.kremlin.ru/events/president/transcripts/21328.
2. President Putin, "Russian President's Statement," September 24, 2001, http://en.kremlin.ru/events/president/transcripts/21338. My colleagues and I organized an April 2002 survey of Russian public opinion on the U.S. war on terror. One question listed a series of motives for establishing military bases in Central Asia (three choices were allowed). Only 17 percent of respondents thought the motive was fighting the ongoing war against terrorism. Almost half thought the motive was to expand the U.S. sphere of influence in the region, with many seeing the U.S. motivation as replacing Russian influence in the area. See John O'Loughlin, Gearóid Ó Tuathail, and Vladimir Kolossov, "A 'Risky Westward Turn'? Putin's 9-11 Script and Ordinary Russians," *Europe-Asia Studies* 56, no. 1 (2004): 3–34.
3. The phrase had an ironic meaning during the Soviet period, which satirized rhetoric calling on Russians to sacrifice for "socialist comrades" in the near abroad. For a discussion of this and other early alternative phrases—"abroad close at hand" and "nearby foreign lands"—see William Safire, "On Language: Near Abroad," *New York Times*, May 22, 1994.
4. For some even the phrase "post-Soviet space" is unacceptable as it defines these countries by a past Soviet inheritance centered on Moscow rather than a future they choose. See David Miliband, "There Is No Such

Thing as Post-Soviet Space," *Moscow Times*, September 2, 2008. The U.S. State Department division addressing the region is called the Bureau of European and Eurasian Affairs, though it does not encompass Central Asia. Few if any of the region's peoples see themselves as Eurasians.

5. Jim Nichol, "Armenia, Azerbaijan, and Georgia: Political Developments and Implications for US Interests" (Washington, DC: Congressional Research Service, 2014), Table 3, 64, https://www.fas.org/sgp/crs/row/RL33453.pdf. See also Charles King, "Potemkin Democracy: Four Myths about Post-Soviet Georgia," *National Interest* (2001): 93–104.
6. Roman Solchanyk, "The Politics of State Building: Centre-Periphery Relations in Post-Soviet Ukraine," *Europe-Asia Studies* 46, no. 1 (1994): 47–68.
7. George H. W. Bush, "A Europe Whole and Free. Remarks to the Citizens in Mainz, Federal Republic of Germany," May 31, 1989, http://usa.usembassy.de/etexts/ga6-890531.htm. Jim Hoagland, "Europe's Destiny," *Foreign Affairs* 69, no. 1 (1990): 33–50.
8. Strobe Talbott, *The Russia Hand* (New York: Random House, 2002), 220.
9. James M. Goldgeier, *Not Whether but When: The U.S. Decision to Enlarge NATO* (Washington, DC: Brookings Institution Press, 1999).
10. Interview with Robert Hunter, February 5, 2014. See Robert Hunter, "The West Has Failed to Find a Constructive Role for Moscow," *Financial Times*, February 17, 2015.
11. NATO, The Alliance's Strategic Concept, April 24, 1999, http://www.nato.int/cps/on/natohq/official_texts_27433.htm.
12. David S. Yost, *Nato's Balancing Act* (Washington, DC: US Institute of Peace Press, 2014), 10.
13. Goldgeier, *Not Whether but When*, 39.
14. Merje Kuus, *Geopolitics Reframed: Security and Identity in Europe's Eastern Enlargement* (New York: Palgrave, 2007).
15. NATO, Bucharest Summit Declaration, April 3, 2008, http://www.nato.int/cps/en/natolive/official_texts_8443.htm.
16. This definition is a paraphrase of the definition of literary criticism offered by Northrop Frye, *Anatomy of Criticism: Four Essays* (Princeton: Princeton University Press, 2000), 5.
17. On classic geopolitics see Gerry Kearns, *Geopolitics and Empire: The Legacy of Halford Mackinder* (Oxford: Oxford University Press, 2009). My notion of geopolitical field here is also inspired by the work of the French sociologist Pierre Bourdieu. See Patricia Thomson, "Field" in *Pierre Bourdieu: Key Concepts*, ed. Michael Grenfell (Stocksfield: Acumen, 2008), 67–81.
18. Juliet Fall, "Artificial States? On the Enduring Geographical Myth of Natural Borders," *Political Geography* 29 (2010): 140–147.
19. Arsène Saparov, "Why Autonomy? The Making of the Nagorno-Karabakh Autonomous Region, 1918–1925," *Europe-Asia Studies* 64, no. 2 (2012):

99–123. Mark Beissinger, *Nationalist Mobilization and the Collapse of the Soviet State* (Cambridge: Cambridge University Press, 2002).
20. John O'Loughlin, Vladimir Kolossov, and Gerard Toal, "Inside the Post-Soviet De Facto States: A Comparison of Attitudes in Abkhazia, Nagorny Karabakh, South Ossetia, and Transnistria," *Eurasian Geography and Economics* 55, no. 5 (2014): 425–456.
21. For an early discussion see Gearóid Ó Tuathail, "Geopolitical Structures and Geopolitical Cultures: Towards Conceptual Clarity in the Critical Study of Geopolitics," in *Geopolitical Perspectives on World Politics*, ed. Lasha Tchantouridze, Bison Paper 4, Winnipeg: Centre for Defence and Security Studies, November 2003, 75–102. A key inspiration is Gertjan Dijkink, *National Identity and Geopolitical Visions: Maps of Pride and Pain* (London: Routledge, 1996).
22. The arguments made here are very different from the naturalization of socially produced power configurations and geopolitical cultures found in contemporary journalistic accounts. See Robert D. Kaplan, *The Revenge of Geography* (New York: Random House, 2012); Tim Marshall, *Prisoners of Geography* (New York: Scribner, 2015).
23. Ambrose Gwinnett Bierce (1842–1913) was an American writer best known for *The Devil's Dictionary* (1912), a collection of satirical definitions that mock political cant and double talk.
24. Among the many influential works are Antonio R. Damasio, *Descartes' Error: Emotion, Reason, and the Human Brain* (New York: Putnam, 1994); Daniel Kahneman, *Thinking, Fast and Slow*, 1st ed. (New York: Farrar, Straus & Giroux, 2011); Nigel Thrift, *Non-Representational Theory: Space, Politics, Affect* (New York: Routledge, 2008); Drew Westen, *The Political Brain: The Role of Emotion in Deciding the Fate of the Nation* (New York: PublicAffairs, 2007).
25. See Armand Matellart, *The Invention of Communication,* trans. Susan Emanuel (Minneapolis: University of Minnesota Press, 1996); and Peter Hugill, *Global Communications since 1844: Geopolitics and Technology* (Baltimore: Johns Hopkins University Press, 1999).
26. I have engaged this theme over the last two decades. See, inter alia, Timothy W. Luke and Gearóid Ó Tuathail, "On Videocameralistics: The Geopolitics of Failed States, the CNN International and (UN) Governmentality," *Review of International Political Economy* 4 (1997): 709–733; Gearóid Ó Tuathail, "Political Geography III: Dealing with Deterritorialization," *Progress in Human Geography* 22 (1998): 81–93; Gearóid Ó Tuathail, "Postmodern Geopolitics? The Modern Geopolitical Imagination and Beyond," in *Rethinking Geopolitics*, ed. Gearóid Ó Tuathail and Simon Dalby (London: Routledge, 1998), 16–38; Gearóid Ó Tuathail, "Borderless Worlds: Problematizing Discourses of Deterritorialization," *Geopolitics* 4 (1999): 139–154; Gearóid Ó Tuathail, "A Strategic Sign: The Geopolitical Significance

of 'Bosnia' in U.S. Foreign Policy." *Environment and Planning D: Society and Space* 17 (1999): 515–533; Timothy W. Luke and Gearóid Ó Tuathail, "Thinking Geopolitical Space: The Spatiality of War, Speed and Vision in the Work of Paul Virilio," in *Thinking Space*, ed. Mike Crang and Nigel Thrift (London: Routledge, 2000), 360–379.
27. Judith Butler, *Precarious Life: The Powers of Mourning and Violence* (London: Verso, 2004).
28. On this see the edited collection *Countdown to War in Georgia: Russia's Foreign Policy and Media Coverage of the Conflict in South Ossetia and Abkhazia* (Minneapolis: East View Press, 2008); and Sergey Markedonov, *The Big Caucasus: Consequences of the "Five Day War," Threats and Political Prospects*, Athens, Greece: International Center for Black Sea Studies, Xenophon Paper Number 7, May 2009, kms2.isn.ethz.ch/serviceengine/Files/RESSpecNet/.../XENOPHON_PAPER_7.pdf.

Chapter 1

1. "Text of the Secret Additional Protocol of the Nazi-Soviet Nonaggression Pact," in Roger Moorhouse, *The Devil's Alliance: Hitler's Pact with Stalin, 1939–1941* (New York: Basic Books, 2014), 302.
2. Text of the Charter of Paris for a New Europe, Paris 1990, http://www.osce.org/node/39516.
3. See Mark Beissinger, *Nationalist Mobilization and the Collapse of the Soviet State* (Cambridge: Cambridge University Press, 2002).
4. A foundational text in the establishment of these terms is George F. Kennan, *American Diplomacy, 1900–1950* (Chicago: University of Chicago Press, 1951).
5. For instances of this work, review the past issues of the journals *Geopolitics* and *Political Geography*. See also Klaus Dodds, Merje Kuus, and Joanne Sharp, eds., *The Ashgate Research Companion to Critical Geopolitics* (Farnham, Surrey: Ashgate, 2013); Gearóid Ó Tuathail and Simon Dalby, *Rethinking Geopolitics* (London: Routledge, 1998); and Gearóid Ó Tuathail, *Critical Geopolitics: The Politics of Writing Global Space* (Minneapolis: University of Minnesota Press, 1996).
6. Richard Ashley, "The Geopolitics of Geopolitical Space: Towards a Critical Social Theory of International Politics," *Alternatives* XII (1987): 403–437.
7. David S. Foglesong, *The American Mission and the "Evil Empire": The Crusade for a "Free Russia" since 1881* (New York: Cambridge University Press, 2007).
8. George F. Kennan, 861.00/2—2246: Telegram, Moscow, February 22, 1946. Available at http://nsarchive.gwu.edu/coldwar/documents/episode-1/kennan.htm. The clipped style reflects its origins as a telegram.

9. George F. Will, "Eastward Ho—and Soon," *Washington Post*, June 13, 1996. The image of Russia as a relentlessly expansionist power is a longstanding trope in European diplomacy. One frequently cited documentary basis for it was the Testament of Peter the Great, a forgery drafted in the early eighteenth century by various enemies of the tsar that ended up in the French archives. In his history of the Crimean War, Figes argued that Russophobia "was arguably the most important element in Britian's outlook on the world abroad. Throughout Europe, attitudes to Russia were mostly formed by fears and fantasies, and Britain in this sense was no exception to the rule." Orlando Figes, *The Crimean War: A History* (New York: Picador, 2010), 70–71.
10. White House, "President Bush Concerned by Escalation of Violence in Georgia," August 9, 2008, https://georgewbush-whitehouse.archives.gov/news/releases/2008/08/20080809-2.html.
11. White House, President Bush Discusses Situation in Georgia, August 11, 2008, https://georgewbush-whitehouse.archives.gov/news/releases/2008/08/20080811-1.html.
12. White House, Remarks of the Situation in Georgia, August 15, 2008, https://georgewbush-whitehouse.archives.gov/news/releases/2008/08/20080815.html.
13. Robert Kagan, "Putin Makes His Move," *Washington Post*, August 11, 2008.
14. White House, Statement by the President on Ukraine, February 28, 2014, https://www.whitehouse.gov/the-press-office/2014/02/28/statement-president-ukraine.
15. White House, Statement by the President on Ukraine, March 6, 2014, https://www.whitehouse.gov/the-press-office/2014/03/06/statement-president-ukraine.
16. Kerry's statements were on two television programs on the same day: *Face the Nation*, March 2, 2014; and *Meet the Press*, March 2, 2014.
17. White House, Remarks by President Obama at 25th Anniversary of Freedom Day—Warsaw, Poland, June 4, 2014, https://www.whitehouse.gov/the-press-office/2014/06/04/remarks-president-obama-25th-anniversary-freedom-day-warsaw-poland.
18. George W. Bush, Remarks on the Situation in Georgia, August 15, 2008.
19. George W. Bush, Radio Address, August 16, 2008, http://georgewbush-whitehouse.archives.gov/news/releases/2008/08/20080815-5.html.
20. William Kristol, for example, argued that because Georgia had soldiers serving with U.S. troops in Iraq, "we owe Georgia a serious effort to defend its sovereignty. Surely we cannot simply stand by as an autocratic aggressor gobbles up part of—and perhaps destabilizes all of—a friendly democratic nation that we were sponsoring for NATO membership a few months ago." William Kristol, "Will Russia Get Away with It?" *New York Times*, August 10, 2008, A17.

21. White House, Remarks by the President in Address to European Youth, Brussels, March 26, 2014, https://www.whitehouse.gov/the-press-office/2014/03/26/remarks-president-address-european-youth.
22. Peter Baker and Michael Shear, "U.S. Weighs Direct Military Action against Isis in Syria," *New York Times*, August 22, 2014, A1.
23. See, e.g., Derek Gregory, *The Colonial Present: Afghanistan, Palestine, Iraq* (Malden, MA: Blackwell, 2004).
24. For a discussion of what social psychologists frame as "cognitive biases" and their relevance to geopolitical reasoning, see Daniel Kahneman and Jonathan Renshon, "Hawkish Biases," in *American Foreign Policy and the Politics of Fear: Threat Inflations Since 9/11*, ed. Trevor Thrall and Jane K Cramer (New York: Routledge, 2009), 79–96.
25. Mikhail Gorbachev, "A Path to Peace in the Caucasus," *Washington Post*, August 12, 2008, A13.
26. Kennan, *American Diplomacy, 1900–1950*, 97.
27. Ibid.
28. Ibid., 96.
29. Ibid., 98.
30. John J. Mearsheimer, *The Tragedy of Great Power Politics* (New York: Norton, 2001).
31. Ibid., 31.
32. See also the appearance of the president of the Nixon Center, Dimitri Simes, on *The Newshour with Jim Lehrer* in debate with Richard Holbrooke, August 12, 2008, http://washingtonnote.com/rush_news_hour/.
33. Paul J. Saunders, "Georgia's Recklessness," *Washington Post*, August 15, 2008, A21.
34. John J. Mearsheimer, "Why the Ukraine Crisis Is the West's Fault," *Foreign Affairs* 93, no. 5 (2014): 77–89.
35. Ibid., 82.
36. Ibid., 84.
37. Ibid., 88.
38. Ibid.
39. President of Russia, "Press Statement and Answers to Questions at a Joint News Conference with Ukrainian President Leonid Kuchma," May 17, 2002, http://en.kremlin.ru/events/president/transcripts/21598; Russian prime minister Vladimir Putin interviewed by the German ARD TV channel, http://archive.premier.gov.ru/eng/events/news/1758/.
40. Michael McFaul, "Moscow's Choice," *Foreign Affairs* 93, no. 6 (2014): 167–171; Stephen Sestanovich, "How the West Has Won" *Foreign Affairs*, 93, no. 6 (2014): 171–175; John Mearsheimer, "Mearsheimer Replies," *Foreign Affairs* 93, no. 6 (2014): 175–178.
41. David Milliband, "There Is No Such Thing as Post-Soviet Space," *Moscow Times*, September 2, 2008.

42. Viatcheslav Morozov, *Russia's Postcolonial Identity: A Subaltern Empire in a Eurocentric World* (New York: Palgrave, 2015).
43. Rogers Brubaker, *Nationalism Reframed: Nationhood and the National Question in the New Europe* (Cambridge: Cambridge University Press, 1996), 3.
44. Ibid., 5, emphasis in original.
45. Rogers Brubaker, "Nationalizing States Revisited: Projects and Processes of Nationalization in Post-Soviet States," *Ethnic and Racial Studies* 34, no. 11 (2011): 1785–1814.
46. Brubaker, *Nationalism Reframed: Nationhood and the National Question in the New Europe*, p. 5.
47. Pierre Bourdieu and Loïc J. D. Wacquant, *An Invitation to Reflexive Sociology* (Chicago: University of Chicago Press, 1992).
48. Taras Kuzio, "'Nationalising States' or 'Nation Building'? A Critical Review of the Theoretical Literature and Empirical Evidence," *Nations and Nationalism* 7, no. 2 (2001): 135–154; Volodymyr Kulyk, "The Politics of Ethnicity in Post-Soviet Ukraine: Beyond Brubaker," *Journal of Ukrainian Studies* 26, nos. 1–2 (2001): 197–221; David J. Smith, "Framing the National Question in Central and Eastern Europe: A Quadratic Nexus?," *Global Review of Ethnopolitics* 2, no. 1 (2002): 3–16; Gëzim Krasniqi, "'Quadratic Nexus' and the Process of Democratization and State-Building in Albania and Kosovo: A Comparison," *Nationalities Papers* 41, no. 3 (2013): 395–411.
49. There was a long history prior to this. The origins of the contemporary understanding of "terrorism" as violence by nonstate actors and individuals, as opposed to "terror" as something unleashed by the state, has its origins in the context of the Irish nationalist Fenian Brotherhood bombing campaign in mainland Britain in 1881–1886. See Gerry Kearns, "Bare Life, Political Violence and the Territorial Structure of Britain and Ireland," in *Violent Geographies: Fear, Terror, and Political Violence*, ed. Derek Gregory and Allan Pred (New York: Routledge, 2007), 7–35.
50. See, for example, Uğur Ümit Üngör, *The Making of Modern Turkey: Nation and State in Eastern Anatolia, 1913–1950* (New York: Oxford University Press, 2011).
51. It should be noted that this notion of geography as circulation is much older than the cybernetic-informational-age rendition of it by Manuel Castells. Mackinder wrote extensively about spaces of circulation, connection, and mobility, as did Jean Gottmann. See Gerry Kearns, *Geopolitics and Empire: The Legacy of Halford Mackinder* (Oxford: Oxford University Press, 2009). Jean Gottmann, *The Significance of Territory* (Charlottesville: University Press of Virginia, 1973).
52. Gertjan Dijkink, *National Identity and Geopolitical Visions: Maps of Pain and Pride* (London: Routledge, 1996).

53. See Peter J. Taylor, *The Way the Modern World Works: World Hegemony to World Impasse* (London: Wiley, 1996).
54. Michael Mann, *The Sources of Social Power: Volume 1, a History of Power from the Beginning to AD 1760* (New York: Cambridge University Press, 1986).
55. George Schöpflin, *Nations Identity Power* (New York: New York University Press, 2000), 80.
56. Northrop Frye, *Anatomy of Criticism: Four Essays* (Princeton: Princeton University Press, 2000).
57. Michael Calvin McGee, "The 'Ideograph': A Link between Rhetoric and Ideology," *Quarterly Journal of Speech* 66, no. 1 (1980): 1–16; Thomas O. Sloane, ed. *Encyclopedia of Rhetoric* (Oxford: Oxford University Press, 2001), 378–381.
58. Janis L. Edwards and Carol K. Winkler, "Representative Form and the Visual Ideograph: The Iwo Jima Image in Editorial Cartoons," *Quarterly Journal of Speech* 83 (1997): 289–310.
59. Odd Arne Westad, *The Global Cold War: Third World Interventions and the Making of Our Times* (Cambridge: Cambridge University Press, 2005).
60. Thomas Goltz, *Georgia Diary: A Chronicle of War and Political Chaos in the Post-Soviet Caucasus* (Armonk, NY: Sharpe, 2006), 49.
61. Michael Dobbs, "Nationalists, Minority Battle in Soviet Georgia; Moscow Accused of Arming Ossetians," *Washington Post*, March 21, 1991, A31.
62. The leading spokesperson for Russia as a "liberal empire" was Anatoly Chubais, an influential minister within the Yeltsin administration and later head of the Russian energy conglomerate, Unified Energy Systems (UES). See Igor Torbakov, "Russian Policymakers Air Notion of 'Liberal Empire' in Caucasus, Central Asia," *Eurasianet.org*, October 26, 2003, http://www.eurasianet.org/departments/insight/articles/eav102703.shtml.
63. Patrick Thaddeus Jackson, *Civilizing the Enemy: German Reconstruction and the Invention of the West* (Ann Arbor: University of Michigan Press, 2006).
64. Thomas De Waal, *Great Catastrophe: Armenians and Turks in the Shadow of Genocide* (Oxford: Oxford University Press, 2015).
65. A broad equivalent to the Soviet practice was an imagining of the Soviet Union as "Red fascism." See Les K. Adler and Thomas G. Paterson, "Red Fascism: The Merger of Nazi Germany and Soviet Russia in the American Image of Totalitarianism, 1930s–1950s," *American Historical Review* 74, no. 4 (1970): 1046–1064.
66. Timothy W. Luke, *Museum Politics: Power Plays at the Exhibition* (Minneapolis: University of Minnesota Press, 2002).
67. Antonio R. Damasio, *Descartes' Error: Emotion, Reason, and the Human Brain* (New York: Putnam, 1994). Arkady Ostrovsky, *The Invention of Russia: From Gorbachev's Freedom to Putin's War* (New York: Viking, 2015).

68. Daniel Kahneman, *Thinking Fast and Slow* (New York: Farrar, Straus & Giroux, 2013).
69. William E. Connolly, *Neuropolitics: Thinking, Culture, Speed* (Minneapolis: University of Minnesota Press, 2002), 75.
70. Ibid., 61.
71. Rachel Pain and Susan J. Smith, *Fear: Critical Geopolitics and Everyday Life* (Aldershot: Ashgate, 2008); Rachel Pain, "The New Geopolitics of Fear," *Geography Compass* 4, no. 3 (2010): 226–240.
72. Andrew A. G. Ross, *Mixed Emotions: Beyond Fear and Hatred in International Conflict* (Chicago: University of Chicago Press, 2014); Dominique Moïsi, *The Geopolitics of Emotion: How Cultures of Fear, Humiliation, and Hope Are Reshaping the World*, 1st ed. (New York: Doubleday, 2009); Adis Maksic, *Ethnic Mobilization, Violence and the Politics of Affect* (New York: Palgrave, 2017).
73. Anssi Paasi, *Territories, Boundaries, and Consciousness: The Changing Geographies of the Finnish-Russian Boundary* (New York: J. Wiley & Sons, 1996).
74. Alexei Yurchak, *Everything Was Forever, Until It Was No More* (Princeton: Princeton University Press, 2006).
75. Yegor Gaidar, *Collapse of an Empire* (Washington, DC: Brookings Institution Press, 2007). More broadly on this theme see Franck Billé, "Territorial Phantom Pains (and Other Cartographic Anxieties)," *Environment and Planning D: Society and Space* 32 (2014): 163–178.
76. Valerie Sperling, *Sex, Politics, and Putin: Political Legitimacy in Russia* (New York: Oxford University Press, 2015).
77. Steven Lee Myers, *The New Tsar: The Rise and Reign of Vladimir Putin* (New York: Alfred A. Knopf, 2015), 16–18.
78. Ibid., 372.
79. Wendell Steavenson, "Marching through Georgia," *New Yorker*, December 15, 2008, 64. Saakashvili began studies for a Ph.D. in International Law at George Washington University. The proposed topic of his dissertation was the *uti possidetis* doctrine.
80. Clifford Levy, "The Georgian and Putin: A Hate Story," *New York Times*, April 18, 2009. Saakashvili's response predictably intensified the hypermasculinity: "he would not have enough rope." Steavenson, "Marching through Georgia" (p. 64).
81. François Debrix, *Tabloid Terror: War, Culture, and Geopolitics* (New York: Routledge, 2008).
82. Orlando Figes, *The Crimean War: A History* (New York: Picador, 2010), 147–155.
83. Nicholas Cull, *The Cold War and the United States Information Agency: American Propaganda and Public Diplomacy* (Cambridge: Cambridge University Press, 2008).

84. Jean Baudrillard, *Simulacra and Simulation* (Ann Arbor: University of Michigan Press, 1995).
85. Kara Rowland, "Russia Today: Youth Served," *Washington Times*, October 27, 2008; William Dunbar, "They Forced Me out for Telling the Truth about Georgia," *The Independent*, September 19, 2010, 24.
86. Christina Cottiero, Katherine Kucharski, Evgenia Olimpieva and Robert W. Orttung, "War of Words: The Impact of Russian State Television on the Russian Internet," *Nationalities Papers* 43, no. 4 (2015): 533–555.
87. Laura Jones, "The Commonplace Geopolitics of Conspiracy," *Geography Compass* 6, no. 1 (2012): 44–59.
88. Shaun Walker, "'Hitler Was an Anglo-American Stooge': The Tall Tales in a Moscow Bookshop," *The Guardian*, August 14, 2015. The phenomenon of popular conspiracy theories in Russian geopolitical culture is hardly new. *The Protocols of the Elders of Zion* was an anti-Semitic forgery created in the early-twentieth-century Russian Empire.
89. Halford J. Mackinder, "The Geographical Pivot of History," *Geographical Review* 23, no. 4 (1904): 421–444.
90. Gearóid Ó Tuathail, "De-Territorialized Threats and Global Dangers: Geopolitics and Risk Society," *Geopolitics* 3, no. 1 (1998): 17–31.
91. Sandra E. Roelofs, *The Story of an Idealist: The First Lady of Georgia* (Tbilisi: Archipel, 2010), 200.
92. Thomas De Waal, "Georgia's Choices: Charting a Future in Uncertain Times" (Washington, DC: Carnegie Endowment for International Peace, 2011), http://carnegieendowment.org/2011/06/13/georgia-s-choices-charting-future-in-uncertain-times-pub-44553.
93. Thomas De Waal, "No America in the Caucasus: The False Promise of Westernization in Georgia," *Foreign Affairs*, December 5, 2012, https://www.foreignaffairs.com/articles/georgia/2012-12-05/no-america-caucasus.

Chapter 2

1. Vladimir Putin, "Address to the Federal Assembly," April 25, 2005, http://archive.kremlin.ru/eng/speeches/2005/04/25/2031_type70029type82912_87086.shtml.
2. BBC News, "Putin Deplores Collapse of USSR," *BBC News*, 2005; Mike Eckel, "Putin Calls Soviet Collapse a 'Geopolitical Catastrophe'," *Associated Press*, April 26, 2005.
3. John Bolton interview, *Fox News*, March 3, 2014. See Katie Sanders, "Did Vladimir Putin Call the Breakup of the USSR 'the Greatest Geopolitical Tragedy of the 20th Century?'" *Politifact.com*, March 6, 2014.
4. The translation sparked a debate on Johnson's Russia List between Anders Aslund and Patrick Armstrong. See http://russialist.org/re-putin-soviet-geopolitical-disaster-deadly-quotations-part-2/.
5. The phrase "one of the greatest" is perhaps closer to the spirit of Putin's expression than "the greatest," with "catastrophe" instead of "disaster"

justifiable since the Russian word he used was *катастрофа/katastrofa*. That Putin was pandering, as politicians do with superlatives ("greatest generation," "greatest country on earth") to an older domestic audience who felt themselves victims of the Soviet collapse seems evident. Domestic and international audiences constructed meaning differently.
6. Vladimir Putin, "Interview with German Television Channels ARD and ZDF," May 5, 2005, http://archive.kremlin.ru/eng/speeches/2005/05/05/2355_type82912type82916_87597.shtml.
7. White House, "Press Conference by President Bush and Russian Federation President Putin," June 16, 2001, http://georgewbush-whitehouse.archives.gov/news/releases/2001/06/20010618.html.
8. Jackie Calmes, "McCain Sees Something Else in Putin's Eyes," *Wall Street Journal*, October 16, 2007.
9. Robert Gates, *Duty: Memoirs of a Secretary at War* (New York: Alfred A. Knopf, 2014), 169.
10. Mark Beissinger, "The Persisting Ambiguity of Empire," *Post-Soviet Affairs* 11, no. 2 (1995): 160.
11. Terry Martin, *The Affirmative Action Empire: Nations and Nationalism in the Soviet Union, 1923–1939* (Ithaca: Cornell University Press, 2001), 19.
12. Ronald Grigor Suny, "Ambiguous Categories: States, Empires and Nations," *Post-Soviet Affairs* 11, no. 2 (1995): 185–196.
13. Robert Kaiser, *The Geography of Nationalism in Russia and the USSR* (Princeton, NJ: Princeton University Press, 1994).
14. David D. Laitin, *Identity in Formation: The Russian-Speaking Populations in the Near Abroad* (Ithaca: Cornell University Press, 1998). On the Russian Empire in comparative perspective see Dominic Lieven, *Empire: The Russian Empire and Its Rivals* (New Haven: Yale University Press, 2001).
15. Ibid., 44. Brian Silver, "Methods of Deriving Data on Bilingualism from the 1970 Soviet Census," *Soviet Studies* 27 (1975): 574–597.
16. Ibid., 67.
17. See Lawrence Broers, "'David and Goliath' and 'Georgians in the Kremlin': A Post-Colonial Perspective on Conflict in Post-Soviet Georgia," *Central Asian Survey* 28, no. 2 (2009): 99–118.
18. This finding, cited in Laitin, *Identity in Formation*, 65, is somewhat misleading since the Ukrainians were the second most populous group after the Russians. Also Ukraine was the power base of both Khrushchev and Brezhnev, and they tended to appoint local loyalists they knew to high posts to consolidate their power. Furthermore, Laitin cites communication with Pål Kolstø, who argued that western Ukraine did not experience any "most-favored-lord" advantage when incorporated into the Soviet Union but was ruled in "integral" style like the Baltic countries.
19. Francine Hirsch, *Empire of Nations: Ethnographic Knowledge and the Making of the Soviet Union* (Ithaca: Cornell University Press, 2005).

20. Yuri Slezkine, "The USSR as a Communal Apartment, or How a Socialist State Promoted Ethnic Particularism," *Slavic Review* 53, no. 2 (1994): 415.
21. Martin, *Affirmative Action Empire*, 8.
22. Roger Moorhouse, *The Devils' Alliance: Hitler's Pact with Stalin, 1939–1941* (New York: Basic Books, 2014).
23. Serhii Plokhy, *The Last Empire: The Final Days of the Soviet Union* (New York: Basic Books, 2014).
24. Slezkine, "The USSR as a Communal Apartment," 451.
25. Roman Szporluk, *Russia, Ukraine, and the Breakup of the Soviet Union* (Stanford, CA: Hoover Institution Press, 2000).
26. Graham Smith, *The Post-Soviet States: Mapping the Politics of Transition* (London: Edward Arnold, 1999).
27. Serhii Plokhy described the immediate purpose of the Soviet Union created in December 1922 as "to keep the Ukrainians in, the Poles out, and the Russians down," *Last Empire*, 230. Serhii Plokhy, *The Gates of Europe: A History of Ukraine* (New York: Basic Books, 2015).
28. Suny, "Ambiguous Categories," 194.
29. Aleksandr Solzhenitsyn, *Rebuilding Russia: Reflections and Tentative Proposals* (London: Harvill Press, 1991), 11. Also see his *"The Russian Question" at the End of the Twentieth Century* (New York: Farrar, Straus & Giroux, 1995).
30. Thomas De Waal, *Black Garden: Armenia and Azerbaijan through Peace and War*, 2nd ed. (New York: New York University Press, 2013).
31. Charles King, *The Moldovans: Romania, Russia, and the Politics of Culture* (Stanford: Hoover Institution Press, 2000), 140.
32. Conventionally, the term *russkiye* defines Russians linguistically and ethnically, whereas *rossiyanye* defines them by citizenship and residence. This distinction, however, can be misleading. Greater terminological use of the singular *russkii* or plural *russkiye* does not necessarily signal exclusivist ethnonationalism for it has an older meaning that refers to the unity of the peoples of historic Rus, the Eastern Slavic nations: Russians, Ukrainians, and Belarussians. Further, Putin's articulations of Russian nationalism, for example, are usually in the inclusivist key of *rossiyanye* yet delimited by tacit ethnic privileging. See Marlene Laruelle, "Misinterpreting Nationalism: Why *Russkii* Is Not a Sign of Ethnonationalism," *Ponars Eurasia Policy Memo* 416 (2016), http://www.ponarseurasia.org/memo/misinterpreting-nationalism-russkii-ethnonationalism.
33. Cited in Dmitri Trenin, *The End of Eurasia: Russia on the Border between Geopolitics and Globalization* (Washington, DC: Carnegie Endowment for International Peace, 2002), 179.
34. See, inter alia, Smith, *The Post-Soviet States;* Andrei P. Tsygankov, "Mastering Space in Eurasia: Russia's Geopolitical Thinking after the

Soviet Break-Up," *Communist and Post-Communist Studies* 36 (2003); 101–127; Andrei P. Tsygankov, *Russia's Foreign Policy: Change and Continuity in National Identity*, 4th ed. (Lanham: Rowman & Littlefield, 2016); John O'Loughlin, Gearoid Ó Tuathail, and Vladimir Kolossov, "Russian Geopolitical Culture in the Post 9/11 Era: The Masks of Proteus Revisited," *Transactions, Institute of British Geographers* 30 (2005): 322–335; Marlene Laruelle, *Russian Eurasianism: An Ideology of Empire* (Washington, DC: Woodrow Wilson Center Press, 2008).

35. Examples include Article 3 of the Charter of the Commonwealth of Independent States (signed June 22, 1993), the Declaration on Observing the Sovereignty, Territorial Integrity, and Inviolability of Borders (1994), and the Memorandum on the Maintenance of Peace and Stability in the CIS (1995).
36. Text of the Alma Ata Declaration, December 21, 1991.
37. Igor Zevelev, *Russia and Its New Diasporas* (Washington, DC: US Institute of Peace Press, 2001), 139.
38. Andreas Umland, "Zhirinovsky's Last Thrust to the South and the Definition of Fascism," *Russian Politics and Law* 46, no. 4 (2008): 31–46.
39. Anton Shekhovtsov, "Alexander Dugin and the West European New Right, 1989–1994," in *Eurasianism and the European Far Right: Reshaping the Europe-Russia Relationship*, ed. Marlene Laruelle (Latham, MD: Lexington Books, 2015), 35–53.
40. Laruelle, *Russian Eurasianism*.
41. Andreas Umland, "Conceptual and Contextual Problems in the Interpretation of Contemporary Russian Ultranationalism," *Russian Politics and Law* 46, no. 4 (2008): 10.
42. Alan Ingram, "'A Nation Split into Fragments': The Congress of Russian Communities and Russian Nationalist Ideology," *Europe-Asia Studies* 51, no. 4 (June 1999): 691.
43. Alexander Lebed (1950–2002) was a battalion commander of Soviet airborne troops in Afghanistan, and later led airborne units to repress nationalist demonstrations in Azerbaijan and Georgia. Transferred to Moldova in June 1992, he ordered the 14th Army to directly attack the Moldovan forces besieging the Transnistrian rebels in the town of Bendery (Tighina to Romanians). This display of violence, which some view as the first "Russian invasion" of post-Soviet space even though the army were in situ, effectively ended the war and secured the future of the PMR (Transnistria) as a de facto state. Lebed ran as a candidate in the 1996 Russian presidential race and placed third in the first round. He subsequently endorsed Yeltsin after the latter appointed him his national security advisor. He died in a helicopter crash in 2002.
44. Solzhenitsyn, *Rebuilding Russia*; Solzhenitsyn, *"The Russian Question" at the End of the Twentieth Century*.

45. Roman Solchanyk, "The Politics of State Building: Centre-Periphery Relations in Post-Soviet Ukraine," *Europe-Asia Studies* 46, no. 1 (1994): 48.
46. Tsygankov, "Mastering Space in Eurasia."
47. Sergei Stankevich, "A Power in Search of Itself," *Nezavisimaya Gazeta*, March 28, 1992.
48. Gennady Charodeyev, "Moscow Concerned over Human Rights Violations in Estonia," *Izvestia*, April 2, 1992.
49. Stankevich, "A Power in Search of Itself"; Sergei Karaganov, "Problemy Zashchity Interesov Rossiiskogo Orientirovannogo Naseleniya V Blizhnem Zarubezhe [The Problem of Protecting the Interests of the Russian Population in the Near Abroad]," *Diplomaticheskii Vestnik*, November 15–30, 1992.
50. Konstantin Eggert, "Russia in the Role of 'Eurasian Gendarme'? The Chairman of a Parliamentary Committee Has Worked out His Own Conception of Foreign Policy," *Izvestia*, August 7, 1992.
51. Andranik Migranyan, "Real and Illusory Guidelines in Foreign Policy," *Rossiikaya Gazeta*, August 4, 1992.
52. Cited in John Lough, "Defining Russia's Relations with Neighboring States," *RFE/RL Research Report* 2, no. 20 (1992).
53. Cited in ibid.
54. Leon Aron, "The Emergent Priorities of Russian Foreign Policy," in *The Emergence of Russian Foreign Policy*, ed. Leon R. Aron and K. M. Jensen (Washington, DC: United States Institute of Peace, 1994), 17–33.
55. Leslie Gelb, "Yeltsin as Monroe," *New York Times*, March 7, 1993.
56. R. Jeffrey Smith and Barton Gellman, "US Will Seek to Mediate Ex-Soviet States' Disputes," *Washington Post*, August 5, 1993.
57. Ibid.
58. Bohuslav Litera, "The Kozyrev Doctrine—a Russian Variation on the Monroe Doctrine," *Perspectives, Institute of International Relations, NGO* 4 (1995).
59. John Lough, "The Place of the "Near Abroad" in Russian Foreign Policy," *RFE/RL Research Report* 2, no. 11 (1993).
60. Clinton's "not whether but when" statement was January 12, 1994. See James M. Goldgeier, *Not Whether but When: The U.S. Decision to Enlarge NATO* (Washington, DC: Brookings Institution Press, 1999). On the question of a tacit agreement not to expand NATO see Joshua R. Itzkowitz Shifrinson, "Deal or No Deal? The End of the Cold War and the U.S. Offer to Limit NATO Expansion," *International Security* 40, no. 4 (Spring 2016): 7–44.
61. This incident occurred on June 11–12, 1999, immediately after agreement was reached to end NATO's war against Serbia. Russia had sought its own peacekeeping zone in Kosovo but the concluded agreement did not establish one. Russian peacekeeping forces from Bosnia unilaterally drove to Kosovo and occupied Pristina airport ahead of a planned

NATO deployment there. The result was a standoff at the airport. The United States pressured neighboring states to refuse Russia use their airspace to send reinforcements to the airport. Bulgaria, Hungary, and Romania complied and Russia had to accept a lesser role in the Kosovo peacekeeping operation than it desired.

62. John O'Loughlin, "Geopolitical Fantasies, National Strategies and Ordinary Russians in the Post-Communist Era," *Geopolitics* 5, no. 3 (2001): 17–48.
63. Steven Lee Myers, *The New Tsar*, chap. 10.
64. Vladimir Putin, "Russia at the Turn of the Millennium." Web supplement to Vladimir Putin et al., *First Person: An Astonishingly Frank Self-Portrait by Russia's President Vladimir Putin*, translated by Catherine A. Fitzpatrick (New York: Public Affairs, 2000), http://pages.uoregon.edu/kimball/Putin.htm.
65. Boris Yeltsin, "Inuagural Address by the President of the Russian Soviet Federative Socialist Republic," *Associated Press*, July 10, 1991. This is not to suggest Yeltsin was so different from Putin. For a discussion of the dark continuities, including use of terrorist provocations to consolidate power, see David Satter, *The Less You Know, the Better You Sleep: Russia's Road to Terror and Dictatorship under Yeltsin and Putin* (New Haven: Yale University Press, 2016).
66. Some Russians, such as former Kremlin spinmeister Gleb Pavlovskiy, have described Putin's agenda in precisely these terms. Following his lead, so also has Dawisha. See Karen Dawisha, *Putin's Kleptocracy: Who Owns Russia?* (New York: Simon & Schuster, 2014), 9–10, 34, 35, 166, 257.
67. Kimberely Marten, *Warlords: Strong-Arm Brokers in Weak States* (Ithaca: Cornell University Press, 2012).
68. John O'Loughlin et al., "Russian Geopolitical Culture," 323.
69. John O'Loughlin, Gearóid Ó Tuathail, and Vladimir Kolossov, "Russian Geopolitical Storylines and Public Opinion in the Wake of 9-11," *Communist and Post-Communist Studies* 37, no. 3 (2004): 281–318.
70. "A 'Risky Westward Turn'? Putin's 9-11 Script and Ordinary Russians," *Europe Asia Studies* 56, no. 1 (2004): 3–34.
71. A number of these enemies, figures like Boris Berezovsky and his employee Alexander Litvinenko, were former insiders who ended up dying in suspicious and spectacular ways. See Steve LeVine, *Putin's Labyrinth: Spies, Murder and the Dark Heart of the New Russia* (New York: Random House, 2008).
72. Gearóid Ó Tuathail, "Placing Blame: Making Sense of Beslan," *Political Geography* 28 (2009): 4–15.
73. Vladimir Putin, "Address by President Vladimir Putin," September 4, 2004, http://en.kremlin.ru/events/president/transcripts/22589.
74. Solzhenitsyn, *Rebuilding Russia*.

75. Vladimir Putin, "Speech at the 43rd Munich Conference in Security Policy, http://www.washingtonpost.com/wp-dyn/content/article/2007/02/12/AR2007021200555.html.

Chapter 3

1. Interview with Damon Wilson, Washington, DC, April 13, 2011.
2. George W. Bush, "Address at Warsaw University," *Public Papers of the Presidents: George W. Bush*, June 15, 2001.
3. The official U.S. policy, the Welles Declaration of July 23, 1940, was nonrecognition of the military occupation of the Baltic States by the Soviet Union in June 1940. It remained official policy until the Baltic states regained their independence.
4. Georgia and Ukraine were geographic areas within the Russian Empire that, like many other areas that later became states, lacked precise territorial definition. The proclaimed borders of the Democratic Republic of Georgia (1919–1921) were more expansive than those established by Bolshevik rule. The Soviet Union established in December 1922 had four founding members. Soviet Socialist Republics were based on a specified Russia, Ukraine, Byelorussia, and a Transcaucasian Socialist Federative Soviet Republic. The latter lasted until 1936 at which time Armenia, Georgia, and Azerbaijan were recognized as separate Soviet republics. For a map that contrasts the borders of the Democratic Republic with those of contemporary Georgia, see Map 1 in *The Making of Modern Georgia, 1918–2012*, ed. Stephen F. Jones (London: Routledge, 2014).
5. Ronald D. Asmus, *A Little War That Shook the World* (New York: Palgrave Macmillan, 2010), 124.
6. Ibid., 127.
7. Ibid., 129–134.
8. Bucharest Summit Declaration Issued by the Heads of State and Government participating in the meeting of the North Atlantic Council in Bucharest on April 3, 2008, http://www.nato.int/cps/en/natolive/official_texts_8443.htm.
9. The phrase "empire of liberty" was first used by Thomas Jefferson in 1780.
10. David S. Foglesong, *The American Mission and the "Evil Empire": The Crusade for a "Free Russia" since 1881* (New York: Cambridge University Press, 2007).
11. Campbell Craig and Fredrik Logevall, *America's Cold War: The Politics of Insecurity* (Cambridge, MA: Belknap Press of Harvard University Press, 2009).
12. It was also an inaccurate summary of the historical record. See Serhii Plokhy, *Yalta: The Price of Peace* (New York: Viking, 2010).
13. John Foster Dulles, "A Policy of Boldness," *Life*, May 19, 1952.
14. Charles King, "Happy Captive Nations Week!," *Slate*, July 24, 2014.

15. Laszlo Borhi, "Rollback, Liberation, Containment, or Inaction? US Policy and Eastern Europe in the 1950s," *Journal of Cold War Studies* 1, no. 3 (1999).
16. Chris Tudda, *The Truth Is Our Weapon: The Rhetorical Diplomacy of Dwight D. Eisenhower and John Foster Dulles* (Baton Rouge: Louisiana State University Press, 2006).
17. Derek H. Chollet and James M. Goldgeier, *America between the Wars: From 11/9 to 9/11: The Misunderstood Years between the Fall of the Berlin Wall and the Start of the War on Terror* (New York: PublicAffairs, 2008), 101.
18. Anthony Lake, "From Containment to Enlargement," September 21, 1993, http://fas.org/news/usa/1993/usa-930921.htm.
19. Goldgeier, *Not Whether but When*.
20. The Washington Declaration, signed and issued by the Heads of State and Government participating in the meeting of the North Atlantic Council in Washington, DC, on April 23–24, 1999, http://clinton2.nara.gov/WH/New/NATO/statement2.html. For the 1999 Strategic Concept, see http://www.nato.int/cps/en/natohq/official_texts_27433.htm.
21. Cable News Network, Crossfire, "Never Again," April 22, 1993. See Gearóid Ó Tuathail, *Critical Geopolitics: The Politics of Writing Global Space* (Minneapolis: University of Minnesota Press, 1996), 210–211.
22. Cable News Network, *Larry King Live*, South Carolina Republican Debate, February 15, 2000. Transcript available at http://transcripts.cnn.com/TRANSCRIPTS/0002/15/lkl.00.html.
23. See John McCain and Mark Salter, *Worth the Fighting For: The Education of an American Maverick, and the Heroes Who Inspired Him* (New York: Random House, 2003). Matt Stearns, "How a Hemingway Hero Inspired McCain," *McClatchy DC*, November 26, 2007.
24. On McCain's rebellious streak see Robert Timberg, *The Nightingale's Song* (New York: Free Press, 1996). On the Luke Skywalker identification and its subsequent repudiation in McCain's 2007–2008 presidential campaign see Jonathan Chait, "McCain Goes over to the Dark Side," *Los Angeles Times*, March 10, 2007. In April 2016 McCain wrote an admiring obituary of the last veteran of the Abraham Lincoln Brigade, those Americans who volunteered to fight against the fascist-aligned government in the Spanish Civil War. John McCain, "Salute to a Communist," *New York Times*, March 24, 2016. In yet another instance of shared military romanticism, Russian nationalist Eduard Limonov claimed the same Spanish International Brigade heritage for his volunteer movement in the Donbas. See Anna Matveeva, "No Moscow Stooges: Identity Polarization and Guerilla Movements in Donbass," *Southeast European and Black Sea Studies* 16, no. 1 (2016): 35.
25. Norman Kempster, "In a Show of Support, Baker Visits Shevardnadze," *Los Angeles Times*, May 26, 1992.

26. Interview with former U.S. government official.
27. The geopolitics of this moment are complex and do not lend themselves to a simple "sphere of influence" contest. The Yeltsin administration was considering U.S. peacekeepers in Abkhazia because of its own financial constraints.
28. Ronald Grigor Suny, *The Making of the Georgian Nation*, 2nd ed. (Bloomington: Indiana University Press, 1994), 310. The remark was made at the 26th Party Congress in Moscow, and to be fair to Shevardnadze the sun he cited was not a geopolitical center but "the sun of Lenin's ideas."
29. Mikheil Sergeevich Gorbachev, *For a "Common European Home," for a New Way of Thinking: Speech by the General Secretary of the CPSU Central Committee at the Czechoslovak-Soviet Friendship Meeting, Prague, April 10, 1987* (Moscow: Novosti Press Agency, 1987); Eduard Shevardnadze, *The Future Belongs to Freedom* (New York: Free Press, 1991), 112.
30. One example was General Igor Rodionov, who was the commander in charge of the Soviet forces responsible for the murder of protesting Georgians on April 9, 1989. Rodionov's responsibility for the deaths was not established, and he later served for a while as Yeltsin's minister of defense.
31. For discussion of Georgian geopolitical narratives see Lincoln Mitchell, "Georgia's Story: Competing Narratives since the War," *Survival* 5, no. 4 (2009): 87–100.
32. Shevardnadze, *The Future Belongs to Freedom*, 162.
33. Per Gahrton, *Georgia: Pawn in the New Great Game* (London: Pluto Press, 2010), 4–6. Initially Georgia benefited because it was a northern transit outlet for a landlocked Armenia facing closed borders to its west (Turkey), east (Azerbaijan), and somewhat limited border crossings to the south with Iran. Armenian lobby organizations supported Georgia because it offered a route to Yerevan.
34. Fiona Hill, "A Not-So-Grand Strategy: U.S. Policy in the Caucasus and Central Asia since 1991," *Politique étrangère/Brookings Institute* (2001), https://www.brookings.edu/articles/a-not-so-grand-strategy-u-s-policy-in-the-caucasus-and-central-asia-since-1991/.
35. Ibid.
36. James Baker, "America's Vital Interest in the 'New Silk Road,'" *New York Times*, July 21, 1997. The newspaper was later forced to add an editor's note that Baker's law firm represented an oil consortium and other companies in the region.
37. Interview with William Courtney, U.S. Ambassador to Georgia 1995–1997, May 11, 2015.
38. Adrian Brisku, *Bittersweet Europe: Albanian and Georgian Discourses on Europe, 1878–2008* (New York: Berghahn Books, 2013).

39. Donnacha Ó Beachain and Frederik Coene, "Go West! Georgia's European Identity and Its Role in Domestic Politics and Foreign Policy Objectives," *Nationalities Papers* 42, no. 6 (2014): 923–941.
40. Tracey C. German, "The Pankiski Gorge: Georgia's Achilles' Heel in Its Relations with Russia?," *Central Asian Survey* 23, no. 1 (2004): 27–39.
41. Andrew Jack and David Stern, "Georgia Plans to Seek NATO Membership: FT Interview Eduard Shevardnadze," *Financial Times*, October 25, 1999.
42. Jim Hoagland, "Shevardnadze's Balancing Act," *Washington Post*, April 12, 2001; Patrick Cockburn, "We Are Doing Everything We Can to Avoid Getting Involved, Says the Grey-Haired Fox," *The Independent*, February 26, 2000.
43. Charles King, "Potemkin Democracy: Four Myths about Post-Soviet Georgia," *National Interest* (2001): 93–104.
44. In 1951, the U.S. Congress passed the Kertsen amendment, which appropriated $100 million for the recruitment of refugees from the Soviet bloc for military service. In May 1953 the Eisenhower administration approved the establishment of a refugee paramilitary group called the "Volunteer Freedom Corps" for potential covert operations behind the Iron Curtain. Strong resistance from Western European states prevented the program from ever becoming fully operational as envisaged. The United States, nevertheless, did conduct various covert operations within the Soviet bloc in the 1950s. See Borhi, "Rollback, Liberation, Containment, or Inaction? US Policy and Eastern Europe in the 1950s."
45. The scheduled principal for the meeting was Mira Ricardel, then deputy assistant secretary of defense for international security, but she was delayed by the Thanksgiving holiday.
46. Two years later the Turkish government refused to allow the U.S. military to stage a northern invasion of Iraq from its territory. For a discussion of Central Asia and bases see Alexander Cooley, *Great Games, Local Rules: The New Great Power Contest in Central Asia* (New York: Oxford University Press, 2012).
47. U.S. ambassador Miles got a taste of this displeasure when he met with Yevgeny Primakov in Bulgaria before his departure for Georgia. Primakov, who knew Miles from earlier in their careers, declared: "What are you [the United States] doing? Don't you realize how close this is to our territory, to our borders?" Interview with Ambassador Richard M. Miles, May 30, 2013.
48. Vladimir Putin, "President Vladimir Putin Met with Georgian President Eduard Shevardnadze," March 1, 2002, http://en.kremlin.ru/events/president/news/28278.
49. Steven Lee Myers, "Georgia Hearing Heavy Footsteps from Russia's War in Chechnya," *New York Times*, August 15, 2002.

50. Civil Georgia, "Kremlin Says Georgia an 'Enclave of Terrorism,'" *Civil.ge*, August 24, 2002.
51. Interview with U.S. ambassador Richard M. Miles. See also The Association for Diplomatic Studies and Training Foreign Affairs Oral History Project interview with Ambassador Miles, 257, available at http://adst.org/oral-history/oral-history-interviews/.
52. Jean-Christophe Peuch, "Georgia: Shevardnadze Officially Requests Invitation to Join NATO," *Radio Free Europe/Radio Liberty*, 2002.
53. Civil Georgia, "Timeline—2003," *Civil.ge*, 2003.
54. Ibid.
55. This phrase was used by U.S. secretary of the Navy and Marine Corp Ray Mabus during a speech at the Georgian National Independence Day celebration in Washington, June 2, 2015, attended by the author. On May 29, 2012, the Georgian Embassy hosted a reception for its independence day at the Ronald Reagan Building on Pennsylvania Avenue. Guests were welcomed at a receiving line by then ambassador Timuri Yakobashvili and his wife, at the end of which was a war-disabled Georgian soldier unable to shake hands. In this way the invitees from Washington society were confronted with the physical bodily sacrifices made by Georgian soldiers fighting alongside U.S. forces in Iraq and Afghanistan.
56. Lincoln Mitchell, *Uncertain Democracy* (Philadelphia: University of Pennsylvania Press, 2008). The very name "Rose Revolution" was a coinage of CNN's television coverage of the protests. The idea of marching with roses to visually signify nonviolent intent is attributed by some to Mark Mullen, then head of the National Democratic Institute in Georgia. The CNN coinage was quickly appropriated by participants and became the preferred "export brand" of their successful challenge to electoral fraud. See Foreign Affairs Oral History Project interview with Ambassador Richard Miles, 2015, 271; Paul Manning, "Rose-Colored Glasses? Color Revolutions and Cartoon Chaos in Postsocialist Georgia," *Cultural Anthropology* 22, no. 2 (2007): 171–213.
57. Saakashvili had first made this call publicly on February 14, 2002, in front of Shevardnadze at a meeting of the Tbilisi City Council, where Saakashvili served as chair.
58. White House, "President Bush Welcomes Georgian President Saakashvili to White House," February 25, 2004, https://georgewbush-whitehouse.archives.gov/news/releases/2004/02/20040225-1.html.
59. Foreign Affairs Oral History Project interview with Ambassador Richard Miles, 2015, 274, 277.
60. Tom Shankar, "Rumsfeld Visits Georgia to Bind a Partnership with an Ally," *New York Times*, December 6, 2003.
61. Angus Roxburgh, *The Strongman: Vladimir Putin and the Struggle for Russia*, 2nd ed. (London: I. B. Tauris, 2013), 111.

62. From 1990 to 2004 the Georgian state flag—dark red with an upper-left black-and-white inset—was a revival of that used by the first Democratic Republic of Georgia in 1918–1919. Because this government was Menshevik, anti-Communist Georgians agitated for the adoption of a medieval Christian flag associated with Queen Tamar (who ruled from 1184–1213). The opposition party Saakashvili created in late 2001, the United National Movement, adopted this preferred alternative—which had actually been adopted by parliament in 1999 but never signed into law by Shevardnadze—as its party flag, a five-cross flag featuring a central red cross on a white background and four smaller red crosses in each quadrant. After Shevardnadze's resignation and days before Saakashvili's inauguration, this flag became Georgia's new state flag. Georgia displayed the flag of Europe (twelve yellow circles of stars on an azure background) as a member of the Council of Europe, though it is better known as the flag of the European Union. On April 16, 2004, the Georgian parliament passed a law mandating display of the flag of Europe next to the Georgian flag on all government buildings.
63. Text of the official English translation of Saakashvili's inauguration address, January 25, 2004. Mikheil Saakashvili, "Inauguration Address," January 14, 2004, http://www.civil.ge/eng/article.php?id=26694.
64. Civil Georgia, "Saakashvili's Vows Improvements with Drastic Measures," *Civil.ge*, January 25, 2004.
65. Sophie Lambrochini, "Georgia: Moscow Watches Warily as Saakashvili Comes to Power," *RFE/RL*, January 6, 2004.
66. Jean-Christophe Peuch, "Georgia: Saakashvili Sees in 'Wahabbism' a Threat to Secularism," *RFE/RL*, February 18, 2004.
67. In January 2003 Zhvania and Saakashvili joined forces to create a broad opposition to Shevardnadze. In an interview soon after, Zhvania discussed the principles shared by the new opposition. During the course of the interview he said: "I highly appreciate President [Vladimir] Putin's governing style, who made Russia's state policy, both internal and external, much better organized and firm." Saakashvili also publicly praised Putin, and indeed was described by some as Georgia's Putin. Civil Georgia, "Zurab Zhvania Speaks to the Readers of Civil Georgia," *Civil.ge*, February 11, 2003.
68. Cited in Roxburgh, *Strongman*, 114–116.
69. Thomas De Waal, "So Long, Saakashvili. The Presidency That Lived by Spin—and Died by It," *foreignaffairs.com*, October 29, 2013. The source of the observation is Georgia's Foreign Minister Tedo Japaradize.
70. White House, "President Bush Welcomes Georgian President Saakashvili to White House," February 25, 2004, https://georgewbush-whitehouse.archives.gov/news/releases/2004/02/20040225-1.html.
71. The language of paternalism is a commonplace in the history of U.S. foreign policy, starting with the notion of Uncle Sam. Ron Asmus writes

of this period: "Saakashvili soon became a poster child for the Bush Administration's 'freedom agenda' and democracy promotion efforts." See Michael H. Hunt, *Ideology and US Foreign Policy* (New Haven: Yale University Press, 1987).
72. Interview with Matt Bryza, September 16, 2013.
73. Georgia was ranked 94th in the 2004 worldwide index of press freedom issued by the Paris-based lobby group for media rights *Reporters Sans Frontieres* (Reporters without Borders). In 2003 Georgia was ranked 73rd. Civil Georgia, "Georgia Falls Back in Media Freedom Index," *Civil.ge*, October 27, 2004.
74. Ilan Greenberg, "The Not-So-Velvet Revolution," *New York Times Magazine*, May 30, 2004.
75. Mitchell, *Uncertain Democracy*.
76. Civil Georgia, "Pace Approved Resolution on Georgia," *Civil.ge*, January 24, 2005.
77. Interview with Ambassador Richard Miles, May 30, 2013.
78. Interview with F. Stephen Larabee, February 11, 2014.
79. Marc Perelman, "Israel's Military on Display in Georgia," *The Forward*, September 11, 2008. This article claims that Georgia purchased approximately $300 million worth of weapons from Israel between 2000 and 2008. See also Noah Shachtman, "How Israel Trained and Equipped Georgia's Army," *Wired*, August 19, 2008.
80. Arnaud de Borchgrave, "Israel of the Caucasus?," *Washington Times*, September 4, 2008. This claim need to be treated with extreme caution since the quality of de Borchgrave's reporting has been questioned.
81. Saakashvili visited Israel in November 2006 and in a series of speeches drew parallels between Georgia and Israel. At the University of Haifa, where he received an honorary degree, he argued that Georgians and Jews have much in common. "We understand each other well and understand what it means to be persecuted." BBC Worldwide Monitoring, "Saakashvili Draws Parallel between Russia's 'Persecution' of Georgians, Jews," *Imedi TV*, November 1, 2006. In Jerusalem he declared: "The Jews used to say for centuries that they should not forget Jerusalem, because living without it is equal to being without their right hand. The same is true with Abkhazia and Sukhumi for the Georgians. Georgia without Abkhazia is a country without its right hand, while we need that right hand very much." Civil Georgia, "Saakashvili Meets Israeli Leadership," *Civil.ge*, November 2, 2006.
82. Shlomo Avineri, "David and Goliath in the Caucasus," *Haaretz*, May 4, 2010.
83. Sam Schueth, "Assembling International Competitiveness: The Republic of Georgia, USAID, and the Doing Business Project," *Economic Geography* 87, no. 1 (2011): 51–77.

84. European Stability Initiative, "Reinventing Georgia: The Story of a Libertarian Revolution," Brussels 2010, http://www.esiweb.org/index.php?lang=en&id=322&debate_ID=3.
85. George W. Bush, "Second Inaugural Address" (2005), January 20, 2005, https://georgewbush-whitehouse.archives.gov/news/releases/2005/01/20050120-1.html.
86. George W. Bush, "President Addresses and Thanks Citizens in Tbilisi, Georgia," May 10, 2005, https://georgewbush-whitehouse.archives.gov/news/releases/2005/05/20050510-2.html.
87. Communications Office of the President of Georgia, "Remarks by President Bush and President Saakashvili of Georgia in a Joint Press Availability," May 10, 2005.
88. Interview with Ambassador Richard Miles.
89. Interview with Randy Scheunemann, September 12, 2016.
90. Ron Asmus, *A Little War That Shook the World*, 16.
91. See http://www.ontheissues.org/International/Barack_Obama_Foreign_Policy.htm#2008-SR439.
92. Wendell Steavenson, "Marching through Georgia," *New Yorker*, December 15, 2008.
93. One conference was in late October 2007, and Saakashvili waited until its foreign participants had left the country before unleashing the riot police against groups protesting his rule.
94. Interview with Ambassador James F. Jeffrey, May 15, 2015.
95. Asmus, *A Little War That Shook the World*, 134.
96. Gideon Rachman, "Neo-Georgian Architect; President Mikheil Saakashvili Shares Caviar, Kebabs and a Helicopter Ride with Gideon Rachman," *Financial Times*, April 26, 2008.
97. Asmus, *A Little War That Shook the World*, 135. There is no official text of Putin's remarks. A Russian-language transcript was published by the Ukrainian Independent Information Agency on April 16, 2008. I have used this as the record of his remarks. See http://www.unian.net/politics/110868-vyistuplenie-vladimira-putina-na-sammite-nato-buharest-4-aprelya-2008-goda.html.
98. Vladimir Putin, "Press Statement and Answers to Journalists' Questions Following a Meeting of the Russia-NATO Council," April 4, 2008, http://en.kremlin.ru/events/president/transcripts/24903.

Chapter 4

1. The recollections are translated quotations from Pliyev's diary, published soon afterward on a Russian-language website in Ukraine. The broader story is based on the author's interviews with him in South Ossetia in March 2010. Inal Pliyev, "Дневник Осетина: "Город Мертвых Птиц" (An Ossetian's Diary: "City of Dead Birds")," *Segodnya.ua*, August 13,

2008, http://www.segodnya.ua/newsarchive/dnevnik-ocetina-horod-mertvykh-ptits.html.
2. Ibid.
3. Ibid.
4. Ibid.
5. Mikheil Saakashvili, "The War in Georgia Is a War for the West," *Wall Street Journal*, August 11, 2008.
6. John McCain, "We Are All Georgians," *Wall Street Journal*, August 14, 2008.
7. Robert Gates, *Duty* (New York: Knopf), 168.
8. See Marcel H. Van Herpen, *Putin's Wars: The Rise of Russia's New Imperialism* (Lanham, MD: Rowman & Littlefield, 2014); Svante E. Cornell and S. Frederick Starr, *The Guns of August 2008: Russia's War in Georgia* (Armonk, NY: M. E. Sharpe, 2009).
9. An Iranian linguistic connection is the basis of the Ossetian claim to exclusive rights to the Alanian cultural and historical heritage in the Caucasus. Claim to this heritage is contested by the Caucasian-language-speaking Ingush and the Turkic-language-speaking Karachai-Balkars. In 1993 the Republic of North Ossetia added the word "Alania" to its official name to underscore this exclusive claim. See Arthur Tsutsiev, *Atlas of the Ethno-Political History of the Caucasus*, trans. Nora Seligman Favorov (New Haven: Yale University Press, 2014), 141–143.
10. The following draws upon Gearóid Ó Tuathail, "Russia's Kosovo: A Critical Geopolitics of the August 2008 War over South Ossetia," *Eurasian Geography and Economics* 49, no. 6 (2008): 670-705; Gerard Toal and John O' Loughlin, "Inside South Ossetia: A Survey of Attitudes in a De Facto State," *Post-Soviet Affairs* 29, no. 2 (2012): 136–172. For a study of the use of historical narratives in the South Caucasus see Oksana Karpenko and Jana Javakhishvili, eds., *Myths and Conflicts in the South Caucasus: Volume 1: Instrumentalization of Historical Narratives* (London: International Alert, 2003), http://www.international-alert.org/resources/publications/myths-and-conflict-south-caucasus.
11. Arsène Saparov, "From Conflict to Autonomy: The Making of the South Ossetian Autonomous Region 1818–1922," *Europe-Asia Studies* 62, no. 1 (2010): 99–123. See also Cory Welt, "A Fateful Moment: Ethnic Autonomy and Revolutionary Violence in the Democratic Republic of Georgia (1918–1921)," in *The Making of Modern Georgia, 1918–2012*, ed. Stephen F. Jones (New York: Routledge, 2014), 205–231.
12. For a brief memoir of the construction process see Alexander Cheldiev, "Транскам—трудный путь от мечты до воплощения" (A Difficult Path from Dream to Realization), *North Ossetia* (newspaper), May 12, 2009, http://kvaisa.ru/news/show/27/749/.
13. Ghia Nodia, "Political Crisis in Georgia," *Current Politics and Economics of Europe* 2, nos. 1/2 (1992): 32.

14. On the concept of the nation as geobody in the Georgian context see Peter Kabachnik, "Wounds That Won't Heal: Cartographic Anxieties and the Quest for Territorial Integrity in Georgia," *Central Asian Survey* 31, no. 1 (2012): 45–60.
15. Thornike Gordadze, "La Géorgeie Et Ses "Hôtes Ingrates," *Critique internationalie* 10, January (2001): 163–176.
16. Stephen F. Jones, "Georgia: A Failed Democratic Transition," in *Nations and Politics in the Soviet Sucessor States*, ed. Ian Bremmer and Ray Taras (Cambridge: Cambridge University Press, 1993), 288–310.
17. The impetus for Gamsakhurdia's march against South Ossetia was a declaration on November 10 by the SOAO Soviet requesting the Georgian Supreme Soviet to upgrade the region's status to that equivalent to Abkhazia, namely an Autonomous Republic. See Robert English, "'Internal Enemies, External Enemies': Elites, Identity, and the Tragedy of Post-Soviet Georgia," in *Russia and Eastern Europe after Communism*, ed. Michael Kraus and Ronald Liebowitz (Boulder: Westview, 1996), 207–222; Jonathan Aves, "The Rise and Fall of the Georgian Nationalist Movement, 1987–91," in *The Road to Post-Communism: Independent Political Movements in the Soviet Union, 1985–1991*, ed. Geoffrey Hosking, Jonathan Avers, and Peter J. S. Duncan (London: Pinter, 1992), 157–179; Darrell Slider, "The Politics of Georgia's Independence," *Problems of Communism* (November–December 1991): 63–79.
18. Alan Parastayev, "North and South. Ossetia: Old Conflicts and New Fears," in *The Caucasus: Armed and Divided; Small Arms and Light Weapons Proliferation and Humanitarian Consequences in the Caucasus*, ed .Anna Matveeva and Duncan Hiscock (N.p.: Saferworld, 2003), 1–17.
19. Elizabeth Fuller, "The South Ossetian Campaign for Unification," *Report on the USSR*, December 8, 1989.
20. Cory Welt, "Fear and Politics in Tskhinvali: South Ossetia's Late Soviet Secession from Georgia," *Nationalities Papers*, forthcoming.
21. Cited in Tim Potier, *Conflict in Nagorno-Karabakh, Abkhazia and South Ossetia: A Legal Appraisal* (The Hague: Kluwer Law International, 2001), 14.
22. Elizabeth Fuller, "South Ossetia: Analysis of a Permanent Crisis," *Report on the USSR*, February 15, 1991.
23. Ellen Barry, "Soviet Union's Fall Unravelled Enclave in Georgia," *New York Times*, September 6, 2008.
24. Human Rights Watch, "Bloodshed in the Caucasus: Violations of Humanitarian Law and Human Rights in the Georgia–South Ossetia Conflict," ed. Rachel Denber (New York: Human Rights Watch, 1992).
25. South Ossetian Ministry of the Press and Mass Communications, *Несломленные Градом—Unbent by Grad* (Rostov-on-Don: Uzhnaja Alania, 2009). UNHCR numbers estimate that between 40,000 and 100,000 Ossetians fled South Ossetia and Georgia proper at this time,

the bulk to North Ossetia. Some 10,000 ethnic Georgians and others of mixed ethnicity were displaced to Georgia, and 5,000 internally within South Ossetia. UNHCR, "Population Movements as a Consequence of the Georgian–South Ossetian Conflict," updated September 1, 2004. Cited by International Crisis Group, "Georgia: Avoiding War in South Ossetia" (Tbilisi: Brussels 2004), 6.

26. Thomas De Waal, *The Caucasus: An Introduction* (New York: Oxford University Press, 2010), 142–143. For a map of how this violence remade South Ossetia see Tsutsiev, *Atlas of the Ethno-Political History of the Caucasus*, 117.

27. Michael Schwirtz and Ellen Barry, "Russia Sends Mixed Signs on Pullout from Georgia," *New York Times*, August 20, 2008.

28. The Russian Law on Citizenship came into force in February 1992 and was subsequently modified in 1993 and 1995. Formulated as certain successor states promulgated exclusivist citizenship laws rendering resident-ethnic Russians stateless, it allowed stateless persons beyond the territory of the Russian Federation but within the territory of the former USSR to acquire Russian nationality by registration. Thousands of Ossetians, many of whom had been forcefully displaced to North Ossetia and had relatives there, did so though hard numbers on this are elusive. Arthur Tsutsiev, a geo-demography expert in Vladikavkaz, speculated that more than a quarter of South Ossetians already had Russian citizenship documents before the procedures were simplified in 2002 (email correspondence July 10, 2016). Some ethnic Russians, Armenians, and Abkhaz from Abkhazia also acquired Russian citizenship for travel and trading opportunities. The experienced South Ossetian observer and official Kosta Dzugayev estimated that by the end of 2002 over 90 percent of the residents of South Ossetia had become Russian citizens (Email correspondence, August 2016).

29. On May 16, 1996, both sides signed a memorandum on security and confidence-building measures in Moscow. On December 23, 2000, the Russian and Georgian governments reached agreement on economic reconstruction in the Georgian-Ossetian conflict zone and on the return of displaced persons. On December 22, 2001, the OSCE mission in Georgia and the European Commission signed an agreement on a €210,000 grant for measures to settle the Georgian-Ossetian conflict.

30. For a useful English-language summary of this important, neglected, and still murky event see RFE/RL Caucasus Report, October 12, 2001, http://www.rferl.org/content/article/1341915.html.

31. Gelayev had previously fought in 1992–1993 as a volunteer in the Confederation of Mountain Peoples of the Caucasus militia, under the command of Shamil Basayev, for the Abkhazian side. Now he was fighting for the Georgians.

32. Kokoity's 2001 election slogan, a slap at the older Chibirov, was: "Those who defended the republic ought to lead it." Alan Parastayev, "US Deployment in Georgia Angers South Ossetia," *Institute for War and Peace Reporting*, March 21, 2002, http://www.cilevics.eu/minelres/mailing_archive/2002-March/002106.html.
33. Kosta Dzugayev, "South Ossetia's President Clamps Down," *Institute for War and Peace Reporting*, July 4, 2003, https://iwpr.net/global-voices/south-ossetias-president-clamps-down.
34. For a discussion of the legal aspects of "passportization" see the Tagliavini Report, vol. 2, chap. 3. See also Scott Littlefield, "Citizenship, Identity and Foreign Policy: The Contradictions and Consequences of Russia's Passport Distribution in the Separatist Regions of Georgia," *Europe-Asia Studies* 61, no. 8 (2009): 1461–1482; Florian Mühlfried, "Citizenship at War: Passports and Nationality in the 2008 Russian-Georgian Conflict," *Anthropology Today* 26, no. 2 (2010): 8–13.
35. For this point of view see Agnia Grigas, *Beyond Crimea: The New Russian Empire* (New Haven: Yale University Press, 2016). For a different perspective see Vincent Ardman, "Documenting Territory: Passportisation, Territory, and Exception in Abkhazia and South Ossetia," *Geopolitics* 18 (2013): 682–704.
36. Parastayev, "US Deployment in Georgia Angers South Ossetia."
37. Kosta Dzugayev, "South Ossetia: President Builds Power Base," *Institute of War and Peace Reporting: Caucasus Reporting Service*, May 19, 2004, https://iwpr.net/global-voices/south-ossetia-president-builds-power-base.
38. Stephen Jones, *Georgia: A Political History since Independence* (London: I. B. Tauris, 2013).
39. Ghia Nodia, "Georgia's Identity Crisis," *Journal of Democracy* 6, no. 1 (1995): 104–116.
40. It should be noted that some Georgian nationalists still claimed Sochi and its surrounding territories.
41. Mikheil Saakashvili, "Speech at the Meeting with Members of Supreme Council of Abkhazia," September 10, 2004. Unfortunately the press releases and speeches of President Saakashvili are no longer available online. The author compiled a comprehensive collection before they were taken down after he left office. This is the basis for this and subsequent references.
42. "President Saakashvili Visits Zurab Zhvania School of Public Administration in Kutaisi," May 3, 2006.
43. See, for example, his speeches before the United Nations General Assembly, in 2005 for use of "annexation" and 2006 for use of "occupation." While these terms were used before 2008, they became commonplaces thereafter. "Remarks by President Mikheil Saakashvili at the 60th Session of the UN General Assembly," September 15, 2005;

"Remarks of Mikheil Saakashvili, President of Georgia, to 61st Annual United Nations General Assembly," September 22, 2006.
44. Civil Georgia, "President Signs National Accord," *Civil.ge*, January 26, 2004.
45. "Thousands Pay Tribute to the First President," *Civil.ge*, March 31, 2007.
46. Kunin is the son of Madeline May Kunin, a Democrat who was elected governor of the state of Vermont from 1985 to 1991. An exchange student to Georgia, he became political adviser to Zhvania and Saakashvili in December 2003, serving until after the 2008 war. His salary throughout this time was reportedly paid by the U.S. development agency USAID. (This agency also had other U.S. citizens working in various branches of the Georgian government.) Thereafter, he became a lobbyist for Saakashvili's government in Washington, DC. See Kate Weinberg, "Daniel Kunin Interview: Georgia's Alistair Campbell," *The Telegraph*, August 23, 2008; and Kevin Bogardus, "Georgia Builds up Its Lobbying, PR Efforts," *The Hill*, March 19, 2009.
47. See Julie A. George, *The Politics of Ethnic Separatism in Russia and Georgia* (New York: Palgrave Macmillan, 2009).
48. Civil Georgia, "Timeline 2004," *Civil.ge*, 2004. Saakashvili recalled that many people hugged and kissed him, speaking Georgian in response to his Russian. However, one man came up to him and told him he was violating South Ossetian sovereignty and that he should leave. See Mikheil Saakashvili, "President Saakashvili Addresses South Ossetia Conference," *BBC Monitoring*, July 10, 2005.
49. "Inauguration Address," January 14, 2004, http://www.civil.ge/eng/article.php?id=26694.
50. "Remarks on the Occassion of the 59th Session of the UN General Assemby," September 21, 2004, http://www.un.org/webcast/ga/59/21.html.
51. "Speech Delivered by President Mikheil Saakashvili at the Parade Dedicated to the Independence Day of Georgia," May 26, 2004.
52. Civil Georgia, "Saakashvili Says Economic Growth Will Help Restore Territorial Integrity," *Civil Georgia*, December 12, 2003.
53. Mikheil Saakashvili, "Georgian President Outlines Three-Stage Development Strategy at the News Conference," September 9, 2005.
54. Civil Georgia, "Saakashvili Vows Improvements with Drastic Measures," *Civil.ge*, January 25, 2004.
55. In defiance of the consensus of international diplomats, Saakashvili had pushed confrontation with Shevardnadze and gained power. In Adjara, he defied U.S. advice to move slowly and won. Also, Russia agreed to withdraw from its bases in Georgia. Victories seemed to belong to the bold, to risk takers.
56. This data is from the Stockholm International Peace Research Institute's database on world military spending. For a visualization of the rise

and subsequent fall from the 2007 high, see http://militarybudget.org/georgia/.
57. Mikheil Saakashvili, "Georgian President Mikheil Saakashvili Visits Senaki Military Base," April 2, 2007.
58. "Speech of the President of Georgia at a Meeting of the Government," August 15, 2006; "Address by President Saakashvili at the Charity Dinner," May 27, 2006.
59. See Salomé Jashi's film documentary, *The Leader Is Always Right* (2010).
60. International Crisis Group, "Georgia: Avoiding War in South Ossetia."
61. Saakashvili, "Speech Delivered by President Mikheil Saakashvili at the Parade Dedicated to the Independence Day of Georgia."
62. ITAR-TASS, "Georgian Leader Says It Was 'Mistake' to Abolish Breakaway Region's Autonomy," *ITAR-TASS*, June 12, 2004; Nino Khutsidze, "Q&A with MP Giga Bokeria over South Ossetia," *Civil.ge*, July 26, 2004.
63. International Crisis Group, "Saakashvili's Ajara Success: Repeatable Elsewhere in Georgia," in *Europe Briefing* (Tbilisi: n.p., 2004). It is likely that discussions between the United States and Russia at this time facilitated this outcome.
64. Civil Georgia, "Abashidze Flees Georgia," *Civil Georgia*, May 6, 2004.
65. These words were reported by Saakashvili himself and require independent verification. Roxburgh, *Strongman*, 118.
66. International Crisis Group, "Georgia: Avoiding War in South Ossetia"; Civil Georgia, "Russia Calls on Georgia to Pull Out Extra Troops from South Ossetia," *Civil Georgia*, July 13, 2004.
67. International Crisis Group, "Georgia: Avoiding War in South Ossetia."
68. Valeriy Dzutsev, "South Ossetians Fear War," *Institute for War and Peace Reporting*, June 16, 2004.
69. Civil Georgia, "Georgia Gives "Last Chance for Peace," in South Ossetia," *Civil Georgia*, August 19, 2004.
70. International Crisis Group, "Georgia: Avoiding War in South Ossetia," 14. For a comprehensive account of the 2004 violence see Cory Welt, "The Thawing of a Frozen Conflict: The Internal Security Dilemma and the 2004 Prelude to the Russo-Georgian War," *Europe-Asia Studies* 62, no. 1 (2010): 63–97.
71. One observer noted that Saakashvili stirs in Putin the same animus as "Fidel Castro does for U.S. politicians." Robert Legvold, "Introduction," in *Statehood and Security: Georgia after the Rose Revolution*, ed. Robert Legvold and Bruno Coppieters (Cambridge, MA: MIT Press, 2005), 19.
72. For details on this see Gearóid Ó Tuathail, "Placing Blame: Making Sense of Beslan," *Political Geography* 28 (2009): 4–15; and John O'Loughlin, Gearóid Ó Tuathail (Gerard Toal), and Vladimir Kolossov, "The Localized Geopolitics of Displacement and Return in Eastern Prigorodnyy

Rayon, North Ossetia," *Eurasian Geography and Economics* 49, no. 6 (2008): 635–669.

73. Former Ingushetia president and Beslan hostage negotiator Ruslan Aushev warned that an interethnic confrontation between the peoples of Ossetia and Ingushetia might destabilize the situation in the Caucasus on the whole. "This involves Georgia, South Ossetia, the Chechen republic, Dagestan, Kabardino-Balkaria, and so on." Interfax, "Former Ingush President Warns of Ethnic Clashes Following Beslan Crisis," *Interfax*, September 28, 2004. Also International Crisis Group, "Georgia: Avoiding War in South Ossetia," 9.

74. Shamil Basayev, the terrorist mastermind of Beslan, had previously fought against Georgia in the Abkhazian war, and was accused of war crimes there. The general disposition of the Putin administration, however, was to externalize its domestic terrorism problem onto neighboring states. See John Russell, *Chechnya: Russia's "War on Terror"* (New York: Routledge, 2007).

75. Giorgi Sepashvili, "CIS Summit Reveals Rift in Russo-Georgian Relations," *Civil Georgia*, September 17, 2004. A regular bus route between Sochi and Sukhum(i) was opened on September 23, 2004.

76. Saakashvili, "Remarks on the Occassion of the 59th Session of the UN General Assemby."

77. Civil Georgia, "Kokoev Visits Moscow, Warns over 'Georgian Incursion,'" *Civil Georgia*, December 8, 2004.

78. Jean-Christophe Peuch, "Ukraine: Regional Leaders Set up Community of Democratic Choice," *RFE/RL*, December 2, 2005.

79. Civil Georgia, "High-Profile Forum Hails Rose Revolution, Gives Advises [*sic*] to Georgian Leadership," *Civil Georgia*, November 23, 2005.

80. Peuch, "Ukraine: Regional Leaders Set Up Community of Democratic Choice."

81. Mikheil Saakashvili, "Georgian President Addresses Nation Ahead of Bush Visit," March 5, 2005.

82. Communications Office of the President of Georgia, "Remarks by President Bush and President Saakashvili of Georgia in a Joint Press Availability," May 10, 2005, http://www.prnewswire.com/news-releases/remarks-by-president-bush-and-president-saakashvili-of-georgia-in-a-joint-press-availability-54339307.html.

83. Mikheil Saakashvili, "Meeting of President Saakashvili with the Parliamentary Majority," April 18, 2006.

84. Civil Georgia, "Georgia Warns Russia to Improve Peacekeeping Operation," *Civil.ge*, October 11, 2005.

85. Civil Georgia, "South Ossetian Leader Pushes Joint Plan for Conflict Resolution,"*Civil.ge*, December 13, 2005.

86. Civil Georgia, "Georgian Chief Negotiator Comments on JCC Talks," *Civil.ge*, December 25, 2005.

87. Civil Georgia, "Saakashvili Says 'Blackmailer' Russia Sabotages Georgia," *Civil.ge*, January 22, 2006.
88. "Resolution on Peacekeepers Leaves Room for More Diplomacy," *Civil.ge*, February 16, 2006.
89. Tea Gularidze, "Russia Threatens Painful Blow to Georgia's Economy," *Civil.ge*, March 31, 2006.
90. For an account see Kimberely Marten, *Warlords: Strong-Arm Brokers in Weak States* (Ithaca: Cornell University Press, 2012).
91. United Nations Security Council, "Resolution 1716," October 13, 2006. For the text see https://en.wikisource.org/wiki/United_Nations_Security_Council_Resolution_1716.
92. Civil Georgia, "Signs of Status Quo Change in S. Ossetia," *Civil.ge*, November 14, 2006. See International Crisis Group, "Georgia's South Ossetia Conflict: Make Haste Slowly" (Tbilisi: International Crisis Group, 2007).
93. Sanakoev had served as prime minister under Lyudvig Chibirov. When Kokoity took over he was out of power and pursued business interests in North Ossetia before reinventing himself.
94. Civil Georgia, "Tbilisi Willing to Formalize S. Ossetia 'Alternative Government,'" *Civil.ge*, December 1, 2006.
95. International Crisis Group, "Georgia's South Ossetia Conflict: Make Haste Slowly."
96. Civil Georgia, "Tbilisi Allots Gel 6 Mln for S.Ossetia Infrastructure," *Civil.ge*, February 22, 2006. It is possible that some of this money came from international aid like the U.S. Millennium Challenge Corporation. The official U.S. attitude toward the Sanakoev project was not supportive. Deputy Secretary Matthew Bryza indicated to me he met Sanakoev only once in a restaurant and left immediately.
97. Civil Georgia, "Signs of Status Quo Change in S. Ossetia," *Civil.ge*, November 14; Civil Georgia, "Sanakoev Appointed as Head of S. Ossetia Administration," *Civil.ge*, May 10, 2007.
98. Civil Georgia, "Moscow Slams Tbilisi for Backing S. Ossetia Alternative Government," *Civil.ge*, December 5, 2006.
99. According to Georgian Ministry of Internal Affairs figures, from 1992 to August 8, 2008, 150 civilians and policemen died in the Tskhinvali region. Their figures on murders in Abkhazia were much higher: "991 cases of murder, 294 cases of physical injury, 1527 cases of kidnapping, 223 cases of setting fire to houses, 21 cases of rape, 716 attacks against representatives of the law enforcement bodies." See Parliament of Georgia, "Parliamentary Temporary Commission on Investigation of the Military Aggression and Other Acts of Russia against the Territorial Integrity of Georgia" (www.parliament.ge2009), 8.
100. Civil Georgia, "Water Dispute Becomes Political in S. Ossetia," *Civil.ge*, June 1, 2007.

101. RFE/RL, "Putin Urges 'Universally Applicable' Solution for Kosovo," *RFE/RL*, January 30, 2006.
102. Vladimir Putin, "Transcript of the Press Conference for the Russian and Foreign Media," January 31, 2006, http://en.kremlin.ru/events/president/transcripts/23412.
103. Vladimir Putin, "Interview with Newspaper Journalists from G8 Member Countries," June 4, 2007, http://en.kremlin.ru/events/president/transcripts/24313.
104. Ronald D. Asmus, *A Little War That Shook the World* (New York: Palgrave Macmillan, 2010), 106.
105. Vladimir Putin, "Kosovo 'Blows up' International Order, Sets 'Terrible Precedent,'" *Channel One TV*, February 22, 2008. See also "Transcript of Annual Big Press Conference," February 15, 2008, http://en.kremlin.ru/events/president/transcripts/24835. Putin's indignation is evident from the manner in which he delivers his remarks almost as if he were spitting. See the footage in the BBC and Brook Lapping Productions documentary *Putin, Russia and the West: War* (released 2011), minute 17:00 onward.
106. Interviews with General Yuri Baluyevskiy, chief of the Russian General Staff, 2004 to June 2008, and General Valery Zaparenko, acting chief of operations in the General Staff, in the video documentary *Lost Day*. Available at https://www.youtube.com/watch?v=JijZovlIvSk.
107. White House, President George W. Bush, "President Bush Meets with President Saakashvili of Georgia," March 19, 2008, available at https://georgewbush-whitehouse.archives.gov/news/releases/2008/03/20080319-4.html.
108. Richard Weitz, "Georgia: Saakashvili Hears Encouraging Words from United States on NATO Membership," *EurasiaNet.org*, March 20, 2008, available at http://www.eurasianet.org/departments/insight/articles/eav032108a.shtml.
109. Email communication with Fiona Hill, July 8, 2016.
110. International Crisis Group, *Georgia and Russia: Clashing over Abkhazia*, June 5, 2008.
111. For a detailed account of the events preceding the August War see the Tagliavini Report, vol. 2, chap. 5.
112. For the Russian perspective see President of Russia, "Vladimir Putin Had a Telephone Conversation with Georgian President Mikheil Saakashvili," April 21, 2008, http://en.kremlin.ru/events/president/news/44165.
113. For details see Mikhail Zygar, *All The Kremlin's Men: Inside the Court of Vladimir Putin* (New York: Basic Books, 2016), 155.
114. Dmitry Medvedev, "Interview with Russian and Georgian Media," *Voice of Russia*, August 5, 2011; and *Putin, Russia and the West: War*.
115. U.S. Department of State, "Remarks by Secretary Rice and Georgian President Saakashvili," July 10, 2008, http://iipdigital.usembassy.gov/

st/english/texttrans/2008/07/20080710161637gmnanahcub0.3613092 .html#ixzz3gdYPAl5u.
116. Medvedev, "Interview with Russian and Georgian Media."
117. This claim is part of the mysterious *Lost Day* video. See https://www .youtube.com/watch?v=JijZovlIvSk.
118. See Anton Lavrov, "Timeline of Russian-Georgian Hostilities in August 2008," in *The Tanks of August*, ed. Ruslan Puknov (Moscow: Center for the Analysis of Strategies and Technologies, 2010), 37–76.
119. See Gordon M. Hahn, "The Making of Georgian-Russian Five-Day August War: A Chronology, June–August 8, 2008" (2008). www .russiaotherpointsofview.com/files/Georgia_Russian_War_TIMELINE.doc.
120. Marc Champion and Andrew Osborn, "Smoldering Feud, Then War—Tensions at Obscure Border Led to Georgia-Russia Clash," *Wall Street Journal*, August 16, 2008. This report states these were "mostly off-duty policemen out fishing or swimming."
121. See http://www.kavkaz-uzel.ru/articles/139953/. On August 6 the Ukrainian Ministry of Defense acknowledged that it had nineteen Special Force personnel in Georgia conducting military exercises. See www .kavkaz-uzel.ru/newstext/news/id/1226686.html.
122. The number of Russian journalists in Tskhinval(i) before the outbreak of war has fueled a Western conspiracy theory that Russia had planned precisely when it would start the war and had predeployed journalists to cover it. See Cornell and Starr, *The Guns of August 2008*; Van Herpen, *Putin's Wars*.
123. Interview with former Georgian ambassador to the United States and Deputy Defense Minister Batu Kutelia, March 21, 2014.
124. Ibid.
125. Saakashvili interview, *Putin, Russia and the West: War*. Saakashvili later explained to journalists that his military advised that the tank column could only be stopped if Georgia moved in long-range artillery to target a vital bridge in central South Ossetia (the Gupta/Gufta Bridge). Taking Tskhinval(i) and linking up with the Georgian enclave to its north would allow it to confront the column. Saakashvili's critics view the column as a figment of his imagination. Marc Champion, "The Conflict in Georgia: Fighting Raises the Stakes of Embattled US Ally—Russian Assault May Be Designed to Topple President," *Wall Street Journal*, August 11, 2008.
126. Civil Georgia, "Georgia Decided to 'Restore Constitutional Order in S. Ossetia'—Mod Official," *Civil.ge*, August 8, 2008; Charles Clover et al., "Countdown in the Caucasus: Seven Days That Brought Russia and Georgia to War," *Financial Times*, August 27, 2008.
127. Georgia, "Parliamentary Temporary Commission on Investigation of the Military Aggression and Other Acts of Russia against the Territorial Integrity of Georgia," 32.

128. Saakashvili also told the U.S. Embassy he was "provoked by the South Ossetians." There is no mention of the Russians. See U.S. Embassy Cable, Tbilisi, August 8, 2008, https://wikileaks.org/plusd/cables/08TBILISI1341_a.html.
129. Mikheil Saakashvili, "Declaration of Universal Mobilization" (Tblisi: n.p., 2008).
130. Hans Mouritzen and Anders Wivel, *Explaining Foreign Policy: International Diplomacy and the Russo-Georgian War* (Boulder: Lynne Rienner, 2012), 61. For the details on the timeline switch see Nicolai Petro, "Crisis in the Caucasus: A Unified Timeline, August 7–16, 2008." Hahn, "The Making of Georgian-Russian Five-Day August War: A Chronology, June –August 8, 2008." The U.S. State Department began work on creating a timeline to support the Georgian storyline on the war but the project was set aside when the fact-checking process by junior analysts raised too many questions.
131. Georgia, "Parliamentary Temporary Commission on Investigation of the Military Aggression and Other Acts of Russia against the Territorial Integrity of Georgia." The most comprehensive Georgian government chronology (prepared for the Tagliavini-led investigation) is "Major Hostile Actions by the Russian Federation against Georgia in 2004–2007," http://euromaidanpress.com/2014/08/08/major-hostile-actions-by-the-russian-federation-against-georgia-in-2004-2007/. The U.S. State Department began work on a chronology but abandoned the project.
132. On Saakashvili's state of mind prior see U.S. Embassy Cable, Tbilisi, August 4, https://wikileaks.org/plusd/cables/08TBILISI1327_a.html.
133. C. J. Chivers, "Georgia Offers Fresh Evidence on War's Start," *New York Times*, September 15, 2008; Asmus, *A Little War That Shook the World*; Georgia, "Parliamentary Temporary Commission on Investigation of the Military Aggression and Other Acts of Russia against the Territorial Integrity of Georgia."
134. Chivers, "Georgia Offers Fresh Evidence on War's Start."
135. Independent International Fact-Finding Mission on the Conflict in Georgia, "Report. Volume I" (2009), 20.
136. Ibid., 23.
137. Israel ended its military training of Georgian forces two weeks before the conflict, reportedly after Putin called Israeli president Peres and demanded that they leave or else Russian-Israeli relations would suffer.
138. Mouritzen and Wivel, *Explaining Foreign Policy*, 71.
139. Asmus, *A Little War That Shook the World*, 79–80 and 143.
140. Niklas Nilsson, *Beacon of Liberty: Role Conceptions, Crises and Stability in Georgia's Foreign Policy, 2004–2012* (Uppsala: Uppsala Universitet, 2015), 188.
141. Georgia, "Parliamentary Temporary Commission on Investigation of the Military Aggression and Other Acts of Russia against the Territorial Integrity of Georgia."

142. Asmus, *A Little War That Shook the World*.
143. Ibid., 32.
144. See U.S. Embassy Cable, Tbilisi, August 8, https://wikileaks.org/plusd/cables/08TBILISI1337_a.html.
145. *Putin, Russia and the West: War*, 42:30.
146. Email correspondence with former Georgian foreign minister, Eka Tkeshelashvili, September 15, 2016.
147. Interview with Matt Bryza, September 16, 2013.
148. Ibid. See also Nilsson, *Beacon of Liberty*, 183.
149. Email correspondence, Eka Tkeshelashvili, September 15, 2016.
150. One Pentagon official stated: "The Georgians figured it was better to ask forgiveness later, but not ask for permission first. It was a decision on their part. They knew we would say 'no.'" Helene Cooper and Thom Shanker, "After Mixed US Signals, a War Erupts in Georgia," *New York Times*, August 13, 2008.
151. Asmus, *A Little War That Shook the World*, 36; and also *Putin, Russia and the West: War*.
152. Harretz staff, "Georgian Minister Tells Israel Radio: Thanks to Israeli Training, We're Fending Off Russian Military," *Haaretz*, August 11, 2008.
153. Nilsson, *Beacon of Liberty*.
154. U.S. Embassy Cable, Tbilisi, July 8, 2008, https://wikileaks.org/plusd/cables/08TBILISI1204_a.html.
155. Clover et al., "Countdown in the Caucasus"; Dan Fromkin, "From Green Light to Yellow," *Washington Post*, August 13, 2008.
156. U.S. Embassy Cable, Tbilisi, August 8, https://wikileaks.org/plusd/cables/08TBILISI1349_a.html.
157. William Kristol, "Will Russia Get Away with It," *New York Times*, August 10, 2008.
158. During the Crimea crisis of March 2014, Saakashvili's biggest defender, Senator McCain, conceded in an interview on Fox News that Saakashvili did make a mistake in August 2008. See http://www.georgianjournal.ge/politics/26639-in-georgia-mistake-was-made-by-saakashvili-mccain-about-2008-russia-georgia-war.html.
159. Cornell and Starr, *The Guns of August 2008*; Van Herpen, *Putin's Wars*.
160. Secretary Rice in her memoir describes the Russians as moving tanks into the Roki tunnel "thirty minutes after Georgia began its offensive." Condoleezza Rice, *No Higher Honor: A Memoir of My Years in Washington* (New York: Crown, 2011), 687.
161. Russian interests were not served particularly by war since it brought with it the danger of wider instability. The August War also meant it incurred significant reputational costs for its actions in 2008.
162. For an independent Russian assessment see Ruslan Pukhov and David Glantz, eds., *The Tanks of August* (Moscow: Centre for Analysis of Strategies and Technologies, 2008). For one U.S. assessment see Ariel

Cohen and Robert E. Hamilton, "The Russian Military and the Georgia War: Lessons and Implications" (Strategic Studies Institute, 2011), http://www.strategicstudiesinstitute.army.mil/pdffiles/pub1069.pdf.

163. Among others, the respected military analyst Mark Galeotti holds this view: "the Kremlin encouraged its local allies to provoke the Georgians." See Mark Galeotti, *Spetsnaz: Russia's Special Forces* (Oxford: Osprey Publishing, 2015), 39.

164. Mouritzen and Wivel, *Explaining Foreign Policy*, 72.

165. The view of U.S. ambassador Teft was that Russia used a series of "active measures" in Georgia to undermine Saakashvili and the country's geopolitical orientation toward the West. See U.S. Embassy Cable, Tbilisi, August 20, 2007, https://wikileaks.org/plusd/cables/07TBILISI1732_a.html.

166. One existential interest of the South Ossetians was to "lock in" Russian protection for their regime and power. In this sense, they, like the Georgian government, may well have believed that violent conflict would generate benefits for them in forcing the hand of their protector.

167. Interview with Republic of South Ossetia government officials, Tskhinval(i), March 2010. See also the interviews in the *Lost Day* video.

168. On this see U.S. Embassy Cable, Tbilisi, July 8, "Georgia Recalls Ambassador, Asks for Help in Protesting Russian Fighters," https://wikileaks.org/plusd/cables/08TBILISI1204_a.html.

169. For South Ossetian points of view on this period see the video *The Lost Day*, https://www.youtube.com/watch?v=JijZovlIvSk.

Chapter 5

1. The attack had echoes of that by Russian forces against the center of Grozny on October 21, 1999, that killed at least 140 people. See Human Rights Watch, "Evidence of War Crimes in Chechnya" (1999), https://www.hrw.org/news/1999/11/02/evidence-war-crimes-chechnya. Russian president Medvedev, when asked directly about the comparison, declared: "The difference is that Russia was not after the same objectives in Grozny as Georgia was in Tskhinval. We were pursuing a legitimate task of restoring order. We were not set on mass-killing our own people. We were fighting criminals: the people who defied a legitimate government, draping themselves with various slogans, from pseudo-Islamic notions to pure extremist propaganda. There was nothing of the kind in either South Ossetia or Abkhazia, since these two republics had long existed as self-proclaimed independent states which had their own governments and maintained some sort of law and order. These cases are essentially different." Dmitry Medvedev, "Interview with Russian and Georgian Media," *Voice of Russia*, August 5, 2011, http://sputniknews.com/voiceofrussia/2011/08/05/54227630/. The Georgian government told the IIFFMCG that their assault began with their artillery units firing smoke bombs, and fifteen minutes

later with regular ordinance. The interval was supposedly a humanitarian gesture to allow the civilian population enough time to leave the area and find shelter. See the IIFFMCG Report, Chapter 5, Volume II, 209.

2. The former British Army captain who was the senior OSCE representative in Georgia later stated that "the attack was clearly, in my mind, an indiscriminate attack on the town, as a town." C. J. Chivers and Ellen Barry, "Accounts Undercut Claims by Georgia on Russia War," *New York Times*, November 7, 2008.

3. Amnesty International, *Civilians in the Line of Fire: The Georgia-Russia Conflict* (London: Amnesty International, 2008), https://www.amnesty.org/en/documents/EUR04/005/2008/en/.

4. There is no record of what Bush and Putin said in the Great Hall. Bush's memoir suggests he was following protocol in calling Medvedev first, but a press briefing on August 10 by Deputy National Security Adviser Jim Jeffrey stated they did meet, a fact he confirmed to me.

5. George W. Bush, *Decision Points* (New York: Crown, 2010), 435.

6. Frank Rich, *The Greatest Story Ever Sold: The Decline and Fall of Truth from 9/11 to Katrina* (New York: Penguin Press, 2006).

7. Interview with James Jeffrey, May 15, 2015.

8. The phrase "hot-blooded" can be traced back to Shakespeare. It has a long history of use as a dichotomizing trope in gendered and racialized colonial discourses.

9. Lavrov, "Timeline of Russian-Georgian Conflict," 45. For an Ossetian account of the first days of battle by a participant see Kosta Dzugayev, Геноцидная война в Южной Осетии (The Genocidal War in Ossetia), http://iarir.ru/node/107.

10. This time is that revealed by the commander of the 58th Army in Vladikavkaz. Lt. General Anatoly Khrulev in a restrospective interview on events with the Russian military journalist Vladislav Shurigin, posted April 25, 2012, and available at http://shurigin.livejournal.com/347559.html. The initial intervention by Russian forces, thus, was a scenario order that did not involve President Medvedev. Commanders opened envelopes and began to operationalize the preplanned response. Whether these orders were in support of "peacekeeping" is a question that was never seriously debated given what followed.

11. Charles Clover et al., "Countdown in the Caucasus: Seven Days That Brought Russia and Georgia to War," *Financial Times*, August 27, 2008.

12. For accounts of the August War written soon thereafter that see it as a prepared Russian operation see Andrei Illarionov, "The Russian Leadership's Preparation for War, 1999–2008"; and Pavel Felgenhauer, "After August 7: The Escalation of the Russia-Georgia War," both in *The Guns of August 2008: Russia's War in Georgia*, ed. S. Cornell and F. Starr (Armonk: M. E. Sharpe, 2009). See also "Military Events of 2008," Chapter 5, Volume II of the Tagliavini Report. For Russian accounts see

Ruslan Pukhov, ed., *The Tanks of August* (Moscow: Centre for Analysis, Strategies and Technology, 2010); and Anatoliy Dimitrievitch Tsyganok, *Voina 08.08.08: Prinuzhdenie Gruzii k Miru* [*The War of 08.08.08: Forcing Georgia to Peace*] (Moscow: Vetche, 2011). For a more recent account based on extensive interviews with participants but with some questionable assertions see Frederic Labarre, "The Battle of Tskhinvali Revisited," *Small Wars Journal*, October 8, 2014.
13. Lavrov, "Timeline," 47.
14. Khrulev interview with Shurigin.
15. Documentary interview with Major General Marat Kulahmetov, *Lost Day*, 31:05.
16. Michael Gordon, "Pledging to Leave Georgia, Russia Tightens Its Grip," *New York Times*, August 18, 2008.
17. Andrew Kramer, "Despite Yielding Ground, Russia Takes Critical Spots," *New York Times*, August 21.
18. Human Rights Watch, "Clarification Regarding Use of Cluster Munitions in Georgia" (N.p.: Human Rights Watch, 2008).
19. Lavrov, "Timeline," 63.
20. Ibid., 59.
21. Ibid., 67.
22. See the testimonies recorded in "Initial Assessment of the Occupied Villages Adjacent to Tskhinvali Region, September 2008. Special report of the Public Defender of Georgia" (in Georgian), http://ombudsman.ge/old/files/downloads/ge/icmqdpmngfmytuhkqvez.pdf.
23. Lavrov, "Timeline," 66.
24. Email correspondence with Kosta Dzugayev, September 6, 2016.
25. Anne Barnard, Andrew Kramer, and C. J. Chivers, "Russians Push Past Separatist Area to Assault Central Georgia," *New York Times*, August 11, 2008.
26. Anne Barnard, "Georgia and Russia Nearing All-out War," *New York Times*, August 10, 2008.
27. Lavrov, "Timeline," 75.
28. IIFFMCG, Volume II, chapter 5, 216.
29. Human Rights Watch, "Georgian Villages in South Ossetia Burnt, Looted" August 12, 2008, https://www.hrw.org/news/2008/08/12/georgian-villages-south-ossetia-burnt-looted.
30. Human Rights Watch, "Georgia: Satellite Images Show Destruction, Ethnic Attacks," https://www.hrw.org/news/2008/08/27/georgia-satellite-images-show-destruction-ethnic-attacks. For a UNOSAT analysis of the damage to these villages based on a satellite image acquired on August 19, 2008, see http://www.unitar.org/unosat/maps/GEO.
31. See http://www.unitar.org/unosat/node/44/1282.
32. Sabrina Tavernise and Matt Siegel, "Signs of Ethnic Attacks in Georgia Conflict," *New York Times*, August 14, 2008.

33. Sabrina Tavernise, "Survivors in Georgia Tell of Ethnic Killings," *New York Times*, August 20, 2008; C. J. Chivers, "South Ossetian Martial Law Creates a No Man's Land," *New York Times*, August 21, 2008.
34. Jonathan Finer, "The Toll of the War in Georgia," *Washington Post*, August 20, 2008.
35. Phillip P. Pan and Jonathan Finer, "Russia Says Two Regions in Geogia Are Independent," *Washington Post*, August 27, 2008.
36. On primary metaphors see George Lakoff, *Moral Politics: What Conservatives Know That Liberals Don't* (Chicago: University of Chicago Press, 1996).
37. Clifford Levy, "Memories and Messages: Russia Prevailed on the Ground, but Not in the Media," *New York Times*, August 22, 2008.
38. T. J. Holmes, "CNN Newsroom," *Cable News Network*, August 9, 2008.
39. Wolf Blitzer, "Late Edition with Wolf Blitzer," Cable Network News, August 10, 2008.
40. Mikheil Saakashvili, "The War in Georgia Is a War for the West," *Wall Street Journal*, August 11, 2008.
41. Haaretz staff, "Georgian Minister Tells Israel Radio: Thanks to Israeli Training, We're Fending Off Russian Military," *Haaretz*, August 11, 2008.
42. Barnard, "Georgia and Russia Nearing All-out War."
43. John Roberts, "The Situation Room," *Cable Network News*, August 11, 2008.
44. Harry Smith, "The Early Show," *CBS News*, August 13, 2008.
45. Marc Champion, "The Conflict in Georgia: Fighting Raises the Stakes of Embattled US Ally—Russian Assault May Be Designed to Topple President," *Wall Street Journal*, August 11, 2008.
46. Barnard, "Georgia and Russia Nearing All-out War."
47. Tom Parfitt, Helen Womack, and Jonathan Steele, "Russia Brushes Aside Ceasefire Calls after Georgia Withdraws," *The Guardian*, August 11, 2008.
48. Peter Finn, "Georgia Retreats, Pleads for Truce; US Condemns Russian Onslaught," *Washington Post*, August 11, 2008.
49. Michael Abramowitz and Colum Lynch, "Bush, Cheney Increasingly Critical of Russia over Aggression in Georgia," *Washington Post*, August 11, 2008.
50. John Roberts, "American Morning," *Cable Network News*, August 13, 2008.
51. Ibid.
52. Glenn Beck, "Glenn Beck," Cable News Network, August 13, 2008.
53. Tom Foreman, "The Situation Room," *Cable News Network*, August 15.
54. Blitzer, "Late Edition with Wolf Blitzer;" Fareed Zakaria, "GPS," *Cable News Network*, August 31.
55. Steven Lee Myers, *The New Tsar: The Rise and Reign of Vladimir Putin* (New York: Alfred A. Knopf, 2015), 348–349.
56. Ibid., 349.

57. The creators of the documentary *A Lost Day*—officially identified as "Alfa studio in the city of Tver" —remain unknown. A member of the South Ossetian parliament, Kazimir Pliyev, claimed that the video was made by journalists from the Saint Petersburg–based Channel 5, but that channel denied it. A seven-minute version of the video was first posted on August 5, 2012, available at http://www.youtube.com/watch?v=OTXPbA9njCw under the telling headline "Medvedev's cowardice killed 1000 people." A forty-seven-minute version went online on August 7, 2012, at http://www.youtube.com/watch?v=sYQeeFXhOQw. Both postings were via newly created accounts with Ossetian-sounding names "Aslan Gudiyev" and "Alan Biragov," respectively. A version of the video with rough English subtitles was posted April 16, 2014. See https://www.youtube.com/watch?v=JijZovlIvSk.
58. Author's copy of the English-language transcript of *Lost Day*, translated by Emil Sanamyan.
59. These sentiments were expressed at the time. See Tom Parfitt and Alan Tskhurbayev, "War in the Caucasus: Refugees Flee South Ossetia after Georgian Attack," *The Guardian Weekly*, August 15, 2008.
60. The words are those of an unnamed Western diplomat quoting Putin in this meeting. Helen Cooper, "In Georgia Clash, a Lesson on US Need for Russia," *New York Times*, August 10, 2008.
61. Dmitry Medvedev, "Dmitry Medvedev Made a Statement on the Situation in South Ossetia," August 8, 2008, http://en.kremlin.ru/events/president/news/1043.
62. For discussion of the legal arguments cited to justify Russia's invasion see Nicolai N. Petro, "Legal Case for Russian Intervention in Georgia," *Fordham International Law Journal* 32, no. 5 (2008): 1524–1549; Roy Allison, *Russia, the West, and Military Intervention* (New York: Oxford University Press, 2011).
63. "Dmitry Medvedev Has Instructed the Emergency Situations Minister Sergei Shoigu to Fly to South Ossetia," August 9, 2008, http://en.kremlin.ru/events/president/news/1050.
64. Anne Barnard, "Georgia and Russia Nearing All-Out War," *New York Times*, August 11, 2008.
65. BBC Monitoring, "Russian PM Putin Decries 'Genocide' at Meeting with Refugees in North Ossetia," August 9, 2008.
66. Vladimir Putin, "Prime Minister Vladimir Putin Holds a Meeting in Connection with the Events in South Ossetia," August 9, 2008, http://archive.premier.gov.ru/visits/ru/6046/events/1683/. There is no official English translation of this speech. It can be seen at https://www.youtube.com/watch?v=EuHKgMowi-Y.
67. On March 21, 2008, the Duma had adopted a resolution calling on the Russian president and government to grant Abkhazia and South Ossetia recognition. On the extensive debate on the de facto states in Russia prior

to the August War see Ana K. Niedermaier, ed., *Countdown to War in Georgia: Russia's Foreign Policy and Media Coverage of the Conflict in South Ossetia and Abkhazia* (Minneapolis: East View Press, 2008).
68. Putin, "Prime Minister Vladimir Putin Holds a Meeting in Connection with the Events in South Ossetia."
69. Dmitry Medvedev, "Beginning of a Working Meeting with Prime Minister Vladimir Putin," August 10, 2008, http://en.kremlin.ru/events/president/transcripts/1052.
70. "Beginning of a Working Meeting with Chairman of the Russian Federation Prosecutor General's Office Committee of Inquiry Alexander Bastrykin," August 10, 2008, http://en.kremlin.ru/events/president/transcripts/1054.
71. Kremlin, August 10, 2008, http://en.kremlin.ru/events/president/news/1056.
72. They also appeared in the Western media. See Marc Champion and Andrew Osborn, "Georgia Routed as Peace Bid Fails," *Wall Street Journal*, August 12, 2008.
73. Ian Traynor, "Surrender or Else, Russia Tells Georgia," *The Guardian*, August 13, 2008.
74. The de facto Republic of South Ossetia later produced an English-language propaganda book on the war with this title.
75. Dmitry Medvedev, "Opening Remarks at a Meeting with Leaders of the Parliamentary Factions in the State Duma," August 11, 2008, https://www.youtube.com/watch?v=9Dy5QK9ihkA.
76. Vladimir Putin, "Prime Minister Vladimir Putin Chaired a Government Presidium Meeting," May 6, 2008, http://archive.premier.gov.ru/eng/events/news/1648/. Putin misattributes this quote to Reagan. It has long been attributed to Franklin Delano Roosevelt as his description of the U.S.-educated and U.S.-Marines-supported Nicaraguan dictator Anastasio "Tacho" Somoza García (1896–1956). However, there is no documentary evidence FDR actually said this about Somoza.
77. Ben Hall and Quentin Peel, "French President 'Soothed' Putin's Rage," *Financial Times*, November 14, 2008.
78. Dmitry Medvedev, "Press Statement Following Negotiations with French President Nicholas Sarkozy," August 12, 2008, http://en.kremlin.ru/events/president/transcripts/1072.
79. Ramesh Thakur, *The United Nations, Peace and Security: From Collective Security to the Responsibility to Protect* (United Nations University, Tokyo: Cambridge University Press, 2005).
80. Sergey Lavrov, "America Must Choose between Georgia and Russia," *Wall Street Journal*, August 20, 2008.
81. Ibid.
82. Medvedev, "Opening Remarks at a Meeting with Leaders of the Parliamentary Factions in the State Duma." Appeasement was "something

the western countries tried 70 years ago." The Nazi-Soviet pact, unsurprisingly, was not mentioned.
83. The Georgian parliament website was hacked early in the crisis and a picture of Saakashvili as Hitler posted there, presumably by Russian state hackers.
84. Medvedev, "Press Statement Following Negotiations with French President Nicholas Sarkozy."
85. Medvedev, "Statement on the Situation in South Ossetia at the Meeting with Veterans of the Battle of Kursk," August 18, 2008, http://en.kremlin.ru/events/president/transcripts/1121.
86. The gold star Hero of Russia award was established in 1992 as an equivalent to the highest award of the Soviet era, Hero of the Soviet Union. Peacekeepers were also awarded the highest military decoration, the Order of Saint George, first established in 1769 and revived by Putin in 2000.
87. Phillip Kennicott, "Gergiev's Russian Overture: A Symphony of Sympathies," *Washington Post*, August 23, 2008. In May 2016 Gergiev and male members of the Mariinsky Orchestra played in the ancient ruins of Palmyra, Syria, as a soft power face of Russia's military intervention there. See John Thornhill, "Orchestral Maneuvers," *Financial Times*, October 17, 2016.
88. Gerard Toal and John O' Loughlin, "Inside South Ossetia: A Survey of Attitudes in a De Facto State," *Post-Soviet Affairs* 29, no. 2 (2012): 136–172.
89. Dmitry Medvedev, "Interview with CNN," August 26, 2008, http://en.kremlin.ru/events/president/transcripts/1227.
90. "Interview with BBC Television," August 26, 2008, http://en.kremlin.ru/events/president/transcripts/1228.
91. "Interview with TV Channel Russia Today," August 26, 2008, http://en.kremlin.ru/events/president/transcripts/1226.
92. RFE/RL, "Kokoity Says South Ossetia Will Become Part of Russia," *RFE/RL*, September 11, 2008.
93. Dmitry Medvedev, "Interview Given by Dmitry Medvedev to Television Channels Channel One, Rossia, NTV," August 31, 2008, http://en.kremlin.ru/events/president/transcripts/48301.
94. Medvedev's remarks were viewed by critics as another Freudian slip by the Russian leadership, a revelation of the "real motive" behind the August War. See Brian Whitmore, "Medvedev Gets Caught Telling the Truth," *RFE/RL*, November 22, 2011, http://www.rferl.org/content/medvedev_gets_caught_telling_the_truth/24399004.html. But these remarks merely underscored the entwined nature of the affective and strategic in Russian geopolitical thinking. They also came before a presidential election where Putin and not he was the candidate. For Medvedev's remarks (in Russian) see http://kremlin.ru/events/president/news/13605.
95. Stephen Benedict Dyson, "What Russia's Invasion of Georgia Means for Crimea," *Washington Post*, March 5, 2014. The staffer in question, Mark

Simakovsky, served in the U.S. Defense Department until 2015 when he left to become a lobbyist.

96. Kevin Garside, "Beijing Olympics: While Georgia Burned George W. Bush Played Volleyball," *Daily Telegraph*, August 9, 2008. Charles Krauthammer, "Making Putin Pay," *Washington Post*, August 14, 2008.
97. Russia's Move into Georgia, Statement by Secretary Condoleezza Rice, Washington, DC, August 8, 2008, https://2001-2009.state.gov/secretary/rm/2008/08/108083.htm.
98. Medvedev, "Interview with Russian and Georgian Media."
99. Peter Baker, *Days of Fire: Bush and Cheney in the White House* (New York: Doubleday, 2013), 603.
100. Press Briefing by Press Secretary Dana Perino and Senior Director for East Asian Affairs Dennis Wilder and Deputy National Security Advisor Ambassador James F. Jeffrey, August 10, 2008.
101. Interview with James F. Jeffrey.
102. Ben Feller, "Bush Says Violence in Georgia Is Unacceptable," *Washington Post*, August 11, 2008; Michael Abramowitz and Colum Lynch, "Bush, Cheney Increasingly Critical of Russia over Aggression in Georgia," *Washington Post*, August 11, 2008.
103. See also Condoleezza Rice, *No Higher Honor: A Memoir of My Years in Washington* (New York: Crown Publishers, 2011), 688. Rice's move poisoned what was already a fraught relationship with Lavrov.
104. UN Security Council, 63rd year, 5953rd meeting, August 10, 2008, http://www.un.org/press/en/2008/sc9419.doc.htm.
105. Interview with Bob Costas of NBC Sports in Beijing, August 11, 2008, http://www.presidency.ucsb.edu/ws/index.php?pid=78078&st=Bob+Costas&st1=.
106. John Bolton, "After Russia's Invasion of Georgia, What Now for the West," *Telegraph*, August 15, 2008.
107. Jon Ward, "Bush Camp Battles Critics on Georgia Response," *Washington Times*, August 18, 2008.
108. Rice, *No Higher Honor*, 688.
109. Rice records the question differently from Baker: "Are we prepared to go to war with Russia over Georgia?" Ibid., 689.
110. Baker, *Days of Fire: Bush and Cheney in the White House*, 604.
111. Wall Street Journal Editorial, "Bush and Georgia," *Wall Street Journal*, August 13, 2008.
112. Dan Eggen, "Et Tu, Wall Street Journal," *Washington Post*, August 18, 2008. "Setting the Record Straight: President Bush Has Taken Action to Ensure Peace, Security and Humanitarian Aid in Georgia," August 13, 2008.
113. White House, "President Bush Discusses Situation in Georgia, Urges Russia to Cease Military Operations," August 13, 2008, https://georgewbush-whitehouse.archives.gov/news/releases/2008/08/20080813.html.

114. Rice subsequently discovered the French peace implementation proposal had allowed the Russians a military presence in Gori as part of a fifteen-kilometer exclusion zone for their troops.
115. White House, "President Bush Discusses Situation in Georgia," August 13, 2008.
116. Interview with Randy Scheunemann, September 12, 2016.
117. Statement by John McCain on Russia's Aggression in Georgia, August 8, 2008, http://www.presidency.ucsb.edu/ws/?pid=90760.
118. For a discussion of the Obama campaign's response see James Mann, *The Obamians* (New York: Viking, 2012).
119. Interview with Randy Scheunemann, September 12, 2016.
120. McCain campaign statement as read by Jake Tapper, ABC's *This Week*, August 10, 2008.
121. See, for example, the appearance of McCain surrogate Bobby Jindal in ABC's broadcast *This Week*, August 10, 2008.
122. Interview with Randy Scheunemann.
123. See Statement by John McCain on the Crisis in Georgia, August 11, 2008, http://www.presidency.ucsb.edu/ws/?pid=90774; "John McCain Addresses the Crisis in Georgia," Town Hall, York, Pennsylvania, August 12, 2008, http://va4mccain.blogspot.com/2008/08/important-read-john-mccain-addresses.html; John McCain, "We are All Georgians," *Wall Street Journal*, August 14, 2008.
124. See the comments of Governor Bill Richardson (D-NM) on ABC's *This Week*, August 10, 2008, http://archives.politicususa.com/2008/08/10/Richarson-This-Week.html. See also Matthew Mosk and Jeffrey H. Birnbaum, "While Aide Advised McCain, His Firm Lobbied for Georgia," *Washington Post*, August 13, 2008. Scheunemann was ideologically consistent in his lobbying. McCain's national campaign manager was a former aide, Rick Davis, who left his office years earlier to become a lobbyist. He teamed up with Paul Manafort to create Davis-Manafort, a company that came to specialize in rehabilitating the reputations of tarnished figures in the West (like Mobuto Sese Seko, Jonas Savimbi, and Ferdinand Marcos). Among their clients were Rinat Akhmetov and Viktor Yanukovych, who hired them in 2005. Davis and Manafort connected McCain to the Russian oligarch, Oleg Deripaska, a Putin crony with interests in Montenegro. McCain reportedly spent his seventieth birthday on Deripaska's yacht in the Adriatic in August 2006. See Franklin Foer, "The Quiet American," *Slate*, April 28, 2016.
125. McFaul told a reporter later that after he and Obama had talked about why the Russian reaction to Georgia's use of force in South Ossetia was unjustified, Obama said: "It *is* about a big nation attacking a small nation, about sovereignty being violated." Nicholas Lemann, "Worlds Apart: Obama, McCain and the Future of Foreign Policy," *New Yorker*, October 13, 2008.

126. Michael Falcone, "Obama Emerges to Talk About Georgia," *New York Times*, August 11, 2008.
127. For a transcript of the September 26, 2008, debate see http://www.debates.org/index.php?page=2008-debate-transcript. For discussion of these comments in retrospect see Dylan Byers, "'Watch Ukraine... Crimea': Behind the Republican Prescience on Russia," *Politico*, March 4, 2014.
128. Steven Lee Myers, "White House Unveils $1 Billion Georgia Aid Plan," *New York Times*, September 3, 2008.
129. For details see Jim Nichol, "Russia-Georgia Conflict in August 2008: Context and Implications for U.S. Interests" (Washington, DC: Congressional Research Service, 2010), https://www.fas.org/sgp/crs/row/RL34618.pdf.
130. Parliament of Georgia, "Parliamentary Temporary Commission on Investigation of the Military Aggression and Other Acts of Russia against the Territorial Integrity of Georgia" (www.parliament.ge2009). See also Georgian Ministry of Foreign Affairs, "Russian Invasion of Georgia: Facts and Figures," December 5, 2008.
131. For consideration of the struggles and losses of this population in displacement see Elizabeth Cullen Dunn, "Humanitarianism, Displacement and the Politics of Nothing in Postwar Georgia," *Slavic Review* 73, 2 (2012): 287–306.
132. South Ossetian Ministry of the Press and Mass Communications, *Несломленные Градом—Unbent by Grad* (Rostov-on-Don: Uzhnaja Alania, 2009); *Georgian Operation "Clean Field": Genocide against the Ossetians*, 2009.
133. Simon Schuster, "Russia Lost 64 Troops in Georgia War, 283 Wounded," *Reuters*, February 21, 2009.
134. For video of this in December 2008 see https://www.youtube.com/watch?v=T5UoyWletok.
135. Shaun Walker, "Russian 'Borderisation': Barricades Erected in Georgia, Say EU Monitors," *The Guardian*, October 23, 2013; Maxim Edwards, "South Ossetia's Creeping Border," *Open Democracy Russia*, July 20, 2015, https://www.opendemocracy.net/od-russia/maxim-edwards/south-ossetia%27s-creeping-border.
136. "ICC Pre-Trial Chamber I authorises the Prosecutor to open an investigation into the situation in Georgia," January 27, 2016, https://www.icc-cpi.int/en_menus/icc/press%20and%20media/press%20releases/Pages/pr1183.aspx.
137. "The Prosecutor of the International Criminal Court, Fatou Bensouda, requests judges for authorisation to open an investigation into the Situation in Georgia," October 13, 2015, https://www.icc-cpi.int/en_menus/icc/press%20and%20media/press%20releases/Pages/pr1159.aspx.

Chapter 6

1. White House, "Press Conference by President Bush and Russian Federation President Putin," June 16, 2001, http://georgewbush-whitehouse.archives.gov/news/releases/2001/06/20010618.html.
2. Ronald D. Asmus, *A Little War That Shook the World* (New York: Palgrave Macmillan, 2010), 105.
3. *Putin, Russia and the West. Episode 3: War*.
4. RIA Novosti, "Georgia, Ukraine NATO Accession May Cause Geopolitical Shift—FM," *RIA Novosti*, June 7, 2006.
5. The original story, citing the unnamed official, was published by *Komersant* on April 8, 2008. "Блок НАТО разошелся на блокпакеты." Available at http://www.kommersant.ru/doc/877224. Among those who picked it up was James Marson, "Putin to the West: Hands Off Ukraine," *Time*, May 25, 2009. For a Ukrainian translation of Putin's remarks see "То що ж сказав Володимир Путін у Бухаресті? Більше читайте тут," April 18, 2008. Available at http://gazeta.dt.ua/POLITICS/to_scho_zh_skazav_volodimir_putin_u_buharesti.html. A Russian-language transcript was published by the Ukrainian Independent Information Agency the same day. See http://www.unian.net/politics/110868-vyistuplenie-vladimira-putina-na-sammite-nato-buharest-4-aprelya-2008-goda.html.
6. In rhetorical terminology of John Austin, Putin's speech was ostensibly a constative, a didactic description of how things are in the world. However, his constatives were also performative, speech acts that do something to the world, for two reasons. First, his speech was a history and geography lesson that was delivered as a warning. Second, his status as the president of Russia meant he was in a position to make his warning come true. See John L. Austin, *How to Do Things with Words* (Oxford: Oxford University Press, 1976).
7. That son was Lev Davidovich Bronshtein, aka Leon Trotsky, who was born in the small village of Bereslavka in Bobrynets rayon in what is now Kirovohrad oblast, north of Odesa. His phrase was originally about the dialectic.
8. Orest Subtelny, *Ukraine: A History*, 4th ed. (Toronto: University of Toronto Press, 2009), 484.
9. Gwendolyn Sasse, "The 'New' Ukraine: A State of Regions," *Regional and Federal Studies* 1, no. 3 (2010): 67–100; Ivan Katchanovski, "Regional Political Divisions in Ukraine in 1991–2006," *Nationalities Papers* 34, no. 5 (2006): 507–532; Lowell W. Barrington and Erik S. Herron, "One Ukraine or Many? Regionalism in Ukraine and Its Political Consequences," *Nationalities Papers* 32, no. 1 (2004): 53–86.
10. Dominique Arel, "Language Politics in Independent Ukraine—Towards One or Two State Languages?," *Nationalities Papers* 23, no. 3 (1995): 597–622.

11. Lowell W. Barrington and Regina Faranda, "Reexamining Region, Ethnicity, and Language in Ukraine," *Post-Soviet Affairs* 25, no. 3 (2009): 232–256.
12. John O'Loughlin, "The Regional Factor in Contemporary Ukrainian Politics: Scale, Place, Space, or Bogus Effect?," *Post-Soviet Geography and Economics* 42, no. 1 (2001): 1–33.
13. Barrington and Herron, "One Ukraine or Many? Regionalism in Ukraine and Its Political Consequences."
14. The results of Ukraine's 2001 census are available at http://www.ukrcensus.gov.ua.
15. Gwendolyn Sasse, *The Crimea Question: Identity, Transition, and Conflict* (Cambridge, MA: Harvard Ukrainian Research Institute, Harvard University Press, 2007).
16. Michael Specter, "Setting Past Aside, Russia and Ukraine Sign Friendship Treaty," *New York Times*, June 1, 1997.
17. Andrew Wilson, "The Donbas between Ukraine and Russia: The Use of History in Political Disputes," *Journal of Contemporary History* 30 (1995): 267.
18. On this distinctive sense of place as a mobilizable resource see Andrew Wilson, "The Donbas between Ukraine and Russia," and Andrew Wilson "The Donbas in 2014: Explaining Civil Conflict Perhaps, But Not Civil War," *Europe-Asia Studies*, 68, 4 (2016): 631–652.
19. Laada Bilaniuk, *Contested Tongues: Language Politics and Cultural Correction in Ukraine* (Ithaca: Cornell University Press, 2005).
20. Dominique Arel, "Demography and Politics in the First Post-Soviet Censuses: Mistrusted State, Contested Identities," *Population* 57, no. 6 (2002): 801–828; "Interpreting "Nationality" and "Language" in the 2001 Ukrainian Census," *Post-Soviet Affairs* 18, no. 3 (2002): 213–249.
21. Marlene Laruelle, "The 'Russian World': Russia's Soft Power and Geopolitical Imagination" (Washington, DC: Center on Global Interests, 2015).
22. Andrew Wilson, *The Ukrainians: Unexpected Nation*, 3rd ed. (New Haven: Yale University Press, 2009), 299.
23. Zbigniew Brzezinski, "The Premature Partnership," *Foreign Affairs*, 73, 3 (March/April 1994), 67–82.
24. Zbigniew Brzezinski, *The Grand Chessboard: American Primacy and Its Geostrategic Imperatives* (New York: Basic Books, 1997), 121.
25. Zbigniew Brzezinski, "Geopolitical Pivot Points," *Washington Quarterly*, 19, 4 (1996): 209–216.
26. Tatiana Silin, "ПОЛИТИЧЕСКАЯ ВОЛЯ В РАМКАХ ДОЗВОЛЕННОГО" ("Political Will Is Permitted"), *Dzerkalo Tyzhnia,* May 24, 2002, at http://gazeta.zn.ua/POLITICS/politicheskaya_volya_v_ramkah_dozvolennogo.html; Taras Kuzio,

"What Future for Russo-Ukrainian Relations," *Moscow Times*, June 10, 2002.

27. President of Russia, "Press Statement and Answers to Questions at a Joint News Conference with Ukrainian President Leonid Kuchma," May 17, 2002, available at http://en.kremlin.ru/events/president/transcripts/21598.
28. See Anatol Lieven, *Ukraine and Russia: A Fraternal Rivalry* (Washington, DC: U.S. Institute of Peace Press, 1999), 50.
29. Alexander J. Motyl, "Ukraine's Orange Blues, September 11, 2013, http://www.kyivpost.com/article/opinion/op-ed/alexander-j-motyl-ukraines-orange-blues-91009.html.
30. Kate Weinberg, "Daniel Kunin Interview: Georgia's Alistair Campbell," *The Telegraph*, August 23, 2008. Saakashvili's predictions on Russian designs on Crimea are discussed in a number of U.S. Embassy Tbilisi cables. See the report on a lunch with Ambassador Bass, October 19, 2009, https://wikileaks.org/cable/2009/10/09TBILISI1915.html; and the report on his discussion with U.S. assistant secretary of defense for International Security Affairs Alexander Vershbow in November 2009, available at https://www.wikileaks.org/plusd/cables/09TBILISI1965_a.html.
31. Russian prime minister Vladimir Putin interviewed by German ARD TV channel, http://archive.premier.gov.ru/eng/events/news/1758/.
32. For a selective video of his remarks see http://www.jamestown.org/press/events/video-russia-the-north-caucasus-after-the-sochi-olympics/. A full audio recording of his remarks is in the possession of the author.
33. Mikheil Saakashvili, "Ukraine Crisis: My Sense of Déjà Vu as the West Appeases Putin," *Financial Times*, March 3, 2014; "The West Must Not Appease Putin," *Washington Post*, March 6, 2014; "Let Georgia Be a Lesson for What Will Happen to Ukraine," *The Guardian*, March 14, 2014.
34. The term "Euromaidan" began as a Twitter hashtag and was adopted as the name of the protest movement.
35. Saakashvili appeared multiple times, in December 2003, and February 23, and March 1, 2014. McCain addressed the crowd on December 15, 2013, along with Democratic senator Chris Murphy.
36. Richard Sakwa, *Frontline Ukraine: Crisis in the Borderlands* (London: I. B. Tauris, 2015); Andrew Wilson, *Ukraine Crisis: What It Means for the West* (New Haven, CT: Yale University Press, 2014); Andrei P. Tsygankov, "Vladimir Putin's Last Stand: The Sources of Russia's Ukraine Policy," *Post-Soviet Affairs* 31, no. 4 (2015): 279–303; Kimberely Marten, "Putin's Choices: Explaining Russian Foreign Policy and Intervention in Ukraine," *Washington Quarterly* 38, no. 2 (20015): 189–204; Kathryn Stoner and Michael McFaul, "Who Lost Russia (This Time)? Vladimir Putin," *Washington Quarterly* (2015): 167–187.
37. Dmitri Trenin, "Russia's Breakout from the Post–Cold War System: The Drivers of Putin's Course" (Moscow: Carnegie Moscow Center, 2014).

38. Ivan Krastev, "Is Vladimir Putin Trying to Teach the West a Lesson in Syria?," *New York Times*, October 7, 2015; James Mann, *The Obamians: The Struggle inside the White House to Redefine American Power* (New York: Viking, 2012).
39. Some of this legislation was payback for the passage of the Magnitsky Act in the United States in December 2012. This act repealed the longstanding Jackson-Vanik restrictions on Russia but replaced them with individualized sanctions against those held responsible for the death of Sergei Magnitsky in a Moscow prison in 2009. See Bill Browder, *Red Notice: A True Story of High Finance, Murder, and One Man's Fight for Justice* (New York: Simon & Schuster, 2015).
40. Laruelle, "The 'Russian World.'"
41. Kazakhstan's president Nursultan Nazarbayev initially championed the idea of a Eurasian economic bloc in the early 1990s. While various trade and customs agreements were signed, the idea as a geopolitical vision languished until Putin started touting it as an alternative to the European Union in his third term as Russian president.
42. Taras Kuzio, *Ukraine: Democratization, Corruption and the New Russian Imperialism* (Santa Barbara: Praeger, 2015).
43. The role of far-right groups during the Euromaidan protests is highly contentious. See Volodymyr Ishchenko, "Far Right Participation in the Ukrainian Maidan Protests: An Attempt of Systematic Estimation," *European Politics and Society*, in press (2016): 1–20; Yyacheslav Likhachev, "The 'Right Sector' and Others: The Behavior and Role of Radical Nationalists in the Ukrainian Political Crisis of Late 2013–Early 2014," *Communist and Post-Communist Studies* 48, nos. 2–3 (2015): 257–271; Vyacheslav Likhachev, "The Far Right in the Conflict Between Russia and Ukraine," *Russie.Nei.Visions* no. 95 (2016), https://www.ifri.org/en/publications/notes-de-lifri/russieneivisions/far-right-conflict-between-russia-and-ukraine; Anton Shekhovtsov, "The Ukrainian Far Right and the Ukrainian Revolution," *Black Sea Link Yearbook*, 2014–15, https://www.academia.edu/28263649/The_Ukrainian_Far_Right_and_the_Ukrainian_Revolution.
44. Wilson, *Ukraine Crisis: What It Means for the West*, 81. See also Serhiy Leshchenko, "Yanukovych's Secret Diaries," *Euromaidan Press (Ukrainska Pravda)*, March 11, 2014.
45. Serhiy Kudelia, "The Donbas Rift," *Russian Politics and Law* 54, no. 1 (2016): 5–27.
46. Daniel Treisman, "Why Putin Took Crimea," *Foreign Affairs* 92, no. 3 (2016): 52.
47. Andrew Higgins, Andrew E. Kramer, and Steven Erlanger, "As His Fortunes Fell in Ukraine, a President Clung to Illusions," *New York Times*, February 23, 2014.
48. Treisman, "Why Putin Took Crimea," 50.

49. Mikhail Zygar, *All the Kremlin's Men*, 275.
50. Treisman, "Why Putin Took Crimea;" Andrew Weiss, "Putin the Improviser," *Wall Street Journal*, February 15, 2014.
51. Alexander Cooley and Volodymr Dubovyk, "Will Sevastopol Survive? The Triangular Politics of Russia's Naval Base in Crimea," *PONARS Eurasia Policy Memo No. 47*, December 2008.
52. Pavel Baev, "Russia Insists on Treating Sevastopol as an Open Question," *Eurasian Daily Monitor*, June 16, 2008. Taras Kuzio, "Strident, Ambiguous and Duplicitous: Ukraine and the 2008 Russia-Georgia War," *Demokratizatsiya: The Journal of Post-Soviet Democratization* 17, no. 4 (2009); Victor Yushchenko, "Georgia and the Stakes for Ukraine," *Washington Post*, August 25, 2008. Dominique Arel, "Ukraine since the War in Georgia," *Survival* 50, no. 6 (2008): 15–25.
53. When they first met Putin gave Zaldostanov a Russian flag to take with him on his group's ride to Sevastopol. In 2010 Putin went to Sevastopol to ride with the group, then to Novorossiisk (east from Crimea on the Black Sea) in 2011, and to Sevastopol again in 2012. Elisabeth Wood, "A Small Victorious War? The Symbolic Politics of Vladimir Putin," in *Roots of Russia's War in Ukraine*, ed. Elisabeth A. Wood et al. (Washington, DC: Kennan Institute, Woodrow Wilson Center, 2015), 97–129.
54. Zygar, *All the Kremlin's Men*, 279.
55. Ibid., 275
56. Ibid., 281.
57. The documentary had hyped promotion and a peak viewing broadcast slot, Sunday evening March 15. It can be viewed with English subtitles at https://slavyangrad.org/2015/03/26/crimea-the-way-back-home-subtitled-in-english-by-v-p-e/.
58. Yuriy Snegirev, "Russian Filmmaker Interviewed on Crimea TV Documentary," *Rossiyskaya Gazeta*, March 17, 2015.
59. Steven Lee Myers, *The New Tsar: The Rise and Reign of Vladimir Putin* (New York: Alfred A. Knopf, 2015).
60. Vladimir Putin, "Meeting with Designers of a New Concept for a School Textbook on Russian History," January 16, 2014, http://en.kremlin.ru/events/president/news/20071.
61. "Russia-EU Summit," January 28, 2014, http://en.kremlin.ru/events/president/news/20113.
62. "Vladimir Putin Answered Journalists' Questions on the Situation in Ukraine," March 4, 2014, http://en.kremlin.ru/events/president/news/20366.
63. "Address by President of the Russian Federation," March 18, 2014, http://en.kremlin.ru/events/president/news/20603.
64. The creation of certain affective conditions that demobilized and confused the enemy was the aim of U.S. "shock and awe" military campaigns in Afghanistan and Iraq. Russian military thinking sought to

do the same. The much-hyped "Gerasimov Doctrine" (after Russia's chief of general staff, Valery Gerasimov) of "hybrid warfare" was actually an effort to copy what Russia thought the Americans were already doing. See Michael Kofman, "Russian Hybrid Warfare and Other Dark Arts," *War on the Rocks*, March 11, 2016, http://warontherocks.com/2016/03/russian-hybrid-warfare-and-other-dark-arts/.

65. Zygar, *All the Kremlin's Men*, 277.
66. Anton Lavrov, "Russian Again: The Military Operation for Crimea," in *Brothers Armed: Military Aspects of the Crisis in Ukraine*, ed. Colby Howard and Ruslan Pukhov (Minneapolis: East View Press, 2014), 164.
67. It was a source of local resentment that Kyiv appointed Sevastopol's mayor. With Yanukovych disgraced, the existing mayor resigned his post. Chaly, a businessman whose company was headquartered in Moscow, was subsequently affirmed as mayor by an emergency meeting of the city council amid considerable tension in the city. Paul Sonne, "In Crimea, Backlash to Uprising Lifts Pro-Russia Leader," *Wall Street Journal*, February 25, 2014.
68. Myers, *The New Tsar*, 159–162.
69. Putin, "Vladimir Putin Answered Journalists' Questions on the Situation in Ukraine."
70. Belaventsev was a former Soviet naval intelligence officer who was expelled from the United Kingdom in 1985. He subsequently served as deputy of Russia's main arms trading company and then as head of Russia's emergency aid organization, Emercom, a unit within the Ministry for Emergency Situations, administered from 1991 to 2012 by Sergey Shoygu. On March 21, 2014, Belaventsev was appointed Plenipotentiary Representative of the President of the Russian Federation to the newly created Crimean Federal District. A corruption monitoring group has accused him of self-dealing in the distribution of Russian emergency aid. See https://www.occrp.org/en/investigations/4565-ex-spy-turned-humanitarian-helps-himself.
71. Andrew Higgins, "Grab for Power in Crimea Raises Secession Threat," *New York Times*, February 27, 2014.
72. Treisman, "Why Putin Took Crimea," 52.
73. See Natalia Kisileva, Andrei Mal'gin, Vadim Petrov, and Alexander Formanchuk. *Ethnopolticheskie protsess' v Krimu: Istorpricheskiy op't soveremenn'e problem' I perspepektiv' ikh resheniya (Ethnopolitical Processes in Crimea: The Historical Experience, Contemporary Challenges and Solution Prospects)* (Salta: Simferopol, 2015).
74. It states: "Alterations to the territory of Ukraine shall be resolved exclusively by the All-Ukrainian referendum."
75. See the interview with Girkin on Russian NiroMirTV available at https://www.youtube.com/watch?v=Go4tXnvKx8Y.
76. Putin, "Address by President of the Russian Federation."

77. Ukrainian officials were acutely aware of this. In a National and Defense Security Council meeting on February 28, Prime Minister Yatsenyuk argued against using force because: "Right after we do this, there will be a Russian statement 'On defending Russian citizens and Russian speakers who have ethnic ties with Russia.' That is the script the Russians have written, and we're playing to that script." Leonid Bershidsky, "Why Russia Stopped at Crimea," *Bloomberg View*, February 22, 2016, https://www.bloomberg.com/view/articles/2016-02-22/why-russia-stopped-at-crimea.
78. The language law was always contentious and the subject of two separate opinions by the Council of Europe's Venice Commission. See Dominique Arel, "Language, Status and State Loyalty in Ukraine," working paper.
79. Lavrov, "Russian Again: The Military Operation for Crimea," 161.
80. The docudrama *Crimea: The Road Home* reenacts such a scenario.
81. See, for example, the footage filmed in late February by Ruptly where the checkpoint barrier says "fascism not allowed," available at https://www.youtube.com/watch?v=HgPlInTNwGs.
82. Quoted in Myers, *The New Tsar: The Rise and Reign of Vladimir Putin*, 462.
83. Ukrainian prime minister Yatsenyuk described it as "the sovereign right of Ukraine on its choice of foreign policy vector."
84. Mark Beissinger, "Self-Determination as a Technology of Imperialism: The Soviet and Russian Experiences," *Ethnopolitics* 14, no. 5 (2015).
85. Farangis Najibullah, "Crimean Tatar Community Mourns Death of Tortured Local Activist," *RFE/RL*, March 18, 2014.
86. These were the numbers reported by the head of the Crimean parliament's commission on the referendum, Mikhail Malyshev, to the world's media. There were 1,233,002 votes "for" integration, with the total number of those who voted standing at 1,274,096 people. Malyshev stressed that the referendum commission has not received any complaints about the vote.
87. Color schemes, as noted in chapter 1, are important anchors of affective geopolitics. In 2005, in response to the very effective use of color by the Yushchenko presidential campaign to mobilize people against falsified election returns, those in the opposition began to use the black and orange of the Saint George's ribbon to symbolize their cause. A colorful disparaging rhetoric developed thereafter: Yushchenko's movement was the "orange plague" to its virulent opponents, while Yanukovych's forces were derided as *koloradi*, named after the Colorado potato beetle that shared the same black and orange stripes. See Daisy Sindelar, "What's Orange and Black and Bugging Ukraine?," *RFE/RL*, April 28, 2014.
88. Putin, "Address by President of the Russian Federation."
89. President of Russia, "Meeting in Support of Crimea's Accession to the Russian Federation 'We Are Together'!" March 18, 2014. For video of Putin's speech see https://www.youtube.com/watch?v=2BMmp07t1fs.

90. Mikhail Suslov, "'Crimea Is Ours!' Russian Popular Geopolitics in the New Media Age," *Eurasian Geography and Economics* 55, no. 6 (2014): 588–609.
91. Elizabeth Wood, "Performing Memory: Vladimir Putin and the Celebration of WWII in Russia," *Soviet and Post-Soviet Review* 38 (2011): 172–200.
92. Neil MacFarquhar, "From Crimea, Putin Trumpets Mother Russia," *New York Times*, May 10, 2014.
93. On the Belogorsk statue see Daisy Sindelar, "Russia Unveils Monument to 'Polite People' Behind Crimean Invasion," *REF/RL*, Transmission, May 7, 2016, http://www.rferl.org/a/russia-monument-polite-people-crimea-invasion/27000320.html. To view the Simferopol statue's unveiling see https://www.youtube.com/watch?v=dkJggE-ysAA.
94. BBC Monitoring, "Russian Filmmaker Interviewed on Crimea TV Documentary," March 21, 2015. Translation of an interview by Yuriy Snegirev of Andrey Kondrashov in *Rossiyskaya Gazeta*, March 17, 2015.
95. A few of the ethnic Russians among the respondents indicated that they also spoke Ukrainian, but we kept this as a separate ethnolinguistic category because it was a meaningful self-ascriptive identity category to respondents.
96. In SE6 the sample there were 28 who declared themselves Moldovan, 16 as Bulgarian, 10 as Belarusian, 3 as Jewish, and one Hungarian and Romanian. Twenty declared themselves "Other" while 4 did not answer. There were 80 who declared themselves both Ukrainian and Russian. In total, 164 respondents were removed. In the smaller Crimean sample, 18 declared themselves Russian and Ukrainian and 24 as "Other," while 2 did not declare. These are also not represented in the ethnolinguistic category graphs presented here.
97. Neil MacFarquhar, "Crimea in Dark after Power Lines Are Blown Up," *New York Times*, November 22, 2015; Nikolai Petrov, "Crimea: Transforming the Ukrainian Peninsula into a Russian Island," *Russian Politics and Law* 54, no. 1 (2016): 74–95.

Chapter 7

1. For accounts that engage the question of the conflict in eastern Ukraine as a "civil war" see, inter alia, Anton Lavrov, "Civil War in the East: How the Conflict Unfolded before Minsk I," in Colby Howard and Ruslan Pukhov, eds., *Brothers Armed: Military Aspects of the Crisis in Ukraine*, 2nd ed. (Minneapolis: East View Press, 2015), 202–227; Anna Matveeva, "No Moscow Stooges: Identity Polarization and Guerilla Movements in Donbass," *Southeast European and Black Sea Studies* 16, no. 1 (2016): 25–50; Serhiy Kudelia, "The Donbas Rift," *Russian Politics and Law*, 54 (2016): 5–27; Andrew Wilson, "The Donbas in 2014: Explaining Civil

Conflict Perhaps, But Not Civil War," *Europe-Asia Studies* 68 (2016): 631–652; Ivan Katchanovski, "The Separatist Conflict in Donbas: A Violent Break-Up of Ukraine?," *European Politics and Society* (2016): in press; Elise Giuliano, "The Social Bases of Support for Self-determination in East Ukraine," *Ethnopolitics* 14, no. 5 (2015): 313–522; Tetyana Malyarenko, "A Gradually Escalating Conflict: Ukraine from the Euromaidan to the War with Russia," in Karl Cordell and Stefan Wolff, eds. *The Routledge Handbook of Ethnic Conflict*, 2nd ed. (London: Routledge, 2016).

2. For a sense of the early outrage felt in Russian political circles at the violence against the Party of Region headquarters and personnel, see Paul J. Saunders's interview with Alexey Pushkov, chair of the International Affairs Committee of the Russian State Duma, "A Top Russian Lawmaker on Ukraine," *The National Interest*, March 12, 2014, http://nationalinterest.org/commentary/top-russian-lawmaker-ukraine-10032.

3. On the notions of translocality and translocals see, inter alia, Arjun Appadurai, *Modernity at Large: Cultural Dimensions of Globalization* (Minneapolis: University of Minnesota Press, 1996); and Katherine Brickell and Ayona Datta, eds., *Translocal Geographies: Spaces, Places, Connections* (Farnham: Ashgate, 2011).

4. According to a nineteen-volume geographical description of Russia by Semenov-Tyan-Shansky and Lamanskiy (vol. 14, published in 1910), historical Novorossiya comprised six guberniyas: Bessarabskaya, Khersonskaya, Tavricheskaya, Yekaterinoslavskaya, the Region of the Don Army, and Stavropol'skaya. John O' Loughlin, Gerard Toal, and Vladimir Kolossov, "The Rise and Fall of 'Novorossiya': Examining Support for a Separatist Geopolitical Imaginary in Southeast Ukraine," *Post-Soviet Affairs* 32 (2016).

5. These were the Ukrainian People's Republic of Soviets (capital in Kharkov, founded December 25, 1917), the Donetsk–Krivoy Rog Soviet Republic (declared February 12, 1918), and the Odessa Soviet Republic (founded on March 1, 1918).

6. Roman Solchanyk, "The Politics of State Building: Centre-Periphery Relations in Post-Soviet Ukraine," *Europe-Asia Studies* 46, no. 1 (1994): 62.

7. James Meek, "Ukraine; Shaken Kiev Likely to Cling to Nuclear Weapons," *The Guardian*, December 14, 1993.

8. Anatol Lieven, *Ukraine and Russia: A Fraternal Rivalry* (Washington, DC: US Institute of Peace Press, 1999), 50.

9. Mara Belaby, "Crisis in Ukraine Intensifies; President Seeks Blockade End; Opposition Calls for Firing," *Associated Press*, November 29, 2004.

10. Email correspondence with Professor Valeriy Kravchenko, formerly Donetsk National University.

11. Dmitri Trenin, *Post-Imperium: A Eurasian Story* (Washington, DC: Carnegie Endowment for International Peace, 2011), 100.

12. Regina Faranda, "How Ukrainians Viewed the Russia-Georgia Conflict," paper presented at the Association for the Study of Nationalities Conference, Columbia University, New York, 2009.
13. Marlene Laruelle, "The 'Russian World': Russia's Soft Power and Geopolitical Imagination" (Washington, DC: Center on Global Interests, 2015), 12.
14. Ibid.
15. Courtney Weaver, "Ukraine's Rebel Republics," *Financial Times*, December 6/7, 2014.
16. Gubarev had little patience for ideological discussion. See, for example, his discussion with Sergei Kurginyan on July 8, 2014. To this visiting Russian ultranationalist he explained he was for the "Red Orthodox project" (before conversation degenerated into disputation and abuse). Catherine Fitzpatrick, "Novorossiya Theory Meets Novorossiya Reality in Donetsk," *The Interpreter*, July 8, 2014.
17. Marlene Laruelle, "Scared of Putin's Shadow: In Sanctioning Dugin, Washington Got the Wrong Man," *foreignaffairs.com*, March 25, 2015.
18. Nikolay Mitrokhin, "Infiltration, Instruction, Invasion: Russia's War in the Donbass," *Journal of Soviet and Post-Soviet Politics and Society* 1, no. 1 (2015); Courtney Weaver, "Malofeev: The Russian Billionaire Linking Moscow to the Rebels," *Financial Times*, July 24, 2014. Another influential figure was Konstantin Zatulin and his Institute for CIS Countries think tank in Moscow.
19. Oleg Kashin, "From Crimea to the Donbass: The Adventures of Igor Strelkov and Aleksandr Borodai," *Slon*, May 19, 2014. See also Catherine Fitzpatrick, "With Cash and Conspiracy Theories, Russian Orthodox Philanthropist Malofeyev Is Useful to the Kremlin," *The Interpreter*, April 28, 2015.
20. Weaver, "Malofeev: The Russian Billionaire Linking Moscow to the Rebels."
21. In Russian *kurator* refers to a person who functions as adviser, minder, broker, and curator of policy. In highly patrimonial states where there is no formal structure or it counts for little, they are the personal representatives of the center ensuring that its wishes are known and its policies pursued. The Bolsheviks used this system to ensure political correctness and to monitor the local leadership it empowered in ethnoterritories in Central Asia and the Caucasus. Kurators are powerful because they are necessary brokers for resource flows from the center. For a discussion of their role in eastern Ukraine see International Crisis Group, "Russia and the Separatists in Eastern Ukraine," in *Crisis Group Europe and Central Asia Briefing Number 79*, February 5, 2016, http://www.crisisgroup.org/en/regions/europe/ukraine/b079-russia-and-the-separatists-in-eastern-ukraine.aspx. For a brief discussion of them in

Putin's *systema* see Gleb Pavlovsky, "Russian Politics Under Putin," *Foreign Affairs* 95, no. 3 (May/June 2016): 10–17.

22. For the text in Russian see http://www.novayagazeta.ru/politics/67389.html. For an English translation see http://www.unian.info/politics/1048525-novaya-gazetas-kremlin-papers-article-full-text-in-english.html.
23. Neil MacFarquhar, "Early Memo Urged Moscow to Annex Crimea, Report Says," *New York Times*, February 25, 2015; Joshua Keating, "Did This Russian Oligarch Come Up With Putin's Ukraine Strategy?," *Slate*, February 26, 2015.
24. Zygar, *All the Kremlin's Men*, 284. For Glazyev's deeply conspiratorial interpretation of the Ukraine crisis see Dmitri Simes, "An Interview with Sergey Glazyev, *The National Interest*, March 24, 2014, http://nationalinterest.org/print/commentary/interview-sergey-glazyev-10106.
25. For the Prosecutor General's released recordings and approximate English translation of them see https://www.youtube.com/watch?v=nLcHcyeppSU. For a slightly better presentation, translation and arrangement of the same recordings see https://www.youtube.com/watch?v=ow78QuxBUeo. For a second video by the Prosecutor General illustrating its changes against Russian officials see https://www.youtube.com/watch?v=Sb5sbMieLfA. The Russian verb *podnimat* is translated as "raise" but can also be translated as "stir up." See Roman Olearchyk, and Neil Buckley, "Ukraine Gathers Evidence to Try to Force Russia to Court," *Financial Times* September 12, 2016.
26. The term "Banderivtsi" is a Russian collective noun for Ukrainian nationalists and not just followers of the Ukrainian nationalist leader Stepan Bandera. It is loosely used to denigrate all Ukrainian political groups considered threatening to Russian state interests. The cited text can be heard on the taped recording of the call to an unidentified Anatoly from 5:36 to 6:54, https://www.youtube.com/watch?v=ow78QuxBUeo.
27. The recorded call with Yatsyuk can be heard from 9:45 to 12:38, https://www.youtube.com/watch?v=nLcHcyeppSU.
28. Andrew Roth, "From Russia, 'Tourists' Stir the Protests," *New York Times*, March 4, 2014.
29. Besides Gubarev, other figures included Aleksey Khudyakov, leader of Russian Shield.
30. Volodymyr Ishchenko, "Far Right Participation in the Ukrainian Maidan Protests: An Attempt at Systematic Estimation," *European Politics and Society* (2016), 1–20.
31. Volodymyr Ishchenko, "Beyond Strelkov and Motorola: Protest Event Analysis of Anti-Maidan Mobilizations Preceding the War in Donbass," *Ponars Workshop: Analyzing Violence in Eastern Ukraine, Working Paper*, May 17, 2015. I am grateful for permission to cite this preliminary research paper.

32. Daniel McLaughlin, "Fears of Further Russian Incursions into Ukraine Multiply in Wake of Crimea Poll," *Irish Times*, March 17, 2014; Roth, "From Russia, 'Tourists' Stir the Protests."
33. Unfortunately for him Gubarev was soon arrested by the Ukrainian Security Service (SBU) and taken to Kyiv, allowing others to assume key leadership positions. He would be released in a prisoner exchange on May 7, 2014.
34. Anthony Lloyd, "View from Donetsk: A Pro-Russian Opportunist Has Risen to Power in the East," *The Times (London)*, March 5, 2014.
35. James Jones's *Frontline* documentary *The Battle for Ukraine*, first aired May 27, 2014, interviewed an anonymous *Oplot* (Stronghold) leader who confirmed that they had met with Russian intelligence officials before the attack and were paid by the hour to attack pro-Maidan supporters in the building. Available at http://www.pbs.org/wgbh/frontline/film/battle-for-ukraine/. For footage of the violence used see https://www.youtube.com/watch?v=OUbm3BsEHnk.
36. See https://instagram.com/p/lADeEivk5_/.
37. Ukrayinska Pravda, "Pro-Russian Rallies in Southeast Ukraine Call for Secession," *pravda.com.ua*, March 1, 2014.
38. Roth, "From Russia, 'Tourists' Stir the Protests."
39. Anthony Lloyd, "Ukraine's Old Guard Doesn't Flinch in Face of Russian Mob," *The Times (London)*, March 10, 2014.
40. The dead man was Dmitry Cherniavsky, twenty-two, an activist for Svoboda, a far-right Ukrainian nationalist party. Andrew Roth, "Rumors Mushroom in Eastern Ukraine, Fueling Expectations of War," *New York Times*, March 15, 2014.
41. Kathy Lally, "In Eastern Ukraine, Fearing the Worst If Russia Invades," *Washington Post*, March 30, 2014.
42. Andrii Portnov, "How Eastern Ukraine Was Lost," *Open Democracy Russia*, January 14, 2014.
43. A released transcript of Ukraine's National Security and Defense Council meeting on February 28, 2014, revealed former prime minister Tymoshenko arguing that Putin "is just waiting for us to give him an opening. Remember how Saakashvili swallowed his bait and lost! We have no right to repeat his mistake." Leonid Bershidsky, "Why Russia Stopped at Crimea," *Bloomberg View*, February 22, 2016. Transcript available at http://www.pravda.com.ua/articles/2016/02/22/7099911/.
44. Volodymyr Dubovyk, "Odessa: A Local Dimension of Ukraine's Revolution, Crisis, and Conflict," in *PONARS Eurasia Policy Memo No. 390* (2015); Christian Caryl, "Novorossiya Is Back from the Dead," *Foreign Policy*, April 17, 2014.
45. "The Self-Defense of Odessa," *Foreign Policy*, April 16, 2014.
46. *Frontline, The Battle for Ukraine*, May 27, 2014, http://www.pbs.org/wgbh/frontline/film/battle-for-ukraine/.

47. Antony Butts's documentary *The Donetsk People's Republic or the Curious Tale of the Handmade Country* provides some excellent insight into this period.
48. This map is visible in the interview Pushilin gave to Paula Slier of *Russia Today's* English-language news broadcast on May 8, 2014.
49. Alexander Prokhanov, "Кто ты, 'стрелок'?" ["Who are you, 'Shooter'?"], interview with the former Minister of Defense of the Donetsk People's Republic, *Zavtra*, November 20, 2014.
50. Shaun Walker, "Russia's 'Valiant Hero' in Ukraine Turns His Fire on Vladimir Putin," *The Guardian*, June 5, 2016.
51. Andrew Roth, "Ukraine Faces Struggle to Gain Control of Militias, Including Those on Its Side," *New York Times*, May 23, 2014.
52. "Who are you, 'Shooter'?"
53. Ryazanova-Clarke, "The Discourse of a Spectacle at the End of the Presidential Term," in *Putin as Celebrity and Cultural Icon*, ed. Helena Goscilo (London: Routledge, 2013).
54. Vladimir Putin, "Direct Line with Vladimir Putin," *Kremlin.ru*, April 17, 2014.
55. Caryl, "Novorossiya Is Back from the Dead."
56. See Harley Blazer, "The Ukraine Invasion and Public Opinion," *Georgetown Journal of International Affairs*, March 20, 2015.
57. Bershidsky, "Why Russia Stopped at Crimea." For an argument that the Kremlin's goals were always limited see Paul Robinson, "Russia's Role in the War in Donbass, and the Threat to European Security," *European Politics and Society* (2016): 1–16.
58. Luke Harding, "Russia Ready to Annex Moldova Region, NATO Commander Claims," *The Guardian*, March 23, 2014.
59. The "Odessa massacre" has subsequently become a prominent citation in a discourse of victimhood cultivated by Russian media. For an analysis of Russian TV distortions on the first anniversary of the event see Stephen Ennis, "Russian TV's Coverage of Odessa Clashes," *BBC Monitoring*, May 7, 2015, http://www.bbc.co.uk/monitoring/odessa-clashes-russian-tv-distorts-events. For a critical Ukrainian perspective see Halya Cornash, "Odesa 2 May Suspect: We Were Financed by Moscow," July 10, 2015, http://www.khpg.org/en/pda/index.php?id=1444010513.
60. Ian Traynor et al., "Putin Says Eastern Ukraine Referendum on Autonomy Should Be Postponed," *The Guardian*, May 8, 2014.
61. BBC News, "East Ukraine Separatists Seek Union with Russia," *BBC News website*, May 12, 2014.
62. John Reed, "'City of a Million Roses' Wilts as Residents Flee East Ukraine," *Financial Times*, June 7, 2014.
63. Alec Luhn, "Volunteers or Paid Fighters? The Vostok Battalion Looms Large in War with Kiev," *The Guardian*, June 6, 2014; Mark Galeotti, "Russia's Secret Weapon," *Foreign Policy*, July 7, 2014; Andrew Roth and

Sabine Tavernise, "Pro-Russia Troops Take Symbol of Ukraine Uprising," *New York Times*, May 29, 2014.

64. Andrew E. Kramer, "After Neutrality Proves Untenable, a Ukraine Oligarch Makes His Move," *New York Times*, May 20, 2014.
65. Associated Press, "Ukraine's Richest Man Plays Both Sides of War's Frontline," September 22, 2015.
66. Michael Weiss, "All Is Not Well in Novorossiya," *Foreign Policy*, July 12, 2014.
67. "Who are you, 'Shooter'?"
68. Alec Luhn, "Russia to Send Humanitarian Convoy into Ukraine in Spite of Warnings," *The Guardian*, August 11, 2014.
69. Anna Nemtsova, "Putin's Number One Gunman in Ukraine Warns Him of Possible Defeat," *Daily Beast*, July 25, 2014.
70. At the end of June 2014, Alexander Dugin was fired from his professorship at Moscow State University. He saw the hand of Surkov in this act and failure to support Novorossiya. Weiss, "All Is Not Well in Novorossiya."
71. Andrew E. Kramer, "Separatist Cadre Hopes for a Reprise in Ukraine," *New York Times*, August 3, 2014.
72. RAI Novosti, "Borodai Says He Will Leave Post of DPR Prime Minister," August 8, 2014, available (in Russian), http://ria.ru/world/20140807/1019193894.html.
73. Zygar, *All the Kremlin's Men*, 291.
74. For video see https://www.youtube.com/watch?v=yRhUMIPtVmA.
75. Gerard Toal and John O'Loughlin, "Russian and Ukrainian TV Viewers Live on Different Planets," *Washington Post*, February 26, 2015.
76. Luhn, "Russia to Send Humanitarian Convoy into Ukraine in Spite of Warnings."
77. The role of regular and irregular Russian soldiers in Ukraine in 2014–2015 was publicized in three politically charged reports presented at the Atlantic Council in Washington, DC, in 2015, the last the Council's own publication. See An Independent Expert Report, *Putin. War: Based on Materials from Boris Nemtsov*, May 2015, http://www.4freerussia.org/putin.war/; James Miller, Pierre Vaux, Catherine A. Fitzpatrick, and Michael Weiss, *An Invasion by Any Other Name: The Kremlin's Dirty War in Ukraine*, September 17, 2015, http://www.interpretermag.com/an-invasion-by-any-other-name-the-kremlins-dirty-war-in-ukraine/; Maksymilian Czuperski, John Herbst, Eliot Higgins, Alina Polyakova, and Damon Wilson, *Hiding in Plain Sight: Putin's War in Ukraine*, October 15, 2015, http://www.atlanticcouncil.org/publications/reports/hiding-in-plain-sight-putin-s-war-in-ukraine-and-boris-nemtsov-s-putin-war.
78. Vladimir Putin, "President of Russia Vladimir Putin Addressed Novorossiya Militia," August 29, 2014, http://en.kremlin.ru/events/

president/news/46506. Putin would later publicly light a candle at a church on September 10 "for those who had suffered defending the people of Novorossiya." Zygar, *All the Kremlin's Men*, 290.
79. Tim Judah, "Ukraine: A Catastrophic Defeat," *New York Review of Books*, September 5, 2014. Shaun Walker, Oksana Grytsenko, and Leonid Ragozin, "Russian Soldier: 'You're Better Clueless Because the Truth Is Horrible,'" *The Guardian*, September 3, 2014.
80. "ATO Is Over, We Have to Defend against Russia—Defense Head" (in Ukrainian), http://korrespondent.net/ukraine/politics/3412905-ato-zavershena-my-dolzhny-oboroniatsia-ot-rossyy-hlava-mynoborony.
81. O' Loughlin, Toal, and Kolossov, "The Rise and Fall of 'Novorossiya.'
82. Comments during a video interview on YouTube channel of "The Day" (*Den*), posted January 1, 2015, https://www.youtube.com/watch?v=nKfCIFl6ivg (in Russian).
83. Andrei Kolesnikov, "Why the Kremlin Is Shutting Down the Novorossiya Project," *Moscow Carnegie Center*, May 29, 2015.
84. Vladimir Dergachev and Dmitry Kirillov, "Проект «Новороссия» Закрыт (Project "New Russia" Is Closed)," *Gazeta.ru* (in Russian), May 20, 2015.

Chapter 8

1. Thomas Frear, Lukasc Kulesa, and Ian Kearns, "Dangerous Brinkmanship: Close Military Encounters between Russia and the West in 2014" (European Leadership Initiative, 2014); Thomas Frear, Ian Kearns, and Lukasc Kulesa, "Preparing for the Worst: Are Russian and NATO Military Exercises Making War in Europe More Likely?," (N.p.: European Leadership Network, 2015); Michael Kofman and Matthew Rojansky, "A Closer Look at Russia's Hybrid Warfare," Wilson Center, Kennan Institute, *Kennan Cable* 7, April 2015; Michael Kofman, "Russian Hybrid Warfare and Other Dark Arts," *War on the Rocks*, March 11, 2016, http://warontherocks.com/2016/03/russian-hybrid-warfare-and-other-dark-arts/.
2. Barack Obama, "Remarks by President Obama to the People of Estonia," September 3, 2014, https://www.whitehouse.gov/the-press-office/2014/09/03/remarks-president-obama-people-estonia.
3. New York Times Editorial Board, "The Pentagon's Top Threat? Russia," *New York Times*, February 3, 2016; Matthew Rosenberg, "Joint Chiefs Nominee Warns of Threat of Russian Aggression," *New York Times*, July 9, 2015; New York Times Editorial Board, "Who Threatens America Most?," *New York Times*, August 21, 2015.
4. Ashton Carter, "Remarks on 'Strategic and Operational Innovation at a Time of Transition and Turbulence' at the Reagan National Defense Forum ," *US Department of Defense*, November 7, 2015, http://www.defense.gov/News/Transcripts/Transcript-View/Article/628147/remarks-on-strategic-and-operational-innovation-at-a-time-of-transition-and-tur.

5. Mark Lander and Helene Cooper, "U.S. Fortifying Europe's East to Deter Putin," *New York Times*, February 1, 2016.
6. Clinton's remarks were made at a fundraiser for a children's charity in California. She also stated: "[N]obody wants to up the rhetoric. Everybody wants to cool it in order to find a diplomatic solution and that's what we should be trying to do." A local press reporter, however, ran a story with the Hitler analogy; national and international newspapers soon picked it up. Karen Robes Meeks, "Hillary Clinton Compares Vladimir Putin's Actions in Ukraine to Adolf Hitler's in Nazi Germany," *Long Beach Press Telegram*, March 4, 2014. Philip Rucker, "Hillary Clinton Says Putin's Actions Are Like 'What Hitler Did Back in the '30s,'" *Washington Post*, March 5, 2014.
7. Terence McCoy, "Here's 'Putler': The Mash-up Image of Putin and Hitler Sweeping Ukraine," *Washington Post*, April 23, 2014.
8. In accepting the Soviet version of the Great Patriotic War, Russia was ignoring Stalin's complicity in starting it, the regime errors that contributed to the horrific losses suffered, and the rape warfare that characterized some aspects of the Soviet conquest of Europe. The contributions of non-Russian peoples and other Soviet Republics to the victory are implicitly, if understandably, marginalized in Russian celebrations of Victory Day. See Elisabeth A. Wood, "Performing Memory: Vladimir Putin and the Celebration of WWII in Russia," *Soviet and Post-Soviet Review* 38 (2011): 172–200.
9. Dan Deudney and John Ikenberry, "The Unravelling of the Cold War Settlement," *Survival* 51 (2009): 39–62.
10. The thin versus thick distinction here echoes Clifford Geertz's classic notion of "thick description," which he borrows from the British ordinary language philosopher Gilbert Ryle. See Clifford Geertz, *The Interpretation of Cultures* (New York: Basic Books, 1977). However, thin geopolitics *is* a practice of symbolic interpretation but of a hegemonic culture that is largely indifferent to local knowledge and experiences of place. Numerous geographers have used similar distinctions to theorize the concept of place, including its affective qualities. I previously cited the distinction in Gearóid Ó Tuathail, "Re-Asserting the Regional: Political Geography and Geopolitics in World Thinly Known," *Political Geography* 22 (2003): 653–655.
11. In discussing the geopolitical circumstances that led to the Crimean War (1853–1856), the English historian Orlando Figes notes how the fight for freedom in Poland captured the imagination of the British public in the years beforehand. He remarks that this "public" (actually only the "educated classes") "readily assimilated it to the ideals they liked to think of as 'British'—in particular, a love of liberty and the commitment to defend the 'little man' against 'bullies' (the principle upon which the British told themselves they went to war in 1854, 1914 and 1939)." Orlando Figes, *The Crimean War: A History* (New York: Picador, 2010), 81.

12. Frank Ninkovich, *Modernity and Power: A History of the Domino Theory in the Twentieth Century* (Chicago: University of Chicago Press, 1994).
13. Walter A. McDougall, *Promised Land, Crusader State: The American Encounter with the World since 1776* (Boston: Houghton Mifflin, 1997).
14. President Harry S. Truman's Address before a Joint Session of Congress, March 12, 1947, http://avalon.law.yale.edu/20th_century/trudoc.asp.
15. Dean Acheson, *Present at Creation: My Years in the State Department* (New York: W. W. Norton), 365.
16. It should be recalled that Kennan's X article contains a famous "red tide" passage that is completely contrary to his later emphasis on select industrial states as strategic. "Its political action is a fluid stream which moves constantly, wherever it is permitted to move, toward a given goal. Its main concern is to make sure that it has filled every nook and cranny available to it in the basin of world power." X (George Kennan), "The Sources of Soviet Conduct," *Foreign Affairs* (July 1947). For a discussion of Kennan's contradictions see Anders Stephanson, *Kennan and the Art of Foreign Policy* (Cambridge, MA: Harvard University Press, 1992).
17. Ernest R. May, "The Nature of Foreign Policy: The Calculated versus the Axiomatic," *Daedalus* 91, no. 4 (Fall 1962): 653, 667.
18. David C. Unger, *The Emergency State: America's Pursuit of Absolute Security at All Costs* (New York: Penguin, 2012).
19. Campbell Craig and Fredrik Logevall, *America's Cold War: The Politics of Insecurity* (Cambridge, MA: Belknap Press of Harvard University Press, 2009).
20. Timuri Yakobashvili, "The 20% Rule," *The Weekly Standard*, March 3, 2014.
21. Evelyn Farkas, "What the Next President Must Do about Putin," *Politico*, January 24, 2016. See also her testimony before the U.S. House of Representatives, Committee on Armed Services, hearing on "Understanding and Deterring Russia: U.S. Policies and Strategies," February 23, 2016, http://docs.house.gov/Committee/Calendar/ByEvent.aspx?EventID=104473.
22. Liam Stack and Karen Zriack, "Frozen Zones: How Russia Maintains Influence in the Post–Cold War Era," *New York Times*, October 14, 2016.
23. International Crisis Group, "Chechnya: The Inner Abroad," in *Europe Report No. 236,* Brussels, June 30, 2015, http://www.crisisgroup.org/en/regions/europe/north-caucasus/236-chechnya-the-inner-abroad.aspx.
24. Joshua Rovner, "Dealing with Putin's Strategic Incompetence," *War on the Rocks*, August 12, 2015, http://warontherocks.com/2015/08/dealing-with-putins-strategic-incompetence/.
25. Joshua Yaffa, "Putin's Dragon," *New Yorker*, February 8, 2016.
26. Kristin Bakke et al., "Convincing State-Builders? Disaggregating Internal Legitimacy in Abkhazia," *International Studies Quarterly* 58, no. 3 (2014): 591–607.
27. For this tendency see Dawisha, *Putin's Kleptocracy*.

28. Tom Bamforth, "Leading Anticorruption Crusader to March Shoulder to Shoulder with Nationalists," *RFE/RL*, November 3, 2011.
29. John O' Loughlin, Vladimir Kolossov, and Gerard Toal, "Inside the Post-Soviet De Facto States: A Comparison of Attitudes in Abkhazia, Nagorny Karabakh, South Ossetia, and Transnistria," *Eurasian Geography and Economics* 55, no. 5 (2014): 423–456.
30. Joshua Yaffa, "The Unaccountable Death of Boris Nemtsov," *New Yorker*, February 26, 2016.
31. International Crisis Group, "Russia and the Separatists in Eastern Ukraine," 5.
32. Laurence Broers, Alexander Iskandaryan, and Sergey Minasyan, eds., *The Unrecognized Politics of De Facto States in the Post-Soviet Space* (Yerevan: Caucasus Institute International Association for the Study of the Caucasus, 2015).
33. By contrast, major benefactors of Israel in the United States like Sheldon Adelson have made a concerted effort to coach U.S. politicians *not* to use the term "occupied territories" as a description of the West Bank and Golan Heights. See Katie Zezima, "Why Chris Christie's 'Occupied Territories' Remark Matters," *Washington Post*, April 1, 2016.
34. John O'Loughlin, Vladimir Kolossov, and Gerard Toal, "Inside Abkhazia: Survey of Attitudes in a De Facto State," *Post-Soviet Affairs* 27 (2011): Figure 8a, 32.
35. O'Loughlin, Kolossov, and Toal, "Inside the Post-Soviet De Facto States," 15–17. See also John O'Loughlin, Gerard Toal, and Rebecca Chamberlain-Chenga, "Divided Space, Divided Attitudes? A Comparative Analysis of Simultaneous Surveys in the Republics of Moldova and Pridnestrovie (Transnistria)," *Eurasian Geography and Economics* 54 (2013): 227–258. For 2010 to 2014 survey data on attitudes towards Russia see Gerard Toal and John O'Loughlin, "Frozen Fragments, Simmering Spaces: The Post-Soviet De Facto States," in E. Holland, M. Derrek, eds., *Questioning Post-Soviet* (Washington, DC: Kennan Institute, 2017).
36. On the tensions between the needs of displaced persons and states using them as pawns, see Gerard Toal and Magdalena Frichova Grono, "After Ethnic Violence in the Caucasus: Attitudes of Local Abkhazians and Displaced Georgians in 2010," *Eurasian Geography and Economics* 52 (2011): 655–678; and Anik Smit, *The Property Rights of Refugees and Internally Displaced Persons: Beyond Restitution* (London: Routledge, 2012).
37. O'Loughlin et al., "Inside Abkhazia."
38. BBC News, "Moldova's Trans-Dniester Region Pleads to Join Russia," *BBC News Europe*, March 18, 2014.
39. Crisis Group, "Russia and the Separatists in Eastern Ukraine."
40. Mearsheimer, "Why the Ukraine Crisis Is the West's Fault."
41. Paranoid geopolitics, of course, does not need proximity though there is often reference back to the homeland and the domestic: the United States

must fight its enemies "over there" (Vietnam, Iraq, Afghanistan) or else it will have to fight them "here."
42. Rachel Maddow, *Drift* (New York: Broadway Paperbacks, 2012), 30–34.
43. *Public Papers of Presidents of the United States: Ronald Reagan, 1986*, July 23, 1986, 992, https://www.reaganlibrary.archives.gov/archives/speeches/1986/072386a.htm.
44. Ronald Reagan, "Address to the Nation on United States Policy in Central America," May 9, 1984, https://www.reaganlibrary.archives.gov/archives/speeches/1984/50984h.htm. See Gearóid Ó Tuathail, "The Language and Nature of the 'New' Geopolitics: The Case of US–El Salvador Relations," *Political Geography* 5 (1986): 73–85.
45. Kevin Buckley, *Panama: The Whole Story* (New York: Simon & Schuster, 1991).
46. There is also the history of U.S. expansionism at the expense of its neighbor Mexico, and its annexation of the settler-created Republic of Texas in 1845. The United States also facilitated the secession of Panama from Colombia in 1903.
47. For an attempt at a rational choice model of the grand strategy choices of major powers, see Peter Trubowitz, *Politics and Strategy: Partisan Ambition and American Statecraft* (Princeton: Princeton University Press, 2011).
48. Dawisha, *Putin's Kleptocracy*; Andreas Umland, "The Flaws of the Putinversteher's Russian Hermeneutics," *Intersection: Russia/Europe/World*, January 17, 2016.
49. The subsequent revelations of state-sponsored Russian doping during the Winter Olympics at Sochi suggest rules were violated and broken here too. Rebecca Ruiz and Michael Schwirtz, "Russian Insider Says State-Run Doping Fueled Olympic Gold," *New York Times*, May 12, 2016.
50. Russian economic statistics are available from the websites of the World Bank, the *Economist* Intelligence Unit, and the *Financial Times*.
51. Thomas Graham, "Russia Achieves Tactical Success in the Middle East, but No Strategic Victory," *Yale Global Online*, March 1, 2016.
52. Graham Allison and Dimitri Simes, "Russia and America: Stumbling to War," *The National Interest*, April 20, 2015.
53. In running for president in 2008 Obama skillfully negotiated the tacit assurances all U.S. presidential candidates need to provide. In foreign policy this included privileging the security of Israel and its citizens. The privileging of certain conflicts and lives over others—those in the Middle East or on Russia's periphery—over those in Africa (especially the Congo) has been an ethical issue the president, and others in his administration, have reflected upon, sometimes publicly. In a 2013 interview for the re-launch of the *New Republic* magazine Obama asked: "And how do I weigh, tens of thousands who've been killed in Syria versus the tens of thousands who are currently being killed in the Congo?" See Gerard Toal, "In No Other Country on Earth": The Presidential Campaign of Barack

54. Ruth Deyermond, "Assessing the Reset: Success and Failures in the Obama Administration's Russia Policy, 2009–2012," *European Security* 22, no. 4 (2013): 500–523.
55. The budget of the Atlantic Council increased from $2 million to $21 million over the last decade. For a list of the Atlantic Council's sponsors, see http://www.atlanticcouncil.org/support/corporate-membership. Precise details on the funding levels, historic patterns of sponsorship, and conditions of giving are not public. It should be noted that the Atlantic Council is a bipartisan think tank and has affiliates with a diversity of viewpoints. For its alleged role, and others, in corporate lobbying see Eric Lipton and Brooke Williams, "Researchers or Corporate Allies? Think Tanks Blur the Line," *New York Times*, August 7, 2016.
56. Atlantic Council, "Georgia in the West: A Policy Road Map to Georgia's Euro-Atlantic Future" (Washington, DC: Atlantic Council, 2011). Damon Wilson, "Completing Europe: Georgia's Path to NATO," *Issue Brief: Atlantic Council*, February 2014.
57. For footage see https://www.youtube.com/watch?v=jLcHkBv_Umw.
58. Jonathan Marcus, "Ukraine Crisis: Transcript of Leaked Nuland-Pyatt Call," *BBC News*, February 7, 2014.
59. Saakashvili was appointed chair of an International Advisory Council on Reforms in February 2015 by his old friend Poroshenko. He immediately began lobbying U.S. legislators for military aid to Ukraine and testified before the U.S. Senate Foreign Relations Committee on March 4, 2015. Two months later Poroshenko appointed him governor of Odesa and granted him Ukrainian citizenship. In taking this, Saakashvili was legally obliged to give up his Georgian citizenship. He faces numerous criminal charges in Georgia from his time as the country's president. In Odesa, Saakashvili styled himself as an anticorruption fighter before quitting. The details are more ambiguous. See Thomas De Waal, "The Odessa Project," *Carnegie Europe*, February 18, 2016; Ian Bateson, "The Education of Mikheil Saakashvili," *Foreign Policy*, October 14, 2016.
60. Petro Poroshenko, "Remarks by Ukrainian President Poroshenko to the U.S. Congress," September 18, 2014, http://www.cfr.org/ukraine/remarks-ukrainian-president-poroshenko-us-congress/p33470.
61. Derek Chollet, *The Long Game: How Obama Defied Washington and Redefined America's Role in the World* (New York: Public Affairs, 2016), 175. Partisan political divisions, and perhaps as Chollet suggests bad advice from lobbyists, played into these speeches but so also did a consistent lack of respect for Obama's presidency by certain factions within the United States and abroad.
62. White House, "Remarks by President Obama and Ukraine Prime Minister Yatsenyuk after Bilateral Meeting," March 12, 2014, https://www

.whitehouse.gov/the-press-office/2014/03/12/remarks-president-obama-and-ukraine-prime-minister-yatsenyuk-after-bilat.
63. For Yatsenyuk's speech see http://www.ajc.org/site/c.7oJILSPwFfJSG/b.9288277/k.7332/2015_Global_Forum_Highlights.htm.
64. Petro Poroshenko, "We're Making Steady Progress in Ukraine, Despite Putin," *Wall Street Journal*, June 10, 2015.
65. Tom Parfitt, "Ukraine Crisis: The Neo-Nazi Brigade Fighting Pro-Russian Separatists," *The Telegraph*, August 11, 2014. Alec Luhn, "Preparing for War with Ukraine's Fascist Defenders of Freedom," *Foreign Policy*, August 30, 2014. Shaun Walker, "Azov Fighters Are Ukraine's Greatest Weapon and May be its Greatest Threat," *The Guardian*, September 10, 2014; Alec Luhn, "Preparing for War with Ukraine's Fascist Defenders of Freedom," *Foreign Policy*, August 30, 2014.
66. For a sample of the debate see Stephen M. Walt, "Why Arming Kiev Is a Really, Really Bad Idea," *Foreign Policy*, February 9, 2015.
67. For the text of the bill see https://www.congress.gov/bill/113th-congress/senate-bill/2277.
68. Ivo Daalder, Michele Flournoy, John Herbst, Jan Lodal, Steven Pifer, James Stavridis, Strobe Talbott, and Charles Wald, *Preserving Ukraine's Independence, Resisting Russian Aggression: What the United States and NATO Must Do*, Atlantic Council, Washington, DC, February 2015.
69. Jennifer Steinhauer and David M. Herszenhorn, "Defying Obama, Many in Congress Press to Arm Ukraine," *New York Times*, June 11, 2015.
70. See http://www.armed-services.senate.gov/hearings/15-02-04-nomination.
71. Chollet, *The Long Game*, 175.
72. White House, "Remarks by President Obama and Ukraine Prime Minister Yatsenyuk after Bilateral Meeting," https://www.whitehouse.gov/the-press-office/2014/03/12/remarks-president-obama-and-ukraine-prime-minister-yatsenyuk-after-bilat.
73. White House, "Press Conference with President Obama and Prime Minister Rutte of the Netherlands," March 25, 2014, https://www.whitehouse.gov/the-press-office/2014/03/25/press-conference-president-obama-and-prime-minister-rutte-netherlands.
74. Kirit Radai, "Russian Tanks in Ukraine, but US Won't Say 'Invasion,'" *ABC News*, November 13, 2014.
75. Jeffrey Goldberg, "The Obama Doctrine," *The Atlantic*, April 2016.
76. White House, "Press Conference with President Obama and Prime Minister Rutte of the Netherlands."
77. Anatol Lieven, *America Right or Wrong: An Anatomy of American Nationalism*, 2nd ed. (New York: Oxford University Press, 2012).
78. On the dilemmas of democracy promotion see Lincoln A. Mitchell, *The Democracy Promotion Paradox* (Washington, DC: Brookings Institution Press, 2016).

79. Barry R. Posen, *Restraint: A New Foundation for US Grand Strategy* (Ithaca: Cornell University Press, 2014), 65.
80. Christian Davies, "The Conspiracy Theorists Who Have Taken over Poland," *The Guardian*, February 16, 2016; Michael Birnbaum, "Obama Slammed Polish Democracy on Friday. Here's How Polish TV Proved Him Right," *Washington Post*, July 9, 2016.
81. The phrase acquired an add-on—"at peace"—that expressed the wishful thinking of its proponents while softening the fact that it was a vision of peace organized around NATO expansionism. That some advocates of a Europe "at peace" also support the United States providing advanced weapons to Ukraine is indicative of contradictions within the vision. Job C. Henning and Douglas A. Ollicant, "Radically Rethinking NATO and the Future of European Security," *War on the Rocks*, March 24, 2016, http://warontherocks.com/2016/03/radically-rethinking-nato-and-the-future-of-european-security/.
82. The Madrid Principles are six broad principles that the OSCE Minsk Group (Russia, France, and the United States) presented in 2007 as the basis for a peace settlement of the Nagorny Karabakh conflict. For the full text visit http://www.aniarc.am/2016/04/11/madrid-principles-full-text/.
83. George F. Kennan, *American Diplomacy, 1900–1950* (Chicago: University of Chicago Press, 1951), 98.
84. Arkady Ostrovsky, *The Invention of Russia: From Gorbachev's Freedom to Putin's War* (New York: Viking, 2015).
85. Gerard Toal and John O'Loughlin, "Russian and Ukrainian TV Viewers Live on Different Planets," Monkey Cage, *Washington Post*, February 26, 2015.
86. Ken Silverstein, "Pay-to-Play Think Tanks: Institutional Corruption and the Industry of Ideas" (Boston: Edmond J. Safra Center for Ethics at Harvard University, 2014); Janine Wedel, *Unaccountable: How Elite Power Brokers Corrupt Our Finances, Freedom, and Security* (New York: Pegasus Books, 2014); *Shadow Elite: How the World's New Power Brokers Undermine Democracy, Government, and the Free Market* (New York: Basic Books, 20); Lawrence Lessing, *Republic, Lost: How Money Corrupts Congress—and a Plan to Stop It* (New York: Twelve, 2012); Tom Medvetz, *Think Tanks in America* (Chicago: University of Chicago Press, 2012).
87. This book has not addressed cyberattacks but there is evidence of a common Russian connection between cyberattacks against Estonia in 2007, Georgia in 2008, Ukraine in 2015, and U.S. political parties and individuals in 2016. In this aspect of the contemporary geopolitical condition, Russia's near abroad has been the proving ground for attacks that are now scaling up to disrupt competing power centers.
88. Wendy Brown, *Walled States, Waning Sovereignty* (Brooklyn: Zone Books, 2010).

INDEX

Abashidze, Aslan, 114, 147
Abkhazia, 4, 8, 47, 146–47
    Abkhazian Autonomous Soviet Socialist Republic, 67
    adventurism in, 122
    building aid for, 79
    displacement from, 101, 129, 142
    as independent, 182, 188–89, 279, 283, 287
    invasion plan for, 164, 174
    killings in, 110
    leadership, 114
    media attention in, 81
    Putin protecting, 155
    as right hand, 326n81
    survey in, 285–86
    toxic legacy, 282
    unresolved conflict, 140
    vulnerability of, 152–54
    war expectation in, 156–57
Acheson, Dean, 278
Adamkus, Valdas, 160–61
Adamon Nykhaz (Popular Shrine), 133
adviser (*kurator*), 247–48, 282, 359n21
affective geopolitics, 300
    anger and resentment in, 48
    color schemes and, 356n87
    definition and significance of, 45
    gendered ideals and subject-positions, 48
    history of, 44
    noble causes and, 49
    in post-Soviet space, 46
    at work, 13
affective geopolitics, of annexation
    producing glorious historical moment, 226–31
    producing mythic threat, 217–20
    producing scenography of legitimacy, 223–26
    producing surprise and confusion, 220–23
Afghanistan, 2
    Georgia troops to, 119
    Operation Enduring Freedom in, 109
Akhmetov, Rinat, 243, 249, 255, 256–58
Aksyonov, Sergei, 222, 228, 229f, 254, 258
Albright, Madeleine, 5
Aliyev, Heydar, 105
Alma Ata Declaration, 74
Ambartsumov, Evgenii, 82
American Enterprise Institute, 100, 119
Amet, Reşat, 225
anti-imperialism
    imperialism and, 41–42
    Soviet Union ideology as, 59–61
Anti-Terrorist Operation (ATO), 260
Antyufeyev, Vladimir, 267
Armenia
    Azerbaijan border dispute, 9
    diaspora, 3
    discussion of, 15
    persecution memorial in, 43

Asmus, Ron, 123, 159, 160
ASSRs. *See* Autonomous Soviet Socialist Republics
as you possess (*uti possidetis*), 65, 68, 85, 103
Atlantic Council, 292
Atlanticism, 80, 211
ATO. *See* Anti-Terrorist Operation
August War, theory of, 183–84, 189–90, 194
Austin, John, 350n6
Autonomous Soviet Socialist Republics (ASSRs), 131
Avakov, Arsen, 253
Azerbaijan
  Armenia border dispute, 9
  discussion of, 15
  GUAM, 151
  regional rebellions in, 4

Baker, James, 100–101, 105, 116
Baku-Tbilisi-Ceyhan pipeline, 105, 178, 194
Baltic states, 19, 60
  CIS and, 65
  incorporating, 5
  intimidation attempt in, 275
  Saakashvili and, 163
  Scandinavia and, 3
Banderivtsi, 250–51, 360n26
Baran, Zeyno, 116–17
Barkashov, Aleksander, 77
Bay of Pigs invasion, 288
Beck Glenn, 178
Belavenstev, Oleg, 221, 355n70
Belavezha Accords, 74–75
Bendukidze, Kakha, 120–21
Ben-Gurion, David, 48, 120
Bezler, Igor, 247
Biden, Joe, 5, 123
Bierce, Ambrose, 10
Bildt, Carl, 160
Black Hundred, 77
Black Sea Fleet, 7, 18, 203, 205, 215, 227, 287
blue blob of democracy, 6
Bol'shaya Rossiya (Large Russia), 246
Bolsheviks
  kurator, 359n21
  National Bolshevik Party, 77
  power of, 59
  Red Army, 130
  rule, 320n4
  Soviet Union formation, 7, 20, 62–63, 95
  Ukraine and, 237
Bolton, John, 56
Borodai, Alexander, 77, 242, 246–47, 265, 267, 272

Boston marathon bombing, 292
Bourdieu, Pierre, 35
BREXIT campaign, 302
Brezhnev, Leonid, 61, 315n18
Brown, Gordon, 124
Brown, Kent, 101–2
Brubaker, Rogers, 34–35
Bryza, Matthew, 116, 156, 160–61, 190
Brzezinski, Zbigniew, 5, 205–6
Bucharest Declaration, 7–8, 94–95, 124–25, 156, 292
Budapest Memorandum, 24, 202, 293
Burbulis, Gennadi, 73*t*, 74
Burjanadze, Nino, 108
Burns, William, 190
Bush, George W.
  alienating Russia, 90
  on Europe whole and free, 5
  firmness with Putin and Medvedev, 192–93
  Freedom Agenda of, 7, 24, 117, 121
  hegemonic masculinity of, 48
  linking Georgia and Iraq, 122
  MAP to Georgia and Ukraine, 94–95
  National Security Council, 93
  on Putin, 57
  Putin meeting, 166–68
  Putin praised by, 198
  revisionist policies, 287
  on Rose Revolution, 112, 121–22
  Saakashvili meeting, 115, 117, 151, 155–56, 162–63
  Saakashvili on, 124, 177–78
  unilateralism of administration, 2
  on young democracies, 93

captive nations, 5, 52, 96–97, 119
Carpathian Declaration, 207
Carter, Ashton, 275, 293
Carter, Jimmy, 5
casus sui generis, 188
Cato Institute, 121
Caucasus
  beacon of liberty, 112–25
  China and, 3
  David and Goliath in, 100–108
  geography and, 95–100
  Georgia as Christian bastion, 103
  overview, 93–95
  Shalikashvili "this is our neighborhood now" message, 108–11
  Turkey and, 3
Center for Geopolitical Expertise, 78
Center for the National Interest, 29
Chaly, Aleksei, 220, 229*f*, 355n67

Charter of Paris for a New Europe, 19, 74, 189
Chechnya
  as everywhere, 2
  as inner abroad of Russia, 1
  pacified, 298
  secession, 69
Cheney, Dick, 119, 191–92
Chibirov, Lyudvig, 139–40
China, 3, 17, 88, 211, 291
Churkin, Vitaly, 224
CIS. *See* Commonwealth of Independent States
civilization
  Europe and West, 42–43
  as shared common identity, 43
  Yanukovych civilizational choice, 288
Clay, Lucius, 193
Clinton, Bill, 6, 98
  geography acknowledged by, 102
  Russia-first policy and, 104
Clinton, Hillary, 210
  on Crimea seizure, 276, 365n6
Cold War
  alliances, 206
  America, 5
  captive nation discourse, 52
  charges during, 41
  commitments, 298
  competition, 96
  end of, 52–53
  geopolitics of, 22
  global, 278
  occupied territories, 285
  propaganda, 98
  psychological war, 50
  rollback dream of, 6
  systems, 277
  Western powers during, 44
Collective Security Treaty Organization (CSTO), 101, 107
color schemes
  affective geopolitics and, 356n87
  revolutions and, 46
common European home, 5–6
Commonwealth of Independent States (CIS), 60, 65, 205
Communism, 61, 80, 96
Community of Democratic Choice, 4, 151, 207
compatriots abroad, 74–75
competing visions of Russia, 72*t*–73*t*
competitive relational nationalisms, 35
concerts
  Georgia, 2008 war, and, 13
  Georgia, 2008 war, rescue missions, 186–87, 187*f*
  music of celebration, 46
  Ukraine, 2014 seizure, and, 13
Connolly, William, 45
conspiracy theories, 51
containment, 98
Cossacks, 61, 79, 204, 215, 240, 253
Courtney, William, Ambassador, 106
Crimea
  advantages, 251
  contingent road to, 210–14
  Crimean War of 1853-1856, 49
  demographics, 237
  little green men in, 252–53
  reunification in Russian Federation, 225
  Ukraine gift, 201
Crimea Is Ours (*Krym Nash*), 229–30
Crimea: The Way Home docudrama, 216–17, 222, 231
critical geopolitics
  critical analysis offered by, 21
  defined, 8
  discursive practices and, 11
  geopolitical culture analyzed by, 10–11
  rejecting geo-determinism, 8–9
Croatia
  Operation Storm, 145, 149, 160
  sponsored separatism, 145
CSTO. *See* Collective Security Treaty Organization
cyberattacks, 371n87
Czechoslovakia, 34

Dagomys Agreement, 67, 135–36, 138. *See also* Sochi Agreement
Daily Mail, 49
Damasio, Antonio, 45
dark double, 22
David and Goliath, 100–108
Davis, Rick, 348n124
Dayton Peace Accords, 138
DCFTA. *See* Deep and Comprehensive Free Trade Agreement
Debrix, François, 49
decolonization
  India, 38
  nation-building requirement, 42
  post-Soviet space, 34
Deep and Comprehensive Free Trade Agreement (DCFTA), 211–12
de Hoop Scheffer, Jaap, 160–61
democracy
  blue blob of, 6
  Bush on, 93
  Potemkin, 108

Democratic Union of Novorossiya, 4
Deripaska, Oleg, 348n124
derzhavniks (great power), 81, 85–86
de-Sovietization, 64
Direct Line, 260–62
Dnipro Battalion, 255
Donbas, 15, 18
　divergence of, 256–58
　dominated, 249
　Euromaidan protests, 252
　saving, 264–68
Donetsk clan, 239
Donetsk People's Republic (DPR), 265, 267
DPR. *See* Donetsk People's Republic
Dudayev, Dzhokhar, 81
Dugin, Alexander, 76–79, 205, 246–47, 267
Dulles, John Foster, 96

Eastern bloc, 44
Eastern Partnership Association
　Agreement, 292
Edinstvo (Unity), 142, 148
EEU. *See* Eurasian Economic Union
Eisenhower, Dwight, 96
Eisenhower administration, 50
enlargement
　of EU, 27, 276
　McCain on, 99
　of NATO, 6, 30–31, 93–95, 100, 276
　strategy, 98
　Yeltsin introduction of, 98
Erdoğan, Recep Tayyip, 274, 300
ethnic cleansing, 185–86
ethnoterritorial units, 59, 62–63
EU. *See* European Union
Eurasian Economic Union (EEU), 8, 211, 281, 286, 288
Eurasianism, 76–78, 80
Euro-Atlantic integration, 7, 207
Euromaidan protests, 228
　challenging institutions, 212–14
　coining, 352n34
　controlling, 248
　counter storyline for, 217, 223–24
　in Donbas, 252
　McCain on, 292–93
　motivation of, 233–34
　for policy change, 208–9
　supporters, 254
　suppressing, 244
　violent, 238
Europe
　civilization and West, 42–43
　completing, 6
　new, 7
　whole and free, 5, 19
European Association Agreement, 8
Europeanizing, 6, 43
European Reassurance Initiative, 275
European Union (EU)
　alignment with, 4
　Eastern Partnership Association
　　Agreement, 292
　enlargement, 27, 276
　expansionism, 30
evil empire, 41, 52, 179
exceptionalism
　geographic, 103
　in Georgia, 142
　national, 88
　in United States, 95–96, 100, 278
expansionism
　in EU, 30
　by NATO, 5, 22, 27, 85, 207, 281, 288, 298–99, 371n81
　by Russia, 22, 309f9
external normative power, 36, 38*f*

Farkas, Evelyn, 280
fascism, 23
　coup, 2, 18
　genocide and, 43–44
　nationalism and, 49
　Soviet Union and, 44
fast thinking, 45
Fata, Daniel, 191
Federal Security Service (FSB), 246–47, 290
feminism, 46, 47
fictional independent states, 281
Filatov, Boris, 255
Finland, 177
flags, 46, 325n62
flow infrastructures, 39
Foreign Affairs, 30, 205
Forest Brothers, 139–40
For Whom the Bell Tolls (Hemingway), 100
The Foundations of Geopolitics: Russia's Geopolitical Future (Dugin), 77–78
Fox News, 51
freedom, 41
　Operation Iraqi Freedom, 119
　in post-Soviet space, 42
　Ukraine Freedom Support Act, 295
　Volunteer Freedom Corps, 323n44
Freedom Agenda, 7, 24, 117, 121
free world, 6, 41, 298
Fried, Daniel, 116, 158, 190
frozen conflicts, 69, 92, 277, 280

frozen zones, 280
FSB. *See* Federal Security Service

Gaidar, Yegor, 47
Gamsakhurdia, Zviad, 67, 101, 132–34, 142, 329n17
Gates, Robert, 94, 190
Gelayev, Ruslan, 139
Gelb, Leslie, 85
genocide
  fascism and, 43–44
  Holocaust and, 44
  Putin on, 183–84
  in Ukraine, 201
geo-determinism, 8–9
geographic determinism, 33
geography
  Caucasus and, 95–100
  as circulation, 311n51
  Clinton acknowledging, 102
  escaping through globalization, 52–54
  exceptionalism, 103
  Georgia, 2008 war, and, 30
  violence and, 15–16
  war and, 10
geopolitical catastrophe
  collapse of empire, 58–64
  overview, 55–58
  post-Soviet space role, 64–69
  Putin addressing, 91
  Putin's revanchism, 58, 87–92
  Russian geopolitical culture and, 69–87
geopolitical condition
  as core concept, 302
  technological systems and, 13–14
geopolitical culture
  competing visions, 40–41
  critical geopolitics analyzing, 10–11
  great power Russia, 80–82, 85–87
  identity and, 39
  as key concept, 300
  myths and power structure, 40–41
  Putin as articulator, 11–12
  Russia, 11–12, 57, 69–87
  Russian geopolitical catastrophe and, 69–87
  Russian liberal European, 74–75
  security and defense, 39–40
  shared discourses in, 39–41
  spatial identity and, 10
  Ukraine as cause in United States, 291–97
  of United States, 10–11, 15, 57, 95–98, 103, 121–22, 275–79, 286, 291, 296–97, 301–3
geopolitical entrepreneurship, 58, 76, 78, 111, 287, 289

geopolitical field
  near abroad as, 9
  notion of, 8
  post-Soviet space as, 298
  statecraft and, 9
geopolitics. *See also* affective geopolitics; affective geopolitics, of annexation
  background in reason for Russian invasions, 28–54
  as cold, 13
  of Cold War, 22
  emotion and, 46
  as great-power realpolitik, 12
  Kissingerian understanding, 12
  multivector, 211–12
  of mutual antagonism, 2
  Obama and, 297
  post-Soviet space as contested, 33–37, 38–39, 38f, 298
  practice of aggressive powers, 21
  return of past, 22–27
  revisionary geopolitical imaginary, 240–44
  of Russia, 13, 15
  as tabloid storylines and state-produced drama, 49–52
  technology, 6–7
  thick and thin, 277–79, 297, 365n10
  of Ukraine, 199–211
  of United States, 13
Georgia. *See also* North Ossetia; Saakashvili, Mikheil; South Ossetia; Tbilisi
  Berlin compared to, 193
  borders, 320n4
  breakaway territories, 68–69, 150, 152, 156
  Bush linking to Iraq, 122
  chauvinism, 62
  as Christian bastion, 103
  exceptionalism, 142
  frozen conflicts in, 92
  identity, 114
  internal imperialism, 42
  language, 61, 102
  lobbying capability, 122–23
  MAP offer, 94–95, 123–24, 155
  McCain affinity for, 177
  as NATO member, 7
  NATO membership and, 110–11
  as occupied territory, 65
  Operation Iraqi Freedom participant, 119
  post-revolutionary syndrome, 118
  as Potemkin democracy, 108
  regional rebellions, 4
  Saakashvili fleeing, 54
  Sakharov on, 42

Georgia (*Cont.*)
  spiritual, 142
  state flag, 325n62
  territorial integrity of, 114
  troops to Afghanistan, 119
  Turkey and, 3
  United States financial support to, 123
  United States security assistance to, 3–4, 14
  United States train and equip mission in, 109–10
  victimhood in Georgia, 2008 war, 190–97
  victimization of, 104
  wide autonomy proposed, 146
Georgia, 2008 war. *See also* August War theory
  as anti-Russian bias, 51
  comprehending, 20
  concerts and, 13
  contested circumstances, 127
  deaths, 127
  espionage arrests, 152–53
  first attacks, 148–49
  as genocide, 127
  geography and, 30
  getting Georgia back, 141–46
  land seizures, 152
  leadership blamed for, 29–30
  McCain condemning, 127
  military offensive authorized, 158–59, 162
  opening shots, 126
  over South Ossetia territory, 12, 17, 128
  overview, 126–29
  reconnaissance, 156
  Rice's advice on, 161–62
  Russia surprise over, 164
  Saakashvili's responsibility in, 14
  shootouts and landmines, 154
  South Ossetia as contested space, 129–41
  start of, 2
  trap theory, 164
  United States aid anticipated, 162–63
  United States condemning, 23, 166–68
  upping ante, 146–58
  weapons buildup, 160
  who started, 158–65
Georgia, 2008 war, rescue missions
  attack on Russia, 186
  buffer zones, 174
  ceasefire agreement, 173–74
  cleanup operations, 172
  ethnic cleansing accusation, 185–86
  Georgian offensive, 171
  Georgia storyline, attack on West, 175–79
  looting and revenge attacks, 174
  memorial concert, 186–87, 187*f*
  naval attacks, 172

Operation Clear Field, 184
  overview, 166–68
  phases of war, 169–75
  Putin at field hospital, 181–82
  revanchism in, 197
  Russian bombing, 171–72
  Russian ground invasion, 172–73
  Russian military response, 170–71
  Russia storyline, genocide and responsibility to protect, 179–90
  South Ossetia territorial integrity achieved, 175
  United States storyline, Russia aggression and Georgia victimhood, 190–97
Georgia, Ukraine, Azerbaijan, and Moldova (GUAM), 153
Gergiev, Valery, 186, 187*f*
Girkin, Igor, 77, 265–66
  coercion from, 222
  on noble cause, 272
  with special forces, 253, 258–60
  writings and public relations, 245–47
Glavnoye razvedyvatel'noye upravleniye (GRU), 76, 247, 258–59, 265
Glazyev, Sergey, 79, 247, 249–51, 262
globalization
  geography escaping through, 52–54
  time-space compression, 53
Glory to Russia (*Slava Rossiya!*), 229
Gorbachev, Mikhail, 81
  on Caucasus, 27
  on common European home, 5–6
  constitutional change, 64
  intervention, 133
  new thinking foreign policy, 102
  reform policies, 68
Graham, Lindsey, 5
The Grand Chessboard: American Primacy and Its Geostrategic Imperatives (Brzezinski), 206
Great Patriotic War, 19, 43, 56, 186, 237–38
  mythologized memories, 276
great power (*derzhavniks*), 81, 85–86
great power Russia, 80–82, 85–87
Great Silk Way, 104, 106
Grenada, 289
GRU. *See Glavnoye razvedyvatel'noye upravleniye*
GUAM. *See* Georgia, Ukraine, Azerbaijan, and Moldova
Gubarev, Pavel, 245–46, 254, 272
GWOT (Global War on Terror), 119, 163

Hadley, Stephen, 192–93
Harmsworth, Alfred, 49–50

Hearst, William Randolph, 49–50
Heletey, Valeriy, 268
Helsinki Final Accords, 19, 24, 74, 189
Hemingway, Ernest, 100
Heritage Foundation, 292
hidden hand, 51
high-profile events, 45
hired thugs (*Titushki*), 212
historical ideology, 41
Hitler, Adolf, 23, 186, 276
Holbrooke, Richard, 123
Hollande, François, 221
Holocaust, 44
homeland nationalism, 34–35
Human Rights Watch, 170
Hungary, 98
Hunter, Robert, 6
Hussein, Saddam, 100, 186, 210
hypermasculinity, 48

ICC. *See* International Criminal Court
identity
 civilization as shared common, 43
 geopolitical culture and, 39
 in Georgia, 114
 Putin's identity politics, 211
 Russia questions, 70–71
 spatial, 10, 34, 36, 71
illusion of transparency, 27
Immediate Response, 2008, 157
imperialism
 anti-imperialism and, 41–42
 Georgia internal, 42
 Russian geopolitical culture, 75–78
 Russian internal, 42
India, 38
indigenization (*korenizatsiia*), 62
informals (*neformaly*), 133–34
information warfare, 11
Ingram, Alan, 78–79
inner abroad, 38f
 Chechnya as, 1
 of Russia, 86–87
 of Turkey, 38
internal abroad, 35–36
internal imperialism, 42
international affairs, 29
International Criminal Court (ICC), 197
International Crisis Group, 149, 156
International Monetary Fund, 293
international relations, 155. *See also* geopolitics
 geopolitical, 89
 Kissingerian understanding of, 33
 scholars, 46

International Republican Institute (IRI), 99–100
interventionism, 289
Iraq
 Bush linking to Georgia, 122
 United States invasion of, 2
 war debates, 301–2
Ireland
 Home Rule in, 36, 38
 Irish Free State, 38
IRI. *See* International Republican Institute
Israel
 Gaza Strip war, 25
 Saakashvili relationship, 119–20, 326n80
Ivanishvili, Bidzina, 195
Ivanov, Sergei B., 110, 147
Izborsky Club, 245

Jamestown Foundation, 292
Jaresko, Natalie, 293
JCC. *See* Joint Control Commission
Jeffrey, James F., 124, 167
Joint Control Commission (JCC), 137, 148, 150, 152, 155
Joint Peacekeeping Forces (JPKF), 137

Kaczyński, Lech, 160
Kadyrov, Akhmad, 89, 280
Kadyrov, Ramzan, 89, 282–83
Kagan, Robert, 23, 177
Kahneman, Daniel, 45
Karaganov, Sergei, 82
Karasin, Grigory, 158
Kempé, Frederick, 292
Kennan, George, 6, 300
 Long Telegram of, 22, 27
 red tide passage of, 366n16
 on state sovereignty, 28
 theory on Russian invasions, 22, 26
Kernes, Gennady, 257
Kerry, John, 24
Kertsen amendment, 323n44
Kezerashvili, Davit, 119–20
KGB, 57, 115, 148
 backgrounds, 177
 persecution by, 132
 police, 129
 Putin in, 48, 219, 223
 security of, 117
Khakamada, Irina, 260
Khodakovsky, Alexander, 265
Khrulev, Anatoly, 171, 283
Khrushchev, Nikita, 61, 202, 315n18
Kisileva, Natalya, 222

Kissingerian understanding
　of geopolitics, 12
　of international relations, 33
Klitschko, Vitali, 255
Klokotov, Nikolai, 77
Kofman, Aleksander, 273
Kohl, Helmut, 100
Kokoity, Eduard, 133, 134, 140, 141*f*
　election of, 153, 158
　Mamsurov meeting, 169
　playing into hands of, 148
　takeover moves, 142
　three-point plan, 152
　Zhvania meeting, 150
Kolomoyskyi, Igor, 255
Kondrashov, Andrey, 231–32
Kongress russkikh obshchin (KRO), 78–79
Konstantinov, Vladimir, 22, 216, 229*f*
Korban, Gennady, 255
korenizatsiia (indigenization), 62
Kosovo
　independence, 7–8
　NATO intervention, 185
　peacekeeping forces in, 318n61
　Putin on, 154–55
　war, 86
Kozyrev, Andrei, 73*t*, 75, 82
KPRF. *See* Russian Communist Party
krai (land), 208
Kristol, William, 163, 177
KRO. *See Kongress russkikh obshchin*
Krym Nash (Crimea Is Ours), 229–30
Kuchma, Leonid, 202, 206–7
Kulakhmetov, Marat, 180
Kunin, Daniel, 143, 163, 331n46
kurator (adviser), 247–48, 282, 359n21
Kutuzov, Mikhail, 241
Kvitsiani, Emzar, 152
Kyiv crisis
　fascist coup, 2, 18
　Putin response to, 30
　threat story, 51

Laitin, David, 59–60
land (*krai*), 208
language
　in Georgia, 61, 102
　in Russia, 59–60
　in Ukraine, 201–2, 204, 261
Lantos, Tom, 85
Large Russia (*Bol'shaya Rossiya*), 246
Laruelle, Marlene, 243–45
Last Thrust to the South (Zhirinovsky), 75–76

Lavrov, Sergey, 23, 185, 191–92
　on NATO membership, 198
LDPR. *See* Liberal Democratic Party of Russia
Lebed, Aleksandr, 79, 80, 317n43
Lemkin, Raphael, 43
Lenin, Vladimir, 59, 62
Levin, Carl, 5
Liberal Democratic Party of Russia (LDPR), 75
liberal empire, 42, 312n62
liberal European Russia, 74–75
liberty, 41
　Caucasus beacon of liberty, 112–25
　Radio Liberty, 98
Lieberman, Joe, 5, 99, 100, 123
Life magazine, 97
Limonov, Eduard, 73*t*, 77
literary figures (*tsisperkhantselni*), 106
A Little War That Shook the World (Asmus), 159
little green men, 252–53
Little Russia, 204
local elite, 9, 60, 239
Lomaia, Alexander, 194
Long Telegram, 22, 27
Lost Day video, 179–80, 344n57
Luhansk
　census, 203
　fighting in, 232, 244
　rebel stronghold in, 267
　seizures in, 256
　self-determination referendum in, 265
　supporting intervention, 262–63
Luzhkov, Yuri, 73*t*, 79, 202

Mackinder, Halford, 52, 77
Madrid Principles, 371n82
Makarov, Nikolai, 191
Malaysian Airlines Boeing 777 crash, 18, 266, 301
Malofeev, Konstantin, 246–48, 258, 267
Malyshev, Mikhail, 356n86
Mamsurov, Taimuraz, 169
Manafort, Paul, 8, 209
MAP. *See* Membership Action Plan
McCain, John, 5, 109, 123, 348n124
　election defeat, 291
　on enlargement, 99
　on Euromaidan protests, 292–93
　Georgia, 2008 war, condemnation, 127
　Georgia affinity, 177
　hero identification, 100
　hypermasculinity of, 48
　on rollback, 99–100
　on Russian aggression, 191

Shevardnadze and, 106
  supporters, 163
  on Ukraine, 195
  withdrawal demands by, 194
McCain Institute, 292
Mearsheimer, John, 28–29
  power-centrism, 31–32
  on Putin, 30–32
  on Ukraine, 2014 seizure, 30
Medvedev, Dmitri, 156–57
  accusations, 170
  atrocities documented, 183
  on August War, 190
  Bush firmness with, 192–93
  cease-fire agreement, 173–74
  delaying armed response, 179–80
  enforcing peace, 186
  presidency, 210
  recognizing independent states, 188–90
  refugee assistance, 181
  on territorial integrity, 185
Membership Action Plan (MAP), 94–95, 123–24, 155
Memorial Human Rights Center, 1
Merabishvili, Vano, 149, 160
Merkel, Angela, 94, 221, 223
Mesić, Stjepan, 145
metropolitan state, 36, 38*f*
Mexican border wall, 302
Migranyan, Andranik, 82
Military Professional Resources Incorporated (MPRI), 145
Millennium Message, 87, 219
Milošević, Slobodan, 145
Minsk Group, 299
Mirror Weekly, 262
Mitchell, Lincoln, 112
Mladić, Ratko, 185
Mogadishu, 102
Moldova
  discussion of, 15
  frozen conflicts in, 92
  GUAM, 151
  historical fault lines of, 67–68
  regional rebellions in, 4
  republic, 68
  Romania unification and, 69
  separatism in, 241
Monroe Doctrine. *See* Russia's Monroe Doctrine
most favored lord, 61
Mother Russia, 41
MPRI. *See* Military Professional Resources Incorporated

Mubarak, Hosni, 210, 214
Mullen, Mark, 324n56
Mullen, Mike, 191
multivector geopolitics, 211–12
Murphy, Chris, 292
music. *See* concerts

Nagorny Karabakh conflict, 286, 299, 371n82
Nagorny Karabakh Republic (NKR), 67
National Bolshevik Party, 77
National Captive Nations Committee, 96
nationalism
  competitive relational, 35
  fascism and, 49
  homeland nationalism, 34–35
  as national ethos, 41
  nationalist heroics, 52
  nationalizing, 34
  radical, 218
  Saakashvili on, 142–43
  in Ukraine, 218–19
  United States, 301
nationalizing states, 35, 36, 38*f*
national minorities, 35, 36
NATO. *See* North American Treaty Organization
naval base lease, 215
Nazism, 23
  collaborators, 237–38
  Putin and, 44
  Soviet Union pact, 19, 44, 63, 201
  territorial aggrandizement, 33
near abroad
  as geopolitical field, 9
  Russia, 81, 86–87, 277
  Russian geopolitics and, 15
  term emergence, 3
  of Turkey, 38–39
  United States as, for Saakashvili, 4
neformaly (informals), 133–34
Negroponte, John, 190–91
Nemtsov, Boris, 73*t*, 283
neoliberal ideology, 120
News Corporation International, 50–51
New Start Treaty, 210
New York Times, 163, 280
Nicaragua Sandinista government, 289
Nicholas I, 237
Nitze, Paul, 97
Nixon, Richard, 29
Nixon Center, 29
NKR. *See* Nagorny Karabakh Republic
Nodia, Ghia, 132
Noriega, Manuel, 289

normative geopolitics, 96
North American Treaty Organization (NATO)
  alignment with, 4
  encroaching on Russia, 12
  encroachment policies, 287
  enlargement, 6, 30–31, 93–95, 100, 276
  expansionism, 5, 22, 27, 85, 207, 281, 288, 298–99, 371n81
  Georgia membership and, 110–11
  Kosovo intervention, 185
  Lavrov on membership, 198
  MAP offers, 94–95, 123–24, 155
  Partnership for Peace, 107
  Putin on Ukraine joining, 32
  Saakashvili on, 151
  shared values and responsibilities, 99
  as sphere of influence, 27, 291
  Turkey, 105
  Ukraine and Georgia as members, 7
  Ukraine membership proposal, 206–7
  Washington Treaty, 274–75
North Atlantic Cooperation Council, 205
North Ossetia
  Alania added to, 328n9
  evacuations to, 165, 169, 330n28
  flight to, 130, 133–34, 181, 329n5
  gas explosions, 152
  Putin in, 180–82
  republic, 129
  Russian citizens, 339n148
  school siege in, 90–91, 91f
  trade, 132
  turmoil in, 149
  uniting with, 189, 286
no-use-of-force pledge, 156–57
Novaya Gazeta, 248
Novorossiya project, 15
  abandoning Novorossiya, saving Donbas, 264–68
  activating Novorossiya, 244–52
  amplifying, 260–64
  attitudes in Southeastern Ukraine, 268–73, 271f–272f
  attitudes survey, 262–64, 263f
  divergence of Donbas, 256–58
  Izborsky Club, 245
  localized power struggles, 252–62
  overview, 237–40
  Putin naming, 261–62
  as revisionist geopolitical imaginary, 240–44
  Russian filibusters over, 258–60
  separatism and, 271f
NSC-68, 97
nuclear weapons, 96

Nuland, Victoria, 94, 292–93

Obama, Barack, 194, 221
  election of, 291
  European Reassurance Initiative, 275
  geopolitics and, 297
  Hitler and, 276
  Mubarak and, 214
  presidential campaign of, 368n53
  on Putin, 296
  on Russia leadership, 24–25
  on Russian invasion, 195
  on Ukraine, 2014 seizure, 23–24
  Ukraine and, 295–97
  women, 210
occupied territories, 25, 66, 277, 297
  Cold War, 285
  Georgia as, 65
  Russian geopolitical archipelago, 279–86, 284t
Odesa, 204
  activists in, 244, 250
  Democratic Union of Novorossiya formation, 4, 241
  as divided city, 257
  founding of, 240
  history of, 261
  massacre, 264, 270
  People's Republic of Odessa proclaimed, 257
  protests in, 252
  unification call in, 254
Operation Clear Field, 184
Operation Iraqi Freedom, 111, 119
Operation Storm, 145, 149, 160
Oplot, 257, 265
Orange Revolution, 4, 8, 150–51, 207, 212, 242–43
Organization for Security and Cooperation in Europe (OSCE), 6, 137
Organization of Ukrainian Nationalists (OUN), 218
OSCE. See Organization for Security and Cooperation in Europe
Osnovy Geopolitiki: Geopoliticheskoe Budushchee Rossii (Dugin), 77–78
Ottoman Empire, 240–41
OUN. See Organization of Ukrainian Nationalists

Palin, Sarah, 195
Partnership for Peace, 107
Party of Regions, 207, 216, 239, 243, 253–57
passportization policy, 140

paternalism, 204, 326n71
Patrushev, Nicolai, 214, 216
peacekeeping forces
  JPKF, 137
  in Kosovo, 318n61
  in South Ossetia, 151–52
  in Transnistria, 283, 285
People's Front, 254
phantom limb pain, 47
Pliyev, Inal, 126–27, 164, 166
Pliyev, Issa, 131
PMR. *See* Pridnestrovian Moldovan Republic
political realism
  core beliefs in, 28–29
  Ukraine, 2014 seizure, and, 30
  in United States, 28
Popular Shrine (*Adamon Nykhaz*), 133
Poroshenko, Petro, 293, 295
Posen, Barry, 299
post-Soviet space, 37*f*, 305n4
  affective geopolitics in, 46
  contested geopolitical field, 33–37, 38–39, 38*f*, 298
  decolonization, 34
  flow infrastructures, 39
  freedom in, 42
  place-based powers in, 36
  as reminder of imperial power, 33–34
  republican-level sovereignty in, 66–67
  role in geopolitical catastrophe, 64–69
  Russian question, 66
  sovereignty and independence in, 65
  successor states, 42
Potemkin, Gregory, 1
Potemkin democracy, 108
Powell, Colin, 57, 110, 113, 148
Power, Samantha, 210
power-centrism, 31–32
power maximization, 29
Pridnestrovian Moldovan Republic (PMR), 68–70, 79, 241
Primakov, Yevgeni, 86
Project for the New American Century (PNAC), 100
Prokhanov, Alexander, 76, 245
propaganda, 11, 98
psychological warfare, 97
Pushilin, Denis, 257
Putin, Vladimir, 229*f*, 230*f*
  Abkhazia protected by, 155
  appealing to, 250–51
  at Bucharest summit, 124–25
  Bush firmness with, 192–93
  Bush meeting, 166–68
  Bush on, 57
  Bush praising, 198
  Chechnya power reassertion, 81
  on criminal actions in South Ossetia, 182–83
  on *Direct Line,* 260–62
  disenchantment with United States, 90
  disparate dependencies created by, 282
  fictional independent states and, 281
  at field hospital, 181–82
  first political lessons, 79
  force in Chechnya, 90
  on genocide, 183–84
  as geopolitical culture articulator, 11–12
  Hitler and, 276
  hypermasculinity of, 48
  identity politics, 211
  on international law, 91–92
  in KGB, 48, 219, 223
  on Kosovo, 154–55
  Mearsheimer on, 30–32
  military surprise, 221
  Millennium Message, 87, 219
  national anthem for, 47
  Nazism and, 44
  in North Ossetia, 180–82
  Novorossiya project naming, 261–62
  Obama on, 296
  pragmatism of, 147
  protests against, 210
  on radical nationalism, 218
  on reform agenda, 55–56
  response to Kyiv crisis, 30
  responsibility to protect, 187–88
  revanchism of, 58, 87–92
  on Saakashvili, 48
  Saakashvili heated exchange, 150
  Saakashvili meeting, 114–15, 155
  on September 11, 2001 terrorist attacks, 2
  setting attack tone, 180
  shared common identity and, 43
  Shevardnadze meeting, 110
  on statism, 88
  systema, 89
  two-part interest group assertion, 270
  Ukraine, 2014 seizure, decision, 214–17, 245–46, 290
  Ukraine, 2014 seizure, justified, 223
  Ukraine, 2014 seizure, speech and celebration, 226–31
  on Ukraine as state, 199, 207–8, 350n6
  on Ukraine joining NATO, 32
  Yanukovych meeting, 213
  Yanukovych rescued by, 214
Pyatt, Geoffrey, 293

Qaddafi, Muammar, 210

Radio Free Europe, 98
Radio Liberty, 98
Reagan, Ronald
    evil empire and, 41
    Grenada and, 289
    on Nicaragua Sandinista government, 289
realpolitik, 12, 33, 62
reckless driving, 299
Red Army, 63, 130
red blob of totalitarianism, 6
red-brown coalition, 76, 79
regime change, 178, 211
rescue missions, 13
revanchism
    in Georgia, 2008 war, 197
    of Putin, 58, 87–92
    Russia, 189, 276
    Saakashvili and, 128, 208–9
    in Ukraine, 209
Rhodes, Ben, 26, 194
Rice, Condoleezza, 23, 94, 116
    call for ceasefire by, 191
    on defending allies, 157
    Georgia, 2008 war, advice, 161–62
    reports to, 190
    sent to Georgia, 193–95
Rice, Susan, 210
Robertson, George, 111
Rodionov, Igor, 77, 322n30
Rogozin, Dmitry, 73t, 76, 78–79, 183–84, 185
rollback, 97, 99–100
Romania unification, 69
Rose Revolution, 123
    anniversary, 151
    Bush on, 112, 121–22
    coining, 324n56
    Saakashvili to power, 142
    skepticism over, 147
RSFSR. *See* Russian Soviet Federal Socialist Republic
rule of law, 41, 89
Rumsfeld, Donald, 7
Russia. *See also* Soviet Union
    aggression in Georgia, 2008 war, 190–97
    authoritarian state, 290
    Black Sea Fleet, 7
    Bush alienating, 90
    chauvinism, 62
    Chechnya as inner abroad, 1
    competing visions of, 72t–73t
    as dark double, 22
    expansionism, 22, 309f9
    geopolitical culture, 11–12, 57, 69–87
    geopolitics of, 13, 15
    Georgia, 2008 war, attack on, 186
    Georgia, 2008 war, storyline, 179–90
    great power geopolitical culture, 80–82, 85–87
    identity and territory questions, 70–71
    imperialist geopolitical culture, 75–78
    inner abroad, 86–87
    intelligence cooperation offered, 2
    internal imperialism, 42
    language, 59–60
    liberal European, 74–75
    Mother Russia, 41
    national anthem, 46–47
    national interests safeguarded, 30
    NATO encroaching on, 12
    near abroad, 81, 86–87, 277
    new Europe, 7
    new growth model for, 291
    Novorossiya project filibusters, 258–60
    Obama on leadership, 24–25
    occupied territory geopolitical archipelago, 279–86, 284t
    phantom limb pain, 47
    revanchism, 189, 276
    sentiment in Transnistria, 286
    size of, 46
    Sochi Winter Olympics, 17, 167, 180, 209, 290, 368n49
    sphere of influence, 3, 12, 20, 52, 85, 243, 277
    Ukraine invasions, 19
    Ukraine Treaty of Friendship, 203
    United States free Russia crusade, 96
    United States sanctions against, 18
Russian Aggression Prevention Act, 295
Russian Communist Party (KPRF), 80
Russian Federation, 3, 86
    expanse of, 280
    joining, 189
    reunification of Crimea, 225
    rule of law in, 89
    salaries and pensions in, 139
Russian idea (*Russkaya ideya*), 71
Russian invasions, reasons for, 14
    geopolitics background, 28–54
    geopolitics past return, 22–27
    imperialist power storyline, 20
    Kennan's theory, 22, 26
    liberal storyline, 25–26
    overview, 17–21
    political-realist storyline, 21
Russian Law on Citizenship, 139–40, 330n28
Russian National Unity, 77
Russianness, 70, 79, 205, 208

Russian question, 66, 71
Russian Soviet Federal Socialist Republic (RSFSR), 240–41
Russian world (*Russki Mir*), 141, 205, 211, 243–44
Russia's Monroe Doctrine, 82, 83*t*–84*t*
Russia Today, 189
Russification, 60, 131
Russkaya ideya (Russian idea), 71
russkieye, 316n32
Russki Mir (Russian world), 141, 205, 211, 243–44
Russophobia, 309n9
Russophone, 201–2, 204, 220, 244, 288
Rutskoi, Alexander, 85

Saakashvili, Mikheil
  aggressive moves by, 147
  authoritarianism, 117–19
  Baltic states and, 163
  break-ups and, 108
  building military, 145
  on Bush, 124, 177–78
  Bush meeting, 115, 117, 151, 155–56, 162–63
  captive nation and, 119
  Cato Institute and, 121
  cause of, 100
  cease-fire agreement, 173–74
  charisma, 113, 118, 142
  domination, 150–51
  editorials by, 209
  education, 99
  election of, 112
  on Finland, 177
  fleeing Georgia, 54
  Georgia, 2008 war, responsibility, 14
  on getting Georgia back, 144
  government of, 93
  hypermasculinity of, 48
  inauguration speech, 113–14
  Israel relationship, 119–20, 326n80
  leadership style, 146
  lobbying by, 369n59
  on nationalism, 142–43
  on NATO, 151
  neoliberal ideology, 120
  network of, 53
  new cause, 294
  no-use-of-force pledge rejected by, 156–57
  offensive launch by, 158–59, 162, 190
  outrage over, 12
  pushing buttons in media, 176–79
  Putin heated exchange, 150
  Putin meeting, 114–15, 155
  Putin on, 48
  recklessness, 29–30
  revanchism and, 128, 208–9
  rhetoric contradictions, 143–44
  risk taker, 332n55
  Rose Revolution bringing to power, 142
  state-building hero, 48
  supporters, 116, 163–64, 195
  on territorial integrity, 114
  testing limits, 111
  United States as near abroad, 4
  as war criminal, 166–67, 183, 186, 191
  as young reformer, 105–6
  Yushchenko supporting, 243
Sakharov, Andrei, 42
salaries and pensions, 139
Salvation Union of Ossetia, 153
Sanakoev, Dmitri, 153, 157
Sanakoev Project, 153–54
Sarkozy, Nicolas, 173, 184
Saunders, Paul J., 29
SBU. *See* Ukrainian Security Service
Scandinavia, 3
scenography of legitimacy, 223–26
Scheunemann, Randy, 100, 123, 194, 348n124
Sea Breeze exercises, 205
security and defense, 46
  geopolitical culture, 39–40
security corridor, 137–38
self-determination, 299
self-help, 29
separatism
  condemning, 74
  curbing, 249
  Moldova, 241
  Novorossiya project, 271*f*
  pro-Russian, 262
  as Russian lever, 208, 242–43
  sponsored, 145
  territorial fragmentation and, 68
  violent, 150
September 11, 2001 terrorist attacks, 2
Serdyukov, Anatoly, 170
Setting the Record Straight statement, 193
Shalikashvili, Otar Joseph, 108–11
Shelov-Kovedayev, Fyodor, 82
Shevardnadze, Eduard, 3
  aid to, 109
  assassination attempts, 103
  break-ups and, 108
  Dagomys Agreement and, 67
  foreign policy of, 107
  Great Silk Way and, 104, 106
  high-profile policies of, 111
  image in West, 105
  McCain and, 106

Shevardnadze, Eduard (*Cont.*)
  personal insecurity of, 102
  Putin meeting, 110
  resignation of, 112
  tolerance policy, 139
  United States aid to, 107–8
  working relationships of, 100–102
Shevchenko, Taras, 255
Shostakovitch's Seventh Symphony, 44
Shoygu, Sergei, 181, 214, 215, 221, 355n70
Slava Rossiya! (Glory to Russia), 229
slow thinking, 45
SOAO. *See* South Ossetian Autonomous Oblast
Sobchak, Anatoly, 79–80, 202
Sochi Agreement, 135
Sochi Winter Olympics, 17, 167, 180, 209, 290, 368n49
Solzhenitsyn, Alexander, 66, 79, 91
Soros Foundation, 112
South Ossetia, *136*. *See also* North Ossetia
  acculturation, 131
  aid to, 145–46
  campaigns against, 132–33
  capital accumulation, 138
  as contested space, 129–41
  deaths in, 135, 196
  elections in, 139
  ethnic polarization, 134
  Georgia, 2008 war, over, 12, 17, 128
  gutted houses, 196*f*
  independence recognition, 155
  International Crisis Group on, 149
  memorial, 137*f*
  militia groups, 133–34
  passportization policy, 140
  peacekeeping forces in, 151–52
  Putin on criminal actions in, 182–83
  salaries and pensions to, 139
  security corridor, 137–38
  sovereignty of, 129, 133
  territorial integrity achieved, 175
  as threat, 175
  victimization in, 130
  vulnerability of, 135
  war crimes and crimes against humanity in, 197
South Ossetian Autonomous Oblast (SOAO), 131–32, 134
sovereignty
  in post-Soviet space, 65–67
  of South Ossetia, 129, 133
  state, 28
  walled, 302
Soviet Union. *See also* geopolitical catastrophe; Russia
  anti-imperialist ideology, 59–61
  big country, 47
  Bolsheviks forming, 7, 20, 62–63, 95
  collapse of, 34
  containment by, 9
  ethnoterritorial units, 59, 62–63
  as evil empire, 41
  fascism and, 44
  as geopolitical complex, 14
  Great Patriotic War, 19
  liberation from, 47
  Nazi pact, 19, 44, 63, 201
  nostalgia for, 45, 57, 85
  original, 7
  sphere of influence, 97–98
  structural contradictions, 63–64
  titular nations, 63–64
spatial identity, 10, 34, 36, 71
sphere of influence, 297–98
  freedom from, 279
  NATO as, 27, 291
  proximity and paranoia, 286–91
  Russia, 3, 12, 20, 52, 85, 243, 277
  Soviet Union, 97–98
  suppression of, 282
  thinking, 33, 198
  Turkey, 109
sponsored separatism, 145
Stalin, Joseph, 60–63, 114–15
Stankevich, Sergei, 80–81
state-challenging movement, 36
state-produced drama, 49–52
state sovereignty, 28
statism, 88
sui generis, 154, 188
Sukhoi Su-24M aircraft downing, 274
Surkov, Vladislav, 248, 282
Suvorov, Alexander, 241

tabloids
  commercial drive, 50
  storylines, 49–52
  television hyperreality and, 51
Tagliavini, Heidi, 160
Tagliavini Report, 160
Taliban, 2, 109
Taruta, Serhiy, 255
Tbilisi
  Baku-Tbilisi-Ceyhan pipeline, 105, 178, 194
  occupation museum, 65
  pipeline project, 105

protests, 163
rallies, 178
take Georgia back campaign, 129
vision of Georgia, 103
weakness, 147
territorial aggrandizement, 33
terrorism, 1–2, 311n49
  ATO, 260
  cyberattacks, 371n87
  September 11, 2001 terrorist attacks, 2
  war on terror, 4, 110, 305n2
thick and thin geopolitics, 277–79, 297, 365n10
Tishkov, Valery, 74
titular nations, 63–64
Titushki (hired thugs), 212
Tkeshelashvili, Eka, 161–62
totalitarianism, 6
Transnistria, 4, 245, 257
  call for careful policies in, 125
  corridor to, 264
  countermobilization in, 241–42
  media attention in, 81
  peacekeeping forces in, 283, 285
  rebels in, 77, 317n43
  Russia sentiment in, 286
  state of, 279, 281
  as unrecognized, 68, 188
  trap theory, 164
Treaty of Friendship, Cooperation and Partnership with Ukraine, 24
Tretyakov, Vitaly, 70
Truman, Harry, 97, 278
Truman administration, 50
Trump, Donald, 302
tsisperkhantselni (literary figures), 106
Tsutsiev, Arthur, 330n28
Turkey
  Caucasus and, 3
  Georgia and, 3
  inner abroad, 38
  NATO member, 105
  near abroad, 38–39
  sphere of influence, 109
  Sukhoi Su-24M aircraft downing, 274
Tymoshenko, Yulia, 249

Ukraine, 200f. See also Euromaidan protests
  as biggest prize, 56
  Bolsheviks and, 237
  borders, 320n4
  as cause in United States geopolitical culture, 291–97
  Cossacks, 61
  Crimea as gift, 201
  diaspora, 3
  genocide, 201
  geopolitical attitudes, 231–36, 235f
  geopolitics, 199–211
  GUAM, 151
  industrialization, 203
  language, 201–2, 204, 261
  Little Russia, 204
  MAP offer, 94–95
  McCain on, 195
  nationalism, 218–19
  as NATO member, 7
  NATO membership proposal, 206–7
  naval base lease, 215
  Obama and, 295–97
  Orange Revolution, 4, 8
  as power base, 315n18
  as prize, 204, 206
  pro-Russian mobilizations, 69
  Putin on joining NATO, 32
  Putin on Ukraine as state, 199, 207–8, 350n6
  referendum proposal, 225–26, 226f
  regional rebellions, 4
  revanchism in, 209
  Russian invasions, 19
  Russia Treaty of Friendship, 203
  Sea Breeze exercises, 205
  as state, 261
  Ukrainophone and Russophone, 201–2, 204, 220, 244, 288
Ukraine, 2014 seizure, 2. See also Novorossiya project
  Clinton, H., on, 276, 365n6
  comprehending, 20
  concerts and, 13
  contingent road to Crimea, 210–14
  dramaturgy and affective geopolitics of annexation, 217–31
  geopolitical attitudes in contested Ukraine, 232–36, 235f
  improvised rescue mission, 214–17
  Mearsheimer on, 30
  music festival, 230, 230f
  Obama on, 23–24
  overview, 198–99
  polite people monument, 231f
  political realism and, 30
  precipitation of, 14
  protests, 259, 259f
  Putin decision, 214–17, 245–46, 290
  Putin justifying, 223
  Putin public speaking on, 221

Ukraine, 2014 seizure (*Cont.*)
  Putin speech and celebration, 226–31
  reductionist view, 32
  still unfolding, 8
  Ukraine geopolitics, 199–211
Ukraine Freedom Support Act, 295
Ukrainianisation, 80
Ukrainian Security Service (SBU), 255–56
Ukrainophone, 201–2, 204, 220, 244, 288
unilateralism, 2
United Nations Observer Mission in Georgia (UNOMIG), 153
United Russia (*Yedinaya Rossiya*), 142
United States
  aid to Shevardnadze, 107–8
  exceptionalism, 95–96, 100, 278
  free Russia crusade, 96
  geopolitical culture of, 10–11, 15, 57, 95–98, 103, 121–22, 275–79, 286, 291, 296–97, 301–3
  geopolitics of, 13
  Georgia, 2008 war, anticipated aid, 162–63
  Georgia, 2008 war, condemned by, 23, 166–68
  Georgia, 2008 war storyline, 190–97
  Georgia financial support, 123
  Georgia given security assistance, 3–4, 14
  intelligence cooperation offered to, 2
  interventionism, 289
  Iraq invasion, 2
  nationalism, 301
  political realism, 28
  Putin's disenchantment with, 90
  as Saakashvili's near abroad, 4
  sanctions against Russia, 18
  Setting the Record Straight statement, 193
  Taliban war, 109
  train and equip mission in Georgia, 109–10
  Ukraine, 2014 seizure, condemnation, 23–24
  Ukraine as cause in geopolitical culture, 291–97
Unity (*Edinstvo*), 142, 148
UNOMIG. *See* United Nations Observer Mission in Georgia
UNSC, 85, 153
uti possidetis (as you possess), 65, 68, 85, 103

violence
  Euromaidan protests and, 238
  history and geography, 15–16
  separatism, 150
Volunteer Freedom Corps, 323n44

Wahabbism, 114
walled sovereignty, 302

Wall Street Journal, 176, 185, 192
war on terror, 4, 110, 305n2
Warsaw Pact, 5, 22, 85, 102, 205
Washington Declaration, 98, 328n20
Washington Post, 29
Weekly Standard, 279
Weimar Russia, 71
Welles Declaration, 320n3
West, 30
  civilization and Europe, 42–43
  during Cold War, 44
  Georgia storyline and, 175–79
  joining, 6
  Shevardnadze image in, 105
"Why the Ukraine Crisis Is the West's Fault" (Mearsheimer), 30
wide autonomy, 146
Will, George, 22
Wilson, Damon, 93, 292
Wilson, Woodrow, 278
World Bank, 120
World Trade Organization, 210, 292

Yakobashvili, Timuri, 120, 162, 177, 279, 324n55
Yalta Conference, 96–97
Yanukovych, Viktor, 8, 23, 209
  civilizational choice, 288
  Donetsk clan and, 239
  driven from office, 30
  flight of, 244, 247, 252
  impeachment of, 218
  multivector geopolitics, 211–12
  ousting of, 224
  as preferred leader, 281
  Putin meeting, 213
  Putin rescuing, 214
Yatsenyuk, Arseniy, 293, 294
Yatsyuk, Denis, 250
Yedinaya Rossiya (United Russia), 142
Yefremov, Aleksander, 255
Yeltsin, Boris, 64–65, 226
  administration of, 74–75
  attack on parliament, 77
  Dagomys Agreement and, 67
  election win, 80, 81
  enlargement introduced to, 98
  government of, 101
  inaugural declaration, 88
  Japan and, 81–82
  resignation of, 87
  shared common identity and, 43
  as unprepared for collapse, 69–70
  UNSC called upon, 85

Yugoslavia
    collapse of, 34
    dissolution of, 5, 20, 98
Yushchenko, Viktor, 4, 151, 207
    naval base lease and, 215
    Saakashvili supported by, 243

Zakharchenko, Aleksander, 265

Zaldostanov, Alexander, 215, 354n53
Zatulin, Konstantin, 250
Zhirinovsky, Vladimir, 75–76, 242
Zhvania, Zurab, 105–6, 108, 112, 143
    death of, 150
    Kokoity meeting, 150
Zygar, Mikhail, 216, 249–50
Zyuganov, Gennady, 73t, 76, 80

Lightning Source UK Ltd.
Milton Keynes UK
UKHW040844221222
414079UK00046B/332